Motorcycle Workshop Practice Manual

by Pete Shoemark

(1454-1Q1) ABCDE FGHIJ

Haynes Publishing Group
Sparkford Nr Yeovil
Somerset BA22 7JJ England

Haynes Publications, Inc
861 Lawrence Drive
Newbury Park
California 91320 USA

Acknowledgements

Our thanks are due to all who have given assistance in the origination of this manual, particularly the Avon Rubber Company who supplied tyre information, Bearing Services Ltd of Bath who supplied bearing and seal information, the Cross Manufacturing Company (1938) Ltd who supplied wire thread insert information. European Industrial Services – Manufacturers of Nettlefolds Fasteners – supplied fastener information and gave permission to reproduce selected line drawings, Grand Union Products (International) Ltd supplied the Lumiweld kit and Heron Trading Ltd gave permission to reproduce their line drawings. HMSO gave permission to reproduce extracts from the MOT Inspection Manual for Motorcycle Testing, Kawasaki Motors (UK) Ltd gave permission to reproduce their line drawings, Mitsui Machinery Sales (UK) Ltd gave permission to reproduce their line drawings, Nigel Carpenter of Carpenters Engine Services kindly allowed us to photograph the workshop equipment featured in many of the photographs and Paul Branson Motorcycles of Yeovil supplied some of the items photographed. Shell UK Ltd supplied the fuel and oil information, Tran Am Ltd provided chain information and gave permission to reproduce line drawings, Tony Tranter gave permission to reproduce illustrations from his book, the *Automobile Electrical Manual* published by the Haynes Publishing Group, and Yeovil Motorcycle Services supplied some of the other items photographed.

We would also like to thank Bruce Main-Smith for his permission to reproduce extracts from his book, the *Motorcyclists' Encyclopedia*, which is now out of print.

A book in the **Haynes Owners Workshop Manual Series**

Printed by J. H. Haynes & Co. Ltd, Sparkford, Nr Yeovil, Somerset BA22 7JJ, England

---

ISBN 1 85010 454 9

Library of Congress Catalog Card Number 87-83502

British Library Cataloguing in Publication Data
Shoemark, Pete, 1952–
Motorcycle workshop practice manual
1. Maintenance & repair – Amateurs' manuals
I. Title
629.28'775
ISBN 1-85010-454-9

Contents

Introduction

Over the years many books have been published about workshop practice, often under the impression that the average D-I-Y mechanic will have a clean, well-lit and adequately heated workshop equipped with a stout workbench, a vice and a wide selection of good quality tools. Whilst this may well be the ambition of those who carry out anything from basic routine maintenance to major overhauls on their motorcycle or car, comparatively few are ever able to realise this objective. Only too often the working conditions are hopelessly inadequate, simply because there is no other alternative available. So, although this latest addition to the Haynes series of 'special' manuals does stress the advantages of being able to work under near ideal conditions, the author is only too aware of the need to improvise whenever necessary, having worked in decidedly primitive conditions himself for much of his motorcycling career.

Intended more as a reference book than one to be read from beginning to end, this manual has been designed to give a basic introduction to the selection and use of tools and how to employ them to their best advantage. Advice is given on building up a basic kit of tools that will deal satisfactorily with most problems, making the use of special service tools unnecessary unless they are considered essential to remove and replace parts without risk of damage. Today, chemicals play their part too, so information is included about sealants, thread locking compounds, adhesives and solvents.

Listed amongst the more specialized tasks is an introduction to riveting, soldering, welding and brazing and working with plastics. Refinishing techniques are also discussed in view of the many specialist paints and finishes available that provide professional results if applied correctly. Even home plating is no longer beyond the means of the amateur, although admittedly not for chrome which necessitates the use of noxious and highly poisonous chemicals.

How to repair accident damage is given attention so that many components that may otherwise have been scrapped can be reclaimed so that they will continue to give satisfactory service. Many of the problems encountered when rebuilding or restoring a machine are included too, such as the removal of broken studs and the reclaiming of stripped threads. Tyres, tubed and tubeless, also receive adequate mention, along with how to repair punctures. As for the electrics – always a bogy amongst amateur mechanics – suggestions are given on how to locate faults and in some cases repair them, to help avoid the often unnecessary renewal of components until the source is found by substitution.

The appendices contain a wealth of information and of particular interest to many will be the list of requirements that have to be met if a machine is to be issued with an MOT Test Certificate.

When using this manual note that the photographs are numbered sequentially in each text section, eg photo 8.4 relates to the fourth photograph in Section 8 of that Chapter. All line drawings are numbered sequentially in each Chapter, eg Fig. 2.3 relates to the third line drawing in Chapter 2.

Chapter 1 The workshop

Contents

1 The workshop building

The shape and size of a workshop building is almost invariably dictated by circumstance rather than personal choice. In an ideal world, every motorcyclist would have a spacious, clean and well-lit building, specially designed as a motorcycle workshop. In the reality, however, many of us must content ourselves with a corner of the garage at best, and few will have escaped the indescribable joys of working on a motorcycle outside on a cold, wet, winter morning.

Whilst the 'al fresco' approach is acceptable for odd maintenance jobs in fine weather, anything more adventurous than this requires some sort of shelter. The best readily-available building would be a normal car-sized garage. This provides ample working and storage space and room for a large workbench. That said, I know of enthusiasts who have completed full restoration jobs on vintage machines within the confines of a small timber shed. In the end you will have to make do with whatever shelter you can find, and adapt your workshop and working methods to this.

Whatever the limitations of your own proposed or existing workshop area, it is well worth spending a little time considering its potential and drawbacks, and even a well-established workshop will benefit from occasional reorganisation. Most users of home workshops find that lack of space causes problems, and this can be overcome to a great extent by giving careful consideration to the layout of benches and storage facilities. In the rest of this Section we take a look at some of the options available when setting up or reorganising a workshop. Perhaps the best approach in designing the workshop is to look at how others do it. Try approaching a respected local dealer and asking to see his professional workshop; take note of how working areas, storage and lighting are arranged, and then try to scale this down to suit your own workshop area, finances and needs.

The building in general

A solid concrete floor is probably the best prospect for any workshop area. The surface should be as even as possible, and must be dry. The surface can be improved further by applying a coating of paint or sealer suitable for concrete surfaces. This will make oil spillages and general dirt easier to remove and will lay the dust, always a problem with concrete. A wooden floor is less desirable and may be damaged by the localised weight of a motorcycle centre stand. It can be reinforced by laying sheets of thick plywood or chipboard over the existing surface. A dirt floor, whilst having a certain rustic charm, is a positive menace, providing an unending source of abrasive dust which will prove impossible to keep away from engine internals. It is second only to gravel or grass as a swallower of tiny dropped parts such as bearing balls and small springs.

Walls and ceilings should be as clean as conditions allow. It is a good idea to clean down the walls and, if possible, to apply a couple of coats of white paint. The paint will minimise dust and reflect light inside the workshop. On the subject of light, the more natural light there is the better. Artificial lighting will be needed as well, but you will need a surprising amount of it to equal ordinary daylight.

A normal domestic doorway is just wide enough to allow all but the biggest machines through, but not wide enough to allow it through easily. If possible, a full-size garage door is preferable. Steps, even one of them, are difficult to negotiate with the dead weight of most machines; make up a ramp to allow easier entry if the step cannot be removed or replaced by a permanent concrete ramp.

Make sure that there is adequate ventilation in the building, particularly during winter. This is essential to prevent condensation problems, and is also a vital safety consideration where solvents and volatile liquids are likely to be used. It should be possible to open one or more of the windows for this purpose, and in addition, opening vents in the walls are a desirable feature.

The workbench

This should be as large and as robust as space and finances allow. Many people prefer a unit constructed from slotted angle and topped with flooring-grade chipboard sheet. The use of a slotted angle construction allows a bench to be built to suit the available space quickly and easily, and the chipboard top can be renewed when it becomes badly damaged. One disadvantage of using chipboard is that it is somewhat absorbent, allowing oil spills to soak into the surface. Some users may prefer to prevent this by covering the bench surface with thin steel or aluminium sheet.

The main drawback of this arrangement is the generally lightweight and springy nature of the construction; a stout timber version, fixed firmly to a wall, provides a more satisfactory basis for heavier work, and will offer a strong mounting point for a vice and other fixed tools. Fixed shelving below the bench surface will make the construction more rigid and will provide useful storage space.

Motorcycle ramps

If funds and space allow, some form of ramp is a very useful refinement in the motorcycle workshop. Most professional ramps are operated by hydraulic rams. The machine is wheeled onto the ramp which is then raised hydraulically until it is at a convenient working height. Less expensive versions are available which raise the working platform mechanically.

Cheapest of all is a home-made equivalent consisting of a stout low bench about two feet high, and with a platform measuring about six feet by two feet. The machine can be wheeled onto the platform using a strong plank. Although this arrangement is less convenient than the professionally made equivalents, it will prove adequate for occasional home use. Do note, however, that motorcycles are heavy; you will need several pairs of hands to manoeuvre a large machine onto the platform safely.

Engine stands

Many manufacturers recommend the use of a special fixture to hold the engine during dismantling and reassembly. Equipment of this type is undoubtedly very useful, but outside the scope of most home workshops. In practice, most owners will have to make do with a selection of wooden blocks which can be used to prop the engine unit on the bench. These can be arranged as required so that the engine is supported in almost any position. It is very helpful when using this approach to have an extra pair of hands to assist in steadying the unit while fasteners are removed or tightened. In some situations, the unit can be held by one of the mounting bosses clamped in a vice, but care must be taken to avoid damage to the castings.

The popular adjustable workbenches, like the *Black & Decker Workmate* and its derivatives, can be very useful for holding engine/gearbox units in position during an overhaul. Note that a large engine may be too heavy for the lightweight versions, however, and it is important to check that the unit's weight does not exceed the rated capacity of the bench.

Storage and shelving

A motorcycle will occupy a great deal of space when dismantled, and adequate storage space is essential if parts are not to be mislaid. In addition, storage space for oils, greases and assorted consumables will be required, as well as for tools and equipment.

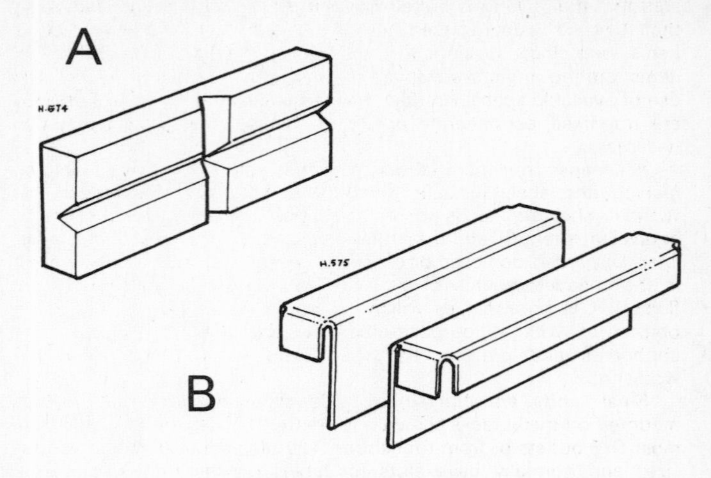

Fig. 1.1 Vice jaw protectors

Some jobs will require the work piece to be protected while it is held in the vice jaws. Alternative jaws (A) in fibre or plastic, often with a notched face to allow small round items to be held, may be available. Another method is to make up soft jaws from aluminium sheet (B). These can be fitted over the standard steel jaws

1.1 A stout bench surface is required to provide a secure fixing for a vice and any similarly-mounted equipment

1.2 Where finances permit, a motorcycle ramp makes working a lot easier

1.3 Note how the machine is secured using tie-down straps

1.4 A simple wooden construction like this will usually suffice in supporting the engine on the workbench

If space permits, fit slotted angle racking around the walls, making sure that the racking is bolted firmly to them. Arrange the shelves so that they are widely spaced near the bottom to take heavy or bulky items, with closer spacings towards the top to take smaller and lighter items. Slotted angle systems are expensive to buy, but make the best use of available space. An added advantage is that the shelf positions are not fixed permanently and can thus be altered as and when necessary.

A cheaper (but more labour intensive) solution is to use wooden racking and shelving built to suit the workshop. Remember that wooden shelving needs to be much bulkier than metal systems of equivalent strength, and that the shelf spacings are less easily changed later. Discarded domestic bookshelves, cupboards and other furniture is the cheapest source of workshop storage, but is also the least flexible. Old bookshelves will not be able to cope with very heavy objects as well as purpose-made storage systems. Similarly, old cupboards and drawers are soon distorted and broken in the workshop.

Small parts are conveniently stored in moulded plastic bins mounted on metal racks screwed to the wall. These are available from most DIY outlets or from tool shops. The bins are available in various sizes and normally have slots for labels on the front lips. Larger free-standing bins and crates are also available and are ideal for larger parts and assemblies. These are also available in a range of sizes, and will normally stack securely. Alternatively, the shelving can be arranged to take the various sizes of bin. There are numerous bin and racking systems available, and all do a similar job. Bear in mind that those produced by one manufacturer will not normally fit rival systems; choose one manufacturer and stick to his products.

Other containers can be used in the workshop to keep storage costs down, but try to avoid round tubs which waste a considerable amount of space. Glass jars are often advocated as cheap storage containers, but invariably get broken. Cardboard boxes are adequate for temporary use, but eventually the bottoms tend to drop out of them, especially if they get damp. Most plastic containers are useful, however, and large ice cream tubs are invaluable for keeping small parts together during a rebuild or major overhaul.

Finally, where space is at a premium, look at the unused areas of the workshop as potential storage areas for larger or awkwardly-shaped items. Where there is sufficient ceiling height, part or all of this space can be made into a loft area. Alternatively, fix hooks or racking to carry infrequently used parts or materials, removing them from the work area.

Electrical power

Of all the facilities useful in the workshop, a supply of mains electricity is by far the most essential. This is relatively easy to arrange where the building is near to or part of the house, but can be difficult and expensive if it is some distance away.

It must be stressed that safety is of prime importance here, and unless you have a very good working knowledge of electrical installations, the work should be left to an electrician. Remember that installing a power supply may mean that local bylaws have to be complied with. The notes below are intended to help you in considering any proposed electrical installation, but are by no means definitive; a skilled electrician will be able to advise on the best choice of equipment and materials for a given installation, and will be able to carry out the work safely. Note that in areas outside the UK, legal requirements and working methods may vary from those mentioned here.

You will need to consider the total electrical requirements of the workshop, making due allowance for later additions of equipment. The power cable can be run from a separate fused connection from the house supply. The cable should preferably be run underground, in which case special armoured cable will be needed. Short overhead runs can be made using conventional PVC insulated cable of the appropriate size, supported by a catenary wire; a steel cable fixed across the area to be spanned, allowing the electrical cable to be suspended from it to prevent damage through stretching.

Inside the workshop, the incoming power cable should be connected to an approved consumer unit (fuse box) from which the lighting and power circuits can be run. It is strongly recommended that the power circuit is protected by residual current circuit breakers, rather than by fuses. These devices have the great advantage that they will break the supply very rapidly in the event of a short circuit, a feature which could save your life if something goes wrong. Power sockets should be positioned well away from areas where they are likely to be damaged while working, and should be of a good quality industrial type housed in surface mounted metal boxes. The wiring connections should be made using armoured cable, or by running PVC cable through metal conduit.

Careful consideration should be given to lighting the workshop. As a general rule, fluorescent lamps are probably the best choice, giving even and shadow-free illumination. The position of the lamps is important; for example, do not position a fitting immediately above the area where the machine will stand during work, because this will cause shadowing even with fluorescent lamps. Fix the lamp at right angles to the machine, or fit one lamp slightly to each side of the area to give even illumination. Note that where solvents or flammable liquids are likely to be used, flameproof fittings should be employed to minimise the fire risk. Do not use fluorescent lighting above machine tools. The flicker produced by alternating current is especially pronounced with this type of lamp, and can make a rotating chuck appear stationary at certain speeds. This can be avoided by using a special type of twin-tube fitting, or by using tungsten lighting.

2 Safety measures

Like it or not, the workshop is inevitably a dangerous place. Electricity, itself a risky commodity if mistreated, becomes doubly so in an environment which is often damp. Hand and power tools offer yet more opportunity for disaster, and stored fuel and solvents present a very real fire risk.

There is no way to make the workshop itself safe, and the topic of safety really relates to minimising the risk of accidents occurring by employing safe working practices (primary safety), and to using the correct clothing and equipment to minimise injury if an accident should happen (secondary safety). The subject of safety is a large one, and could easily fill a chapter on its own. To keep the subject within reasonable bounds – and because few will bother to read an entire chapter on safety – this topic has been confined to one section, with further notes elsewhere in the main text where necessary.

The rest of this section covers some of the more important and relevant safety topics, but is not intended to be definitive. Read through it, even if you have run a workshop for years without grazing as much as a knuckle. Most of all, it must be stressed that the best piece of safety equipment of all is the human brain. Try to get into the habit of thinking about what you are intending to do, and what might go wrong. A little common sense and foresight can prevent the majority of workshop accidents.

A safe workshop

Take a regular look around you workshop, actively checking for potential dangers. The working area should always be kept tidy, and all debris should be swept up and disposed of after each working session. Be particularly careful about oily rags. If left in a pile it is not unknown for spontaneous combustion to occur, so dispose of them before this can take place.

Check all fixtures and fittings for security, and make any necessary repairs or alterations as soon as a fault is noticed – don't wait for a shelving unit to collapse on top of you before you take action.

Electricity

Whilst a supply of electricity in the workshop can be considered essential, do remember that it poses even greater dangers here than the normal domestic environment. Workshop buildings can often become damp, and condensation forming on outlets and electrical equipment could allow normally safe installations to become live. This risk can be minimised by using residual current circuit breakers, either in the incoming supply point, or as individual units plugged between the socket and power tool. These devices are able to react to a short circuit very quickly, shutting off the supply before it can do any real damage.

Be aware of more obvious dangers too, checking the condition of power tool leads regularly for signs of damage, and renewing them at once if there is any doubt about their condition. The mains plugs used in the workshop should always be of the moulded rubber or nylon type; *never* hard plastic which can break if dropped.

Where extension leads have to be used, be aware of any load limitations applicable to them. Most recently-produced extension leads are marked to indicate the maximum rating, both fully extended and when coiled around the reel. In the latter case, the rating is significantly reduced. This is because when coiled, inductance in the lead allows heat to build up. Even though most portable power tools do not draw a great deal of current, it is preferable to uncoil any unmarked extension lead fully to avoid this risk. Equipment such as arc welders draw a much heavier current, and with these it is best not to use an extension lead at all. If it is unavoidable, check that the lead is as short as possible, and that it is sufficiently heavy to cope with the demands on it.

Any electrical equipment used in the workshop poses a fire risk when the inevitable solvents and fuel present are considered. If fuel vapour is present, the slight arcing from a power tool can be enough to ignite it. Bear this in mind before using power tools, and minimise the risk of a build-up of flammable vapour by storing solvents and fuel safely, preferably outside the main workshop area, and ensuring adequate ventilation.

Protective clothing and equipment

Starting from the ground up, use suitable footwear whenever you use the workshop. The best form of foot protection is provided by safety boots with steel toecaps. Safety boots are not the most comfortable or elegant of footwear, but can prevent serious injury if a heavy object should fall. The soles of the boots are also able to withstand oils and fuel. Ordinary shoes or trainers offer almost no protection at all – you wear them at your peril.

Use a full overall or a boiler suit over your normal clothing. This has the obvious function of keeping you and your clothing relatively clean. Less obvious, but more important, the overall will offer less loose material which might otherwise become caught while working. You should be constantly aware of the risk of catching clothing, rings, wrist watches or long hair on projections or moving machinery. Sadly, even in these safety-conscious days, horrific accidents occur in machine shops and similar working environments, and all too often the accident will have been caused by one of the above items getting caught in a moving chuck or similar.

The home workshop may seem a relatively safe place, but a ring caught by the chuck of your bench drill will cause the same sort of injuries as one caught in a lathe. Always remove all jewellery and watches before starting work. Any metal ring, bracelet or watch strap can also cause electrical short-circuits if it accidentally bridges the gap between a live terminal and earth. Do not wear a tie while working (if you must do so, make sure that it cannot work free of your overall). If you have long hair, tie it back or use some sort of hairnet to keep it well away from danger.

Eye protection is of vital importance during many workshop procedures. Filing, drilling, sawing or grinding metal all produce swarf or fine particles which can cause serious injury if they enter the eyes. It is a sound precaution to wear safety glasses during this type of operation, but when carrying out any form of grinding the risk of eye injury is much greater due to the small size of the particles and their high speed. For this type of work, use approved grinding goggles with full side protection. Welding work of any sort requires more specialised eye protection in the form of masks or goggles designed to filter out the harmful ultra violet radiation produced, and this is discussed in Chapter 3.

Hand protection, in the form of barrier cream, should be used at all times. This will prevent dirt becoming ground into the skin, making it much easier to remove when the work is complete. More importantly, barrier creams are formulated to protect the skin from dermatitis, a condition often provoked by continuous contact with oils and similar products.

Physical hand protection is offered by gloves. Whilst it is almost impossible to manipulate hand tools with gloved hands, they are invaluable when lifting heavy, sharp-edged objects like engines, or when using wire brushes or similar tools. Industrial grade rubber gloves should be used during particularly dirty operations such as degreasing.

Other areas requiring specialist protection are the nose, mouth and lungs, and also the ears. To protect the respiratory system from solvents would require very expensive clothing and an air-fed helmet. Such equipment is used commercially when spraying certain types of paint which release cyanide during application. For home purposes, it is not likely that such dangerous products will be encountered, but even so precautions should be taken to avoid inhaling any volatile solvent product. Many solvents may simply make you feel sick if inhaled in any quantity, but some are more dangerous, especially over a long period. The best approach is to make sure that the workshop is well ventilated, and to minimise exposure to solvents and fuel vapour. If possible, work outside to allow solvent fumes to disperse quickly, and make sure that all solvent containers are kept closed when not in use.

You should also remember that the exhaust gases produced by motorcycle engines contain carbon monoxide, and this should be avoided by running the engine outside only, or by ensuring good ventilation of the work area. It is quite easy to position the machine so that the exhausts face an open doorway before the engine is started, if it must be run inside the workshop.

Dust is another problem in the workshop, and whilst much of it is simply irritating, dust containing asbestos is definitely dangerous if inhaled. The fine fibres of asbestos can produce a serious illness, asbestosis, and there have been connections made between the inhalation of asbestos and lung cancer. The best course of action is to wear a particle mask whenever working on asbestos-based materials, to work outside if possible, and to avoid working methods likely to raise dust.

Asbestos is used in friction materials such as brake linings and shoes, and also in clutch friction linings. Of these, the braking system is the most risky area to deal with, since there will be asbestos dust on all brake parts as a matter of course. The golden rule is **never** to use compressed air to clean brake parts. Use instead a rag soaked in a cleaning solvent to wipe away the dust, and then dispose of the rag immediately and safely.

The ears should be protected from excessive noise, or the hearing may be permanently damaged. This is not likely to be a common problem in the home workshop, since although some power tools are quite noisy in operation, it is unlikely that they will be used for long periods. Even so, ear defenders are inexpensive, and worth having just to make life more comfortable. A lengthy session using an angle grinder may not cause permanent deafness, but it can certainly cause dull hearing and a headache for a while.

Fire!

With all the care in the world, accidents will happen, and you should be prepared to deal with them when they do. Probably the greatest risk of all is fire, which even in the most modest workshop can be dangerous and very expensive. Sit down sometime and add up the value of your workshop, your tools and equipment and your motorcycle, then weigh that up against the cost of a reasonable quality fire extinguisher.

A good extinguisher need not cost the earth, but be wary of the small aerosol types sold in car accessory shops and filling stations. They are better than nothing, but are really too small to be much good in the event of a fire of any consequence. Find a reputable supplier and explain your requirements in details. Make sure that you get an extinguisher suitable for use on fuel and electrical fires – a water based extinguisher would be worse than useless.

Accidents and emergencies

These range from minor cuts and skinned knuckles to serious injuries requiring immediate specialist attention. The former are inevitable, whilst the latter are, hopefully, avoidable or at least uncommon. You should give some thought to what you would do in the event of an accident. Try to get yourself a little first aid training, and have an adequate first aid kit somewhere within easy reach.

Think about what you would do if you were badly hurt and incapacitated. Is there someone nearby who could be summoned quickly if the need arose? If possible, never work alone just in case something goes wrong.

If you had to cope with someone else's accident, would you know what to do? Dealing with accidents is a large and complex subject, and it is all too easy to make matters worse if you have no idea how to tackle the problem. Rather than attempt to deal with this subject in a superficial manner, I would suggest that at the very least you buy a good first aid book and read it carefully.

3 Cleaning and degreasing

This is a universally unpopular job which is necessary before commencing any engine overhaul or major repair job. It is an important part of the process, and any cut corners in this aspect of preparation will cause delays and extra work later. It follows that if the machine is cleaned thoroughly on a regular basis, the job will be that much easier than if dirt and grease are allowed to accumulate.

By far the easiest and least unpleasant method is to use a hot pressure washer. These machines deliver a high pressure stream of hot water, normally with a detergent mix, and this is played over the machine to blast away the mud and dirt. Until fairly recently, these machines were expensive and thus inaccessible to most motorcyclists, but there are now coin-operated machines in many filling stations. In addition, a range of reasonably-priced home machines are now on the market, and whilst these are not normally capable of supplying hot water, they will shift most of the dirt using cold water with a detergent.

Before using a pressure washer, make sure that vulnerable areas, like the carburettor air intakes and exposed electrical components are protected, or they will be swamped during the washing operation. It is a good idea to wear waterproofs while using the pressure washer, and an old riding suit is ideal for this purpose.

In the absence of pressure-washing facilities you will have to resign yourself to the traditional methods, using a proprietary degreaser and brushes. Wear overalls, industrial rubber gloves and some form of eye protection to guard against accidental splashes.

The degreaser should be applied according to the maker's directions, but as a general guide it is preferable to use a screwdriver or similar to dislodge any thick greasy deposits before work commences. Most degreasers work better if they are left for a while to penetrate the grease film, and all are most effective if the degreaser is worked in with a stiff brush to make sure that it has penetrated fully. Once the grease film has been broken down, the machine can be washed off as normal, carrying the dirt and grease with it.

Degreasers are available in liquid or aerosol form, the latter covering the machine in a foam blanket which gradually subsides as the grease film breaks down. Of the two, the aerosol versions seem marginally better, but are a good deal more expensive to use. Whichever type you use, keep it away from the tyres and braking system components, and cover any vulnerable electrical components or wiring connectors to prevent them being soaked when the machine is washed off.

After engine removal

Once preliminary cleaning has been carried out, the engine unit can be removed, should this be required. The operations described above will have shifted the majority of the dirt and grime, though there will be more cleaning to be carried out on previously inaccessible corners and recesses. Only after all external dirt has been removed should dismantling begin, or there is a risk of dirt entering the engine. If this happens it will be necessary to strip and clean the whole unit very carefully, or the residual dirt may cause accelerated wear or block oil passages.

During engine overhauls

If the engine unit is to be stripped for inspection or overhaul, there will be further cleaning to be carried out. The inside of the cases, and the surfaces of the internal parts will be coated in a film of oil, and this usually carries a certain amount of dirt, fine metal particles and carbon. All internal parts should be cleaned meticulously after dismantling has been completed. Clean one part or assembly at a time to minimise the risk of parts getting lost or interchanged.

The traditional method of cleaning internal parts has been to wash them off in a petrol (gasoline) bath, blowing the surfaces and passages dry with compressed air. This method works well and is inexpensive, but it does present a real fire hazard, as well as releasing hydrocarbons into the atmosphere. It is preferable, therefore, to use a conventional degreaser, washing each part off in water and then blowing it dry. Once clean, apply a thin film of clean engine oil to steel or iron surfaces to inhibit rusting (WD40 can be used to good effect here) and cover all parts with clean cloth until they are required during reassembly.

Using contact cleaner

This is a solvent cleaner, normally based on trichloroethane and is used for cleaning electrical switches, contacts and connections. It is a highly volatile product which washes out dirt or grease and then evaporates rapidly to leave a dry surface with no residual grease film.

Although primarily intended for use on electrical parts, it is very useful for any application where a part must be completely grease-free. A good example would be preparation of two surfaces to be glued together. Care must be taken when using this type of product on or near plastic surfaces, some of which may be damaged by it. It should only be used in a well-ventilated area, and care should be taken to avoid inhaling the vapour.

4 Chemicals and lubricants

In addition to specific products mentioned elsewhere in this book, you will find it necessary to build up a stock of general-purpose chemicals, cleaners and lubricants for day-to-day use. These are in addition to the oils and greases which may be specified for use on a particular machine. The range of products is vast, and those shown below represent only a selection of the more commonly-needed types. It will not be necessary to rush out and buy all of these; it is better to buy them as and when required so that you end up buying only those which you find necessary or helpful.

Cleaning products

Carburettor cleaner is used for removing the carbon and gum deposits from carburettor parts. A cheap alternative is cellulose paint thinner, which is fairly effective on carburettor bodies. Beware of using it on plastic parts, since these may be damaged.

Brake system cleaner is another specialised cleaning product for use on hydraulic brake parts. In the absence of this product, parts can be cleaned in clean brake fluid, or in some cases, alcohol. You should **never** use petrol (gasoline) or similar solvents on brake parts, or the seals will be damaged.

Electrical contact cleaner is a highly volatile trichloroethane-based solvent for cleaning electrical contacts and switches, which it leaves completely free of grease or residue. It is also very useful for cleaning other parts which need to be absolutely free of grease.

Maintenance sprays such as *WD40* and similar silicone-based products are invaluable general-purpose lubricants and dewatering fluids. They provide a very light, penetrating lubrication, and will displace water from switches and electrical connectors. Never be without this one.

Oils

Motor oil is used in the engine/transmission units of four-stroke motorcycles and in the transmission of two-stroke models. Motor oils are available in a wide range of grades, those used in current models being almost exclusively multigrade types.

Monograde motor oils are rarely required in current engine designs, being better suited to the ball and roller bearing engines for which they were originally designed. The current multigrade types are designed for use in modern engines, where the plain bearing surfaces require their lower viscosity combined with high load capacity.

The exact viscosity required will be specified by the manufacturer, SAE 10W/40 being a common choice. In this example, the oil has an initial viscosity of 10W, and is able to perform well even at low temperatures. At the same time, additives give the oil the load-bearing capabilities of a monograde SAE 40 oil, allowing it to maintain performance as the engine temperature rises.

In addition to the viscosity rating, many manufacturers will also specify a rating to which the oil must conform, commonly type SE or SF. Again, this will be specified by the motorcycle manufacturer, and these recommendations must be followed carefully.

Gear oils are normally found on shaft-drive models, where the bevel gears in the final drive system demand a lubricant capable of withstanding the high pressures and shear loads generated by the gear teeth. Gear oils are generally heavier than motor oils, typically around SAE 80 or SAE 90.

Ironically, few motorcycle gearboxes actually use gear oil. This is because the gearboxes on four-stroke models are normally built in unit with the engine, and so must share the same lubrication supply. This in turn means that the gearbox assembly must be designed to operate with engine oil, rather than gear oil. Another reason for the use of engine oils in the transmission system is the use of wet clutches; because the clutch runs in oil, it must be of low viscosity or clutch slip and drag would occur. This applies equally to two-stroke models.

Fork oil is a special light oil grade specially designed for use in telescopic forks. Most types contain additives to minimise foaming and consequent cavitation of the damping oil. Some manufacturers specify automatic transmission fluid (ATF) for use in their forks.

Most manufacturers specify a particular grade of fork oil, and these usually range from SAE 5W to SAE 20W. Whilst some oil manufacturers are able to offer a range of fork oils in varying grades, others are more vague and may not even mention the oil viscosity on the container. It is always worth ensuring that the correct grade is used, since this can have an appreciable effect on the damping characteristics of the fork. In some cases, especially on off-road models, altering the grade of fork oil can be used as an easy method of fine-tuning fork performance.

Greases

Multi-purpose high melting-point grease is used for numerous greasing applications on various parts of the motorcycle. These lithium-based greases will withstand fairly high temperatures, such as those generated in the brake hubs, without melting, and are usually chosen for use in wheel bearings for this reason. This grade of grease should be kept in the grease gun for regular maintenance purposes.

Graphite and molybdenum disulphide greases are specialised lubricants containing additives which will provide a degree of dry lubrication. These greases are often specified for use during engine assembly, where they provide initial protection and lubrication until the oil in the lubrication system begins to circulate. They are also useful in areas like spark plug threads, where the graphite or molybdenum is left behind to provide lubrication after the grease base has burnt off.

Silicone grease is a useful product for the wet-weather or off-road rider. Packed into wiring connectors and switches, it is extremely effective at excluding water, and a machine protected in this way will often work quite happily even if half submerged in water. It is not always easy to locate a supplier of silicone grease, but electronics shops are useful sources. It is also possible to obtain silicone grease from builders merchants where it is used to lubricate plastic drainage systems during assembly.

Heat sink grease is a special silicone-based product designed for use in some electronic components, and may be found packed around temperature sensors on some machines with automatic cold-start systems. As its name suggests, this grease is designed to help conduct heat away from the component which it is packed around. Like silicone grease, it may prove difficult to locate a source of supply. An electronics shop may be the best place to try.

Brake greases are specially-formulated products designed for use on hydraulic brake components. Conventional greases must **never** be used in braking system applications; the grease would attack and damage the brake seals.

Copper grease is an anti-seize and anti-squeal product. It can be applied to the backing metal of brake pads to reduce squealing, and is useful on fastener threads during assembly to prevent seizing. It is especially good on fasteners subject to high temperature, such as exhaust mounting nuts and studs. Like graphite and molybdenum greases, it is ideal for use on spark plug threads.

Special lubricants

Chain lubricants are available in aerosol spray form and also hot immersion products. In the case of the former, the lubricant is applied in dilute form along with a solvent. This mixture is designed to penetrate the chain rollers and bushes, the solvent evaporating to leave the lubricant film as a sticky residue. Some of these products incorporate polymers and other non-fling additives to help them adhere to the chain. Care must be taken when lubricating O-ring chains; you should only use a product which is marked to indicate that it is safe for use on these chains or the O-rings may be destroyed.

Immersion-type chain lubricants are usually sold in round containers designed so that the chain can be cleaned and then coiled up and placed in the container. The lubricant is then heated until it melts, the chain sinking into the lubricant which penetrates the rollers and bushes. The chain is then removed and hung up to allow residual lubricant to drain off. Once cooled, the bushes will be filled with the waxy lubricant.

Of the two types, aerosol sprays are far quicker and easier to use, and a lot less messy. Immersion lubricants probably provide better protection for the chain, but only if the process is carried out regularly. In the case of O-ring chains, use either a compatible aerosol product, or gear oil, because the hot immersion type will be unable to penetrate the rollers and bushes and will damage the O-rings.

Penetrating fluids are invaluable for helping in the task of freeing seized or rusted fasteners. The product should be applied to the affected fastener and allowed to work for as long as possible. Where the location and position of the seized fastener permits, the fluid can be contained around a fastener by forming a funnel from modelling clay around the area. Fill the funnel with the penetrating fluid and allow it to stand overnight if possible. If you can move the fastener even fractionally, penetration will be speeded up considerably. Products like *WD40* will act as a penetrating fluid to some extent, though *Plus Gas* and similar products are specifically designed for this job and usually work better.

Hydraulic fluids

Hydraulic fluids are used in the hydraulic systems used on disc brakes (and some clutches). Available in various grades and specifications, it is important that the fluid used is as specified by the manufacturer. Using non-specified fluids or other oils can cause rapid seal failure, a potentially lethal condition. Most manufacturers specify a fluid conforming to SAE J1703 (UK) or DOT 3 or 4 (US), but check this in your owner's handbook or workshop manual before purchasing new fluid.

A peculiarity of most hydraulic fluids is that they are hygroscopic, which means that they absorb moisture from the air. This can lead to internal corrosion of the braking system components, and will also lower the boiling point of the fluid. This can in turn allow the fluid to boil under heavy braking, causing rapid and unexpected brake fade. This is why most manufacturers specify that the fluid should be changed every 1 – 2 years. Note also that hydraulic fluid will mark or damage paint and plastic surfaces.

DOT 5, or synthetic hydraulic fluids avoid most of the normal drawbacks; they do not absorb water, and so the problems of corrosion and reduced boiling points do not occur. It may be possible to convert to a synthetic fluid, but note that this will normally require a complete overhaul of the system first. Before attempting the conversion, consult an authorised dealer for the machine concerned to check whether the conversion is advisable.

Other products

Oil additives are designed to enhance the protection or performance of engine or gear oils, and are often molybdenum-based. It should be noted that few oil or motorcycle manufacturers condone the use of these products, and you could invalidate your warranty by using them. On most motorcycles, a wet clutch arrangement is used. Any additive which makes the oil more slippery will thus do the same for the clutch – a generally undesirable state of affairs.

Fuel additives are often used to boost the octane rating of fuels and can be helpful in reducing pinking (knocking) and pre-ignition problems in high compression engines when using low-lead fuels. Other fuel additives are intended to reduce carbon and gum deposits in the fuel system or carburettor, or to act as an upper-cylinder lubricant.

Like oil additives, fuel additives are not normally recommended by manufacturers, and their use may invalidate any warranty. For normal road use, it is unlikely that the use of any additive is justified. This advice does not apply to fuel additives already present in fuel at the pumps; these can be considered safe to use since they have been extensively tested on a wide variety of vehicles to ensure that no harm will result.

Note: Unleaded fuel is currently being introduced in the UK to reduce the level of pollution from vehicle exhaust systems. In many cases, current models of motorcycle will run quite happily on low-lead or unleaded fuel, but in a few instances, serious engine damage could result. Before using fuels of this type, always check with the manufacturer, importer or dealer for your model.

Waxes and polishes are intended to protect the paint finish from oxidation and to help resist fading due to sunlight. Some of these are based on natural waxes, while more recent developments have seen the introduction of polymers and synthetic waxes. Many polishes contain a fine abrasive to remove oxidised paint, and must be used with discretion.

Cutting compounds and paint restorers are abrasive polishes intended to restore the surface finish of old and weathered paint by removing the top surface. These can be very effective in cleaning up old paint finishes, and will even remove light scratches. Be careful not to use them too often or too hard (especially on the edges of panels and fuel tanks) or you will rub right through the top coat to expose the primer or bare metal.

5 Sealants

The various covers and fittings on motorcycle engines are sealed to prevent the egress of oil and the ingress of dirt and water. This is normally performed by a gasket or an O-ring, and in many cases these are used dry. In other situations, the gasket must be sealed to the joint faces with some sort of sealing or jointing compound. In other applications, no gasket is used, and the metal faces are joined using a sealant alone.

Dry joints must be made in some areas, typically the cylinder head to cylinder barrel joint, where the very high temperatures would destroy a sealant. Many manufacturers use sophisticated (and expensive) composite gaskets, where a copper or aluminium gasket deals with the cylinder sealing and bonded rubber beads seal oilways. Other types include asbestos-based sheet material in place of the metal materials. It is important that the correct gasket is used in these situations, and no attempt at improvisation using home-made gaskets or liquid sealants should be contemplated. In certain applications, a specialist gasket will require the addition of a specific sealant at certain areas only. Where this applies, instructions will be found in the factory service manual or the appropriate Haynes Owners Workshop Manual.

Jointing compounds are used in conjunction with a gasket to form an oil-tight seal between two engine parts. Sometimes the manufacturer will specify a particular product for a given application, in which case you will need to purchase this through a dealer. Where no specific recommendation is given, a good quality general purpose compound should be used. *Hylomar* is a well established and effective compound which will withstand the effects of heat, oil and water, and is easily removed from gasket faces using cellulose thinner. There are numerous similar products available.

Silicone rubber 'Instant Gasket' products are effectively a room-temperature vulcanising (RTV) silicone compound which sets to a flexible rubber-like consistency. They are ideal for sealing outer covers on engines, and will take up small irregularities in the sealing faces. They can be used without a conventional gasket, provided that the gasket does not act as a spacing element, in which case some clearance will be lost.

Although RTV sealants are invaluable, they must be used with caution. **Never** apply more than a thin film, or the excess will be forced out of the joint. It is not unknown for detached strands of RTV sealant to find their way into the lubrication system, and a blocked oilway will soon lead to serious damage. This is not the fault of the product, rather one of misuse. Other problems will be encountered if the sealant is applied to areas subject to immersion in fuel, such as carburettors and fuel taps. The fuel attacks the sealant, turning it into a rubbery sludge.

RTV sealants are used for a wide variety of applications, and whilst those produced for sealing jobs on engines are good, similar products for other purposes do exactly the same job and are less expensive. There may be some difference in the two types, but there seem to be few if any disadvantages. Cheaper sources of RTV sealer include types produced for sealing bathtubs and kitchen worktops (in delightful pastel colours!), a particular favourite is any clear sealer, such as that produced for sealing aquariums.

Exhaust sealant is useful for preventing leakage from joints in the exhaust system. In the UK, *Holts Firegum* is a white paste which is applied to the joints during assembly. In use it hardens, forming a gas-tight joint. It is also possible to purchase repair pastes which can be applied to holes and small splits in the exhaust system. Where larger holes are encountered, exhaust bandages can be used to good effect, the fibrous bandage forming a strong repair after it has set hard under the effects of heat from the system. Note that repairs of this type must be considered temporary solutions, but they are worth considering if it is necessary to order a replacement silencer.

General purpose sealants may be needed for sealing items like windscreens on fairings or lamp lenses. Once again, transparent RTV aquarium sealer reigns supreme, though bear in mind that it is damaged by direct contact with fuel. It is worth noting that RTV has fairly good adhesive properties in some areas. It can be used for sealing and bonding in place loose instrument lenses and can even stick and seal headlamp reflectors to their lenses. Good results can be obtained by using RTV to waterproof electrical connectors, and similar techniques can be employed for projects such as converting an inexpensive car dashboard clock for use on a motorcycle by sealing the face of the LCD display to the case.

Removing gasket sealants

It is worth mentioning at this point the problem of removing old gasket compounds and sealants without risking damage to the gasket face. Injudicious use of scrapers can easily cause damage, and for this reason it is best never to use a scraper made from steel. Screwdrivers appear to make convenient scrapers, but try to resist the temptation to use them. Make up your own tool from a piece of aluminium or brass instead.

Many types of sealant can be removed by dissolving them, the type of solvent being dependent on the sealant encountered. In many cases, cellulose thinner will work quite well, whilst RTV-type products will be softened if soaked in petrol. It is also possible to obtain specialist solvents in aerosol form, and some of these will even soften residual areas of old gaskets which may still be stuck to the gasket faces. When using products of this type, be careful to follow the manufacturer's directions closely.

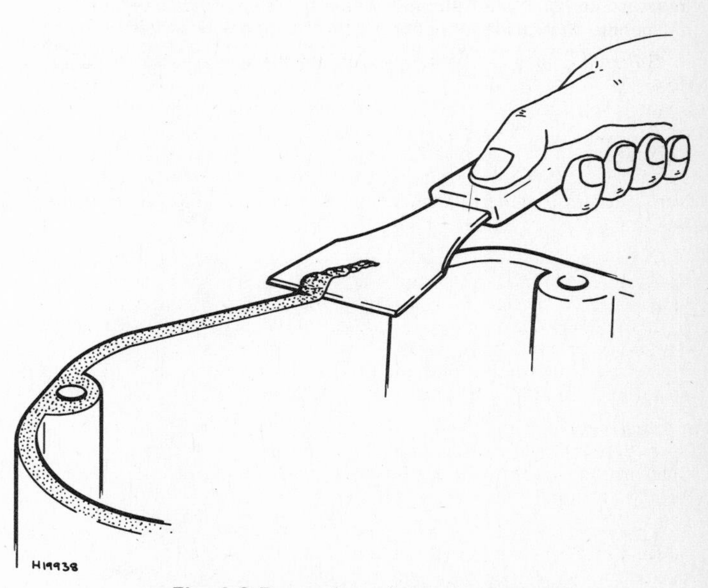

Fig. 1.2 Removing old gasket cement

Use a blunt scraper to remove old gaskets and gasket cement. Take care not to allow the scraper to dig into the gasket face

6 Thread locking compounds

There are a number of types of thread locking compound on the market, the best known being those produced by Loctite. These products are used to ensure that fasteners cannot loosen in service, and are often employed in place of lock washers or spring washers. In some instances, the manufacturer of a particular machine may specify the grade of locking compound to be used on specific fasteners, but for general use there are three basic types.

For small nuts and screws, a low strength compound will be needed. If a medium or high strength type is used it may prove impossible to release the fastener later. For general purpose use, a medium-strength compound is normally used. This gives good resistance to loosening, whilst permitting easy dismantling later. In specific applications, such as anchoring studs into casings, a high-strength product is used. This will make a semi-permanent fitting, and may prove difficult to shift at a later date, and so should never be used on ordinary nuts and bolts.

All of these locking products require the thread to be completely clean and free of grease. It is well worth using aerosol contact cleaner to ensure this. One or two drops of the compound are applied to the cleaned thread, and the fastener can then be fitted and tightened. The compound will begin to harden after it loses contact with air, sealing the threads together. An added advantage is that corrosion is effectively prevented once the fasteners have been assembled. It should be noted that all traces of old thread locking compound should be removed during subsequent overhauls.

7 Adhesives

You may have reason to require one of a number of adhesives during repair or overhaul work on motorcycles. Amongst these are impact adhesives, suitable for sticking trim or fixing seat covers to bases. Another general-purpose adhesive is one of the 'superglue' types, which is very useful for joining O-rings. This can be handy if you find yourself in need of a certain size of O-ring in an emergency.

The technique involves selecting a slightly larger O-ring than the one you need, and carefully cutting out a section using a scalpel or craft knife to give clean ends. These can then be secured with a drop of superglue. This practice is well proven, and you can even buy 'custom' O-ring kits, if required. Be wary of using shortened O-rings where sealing is vital, though, such areas being oilways, where loss of pressure could cause serious damage.

Epoxy resin adhesives are two-pack products which are mixed together to form a sticky paste. This is applied to the parts to be joined and allowed to harden. The curing time can often be reduced by warming the parts in an oven, assuming that this will not cause damage. Once hard, an extremely strong bond is formed, provided that the parts were cleaned carefully before the joint was made. Epoxy adhesives can be used on many non-flexible materials, and are often successful in sealing small cracks in castings.

A development of the epoxy adhesive are epoxy putties, which allow damaged metal to be built up. The hardened putty can be treated much like metal, and can be drilled and tapped, if necessary. Repairs made using this method are strong, but do not rely on them for structural repairs to engine or frame parts. They are, however, ideal for repairing cracks in engine covers or similar damage.

There are numerous other adhesives suitable for specific bonding applications, and the best advice with these is to assess their suitability as the demand arises.

Not strictly speaking adhesives, but related products, are the various products for repairing or restoring metal parts. These range from bearing fitting compounds; a sort of heavy-duty thread locking compound, to plastic metal compounds. Bearing fitting compounds are very useful where a bearing housing has become slack, allowing the bearing outer race to turn. This can occur after a bearing has seized and been forced to spin in the casing. The plastic metal compounds can be used to repair wear in keyways and similar areas where a good fit is essential and play has developed.

Chapter 2 Tools and tool usage

Contents

Introduction

Most motorcycle owners will know only too well that the toolkit supplied by the motorcycle manufacturer is, at best, just about adequate for basic routine maintenance. Anything more demanding than this will require a more comprehensive selection of hand and power tools, together with measuring and test equipment. In this Chapter we will be looking at some of the more commonly needed tools and their uses.

It is tempting to start with the recommendation that you should only ever purchase individual, high quality tools, gradually expanding your toolkit as need and finances dictate. Whilst this is a traditional and very sound approach to the subject, and the route normally suggested in books and magazine articles, it is difficult to follow in practice, given the high costs involved. For a start, a glance through most motorcyclists' tool collection reveals a very mixed assortment. There are the remains of numerous sets of spanners, for example. These can range from top-quality, lifetime warranty items, to cheap sets purchased on the spur of the moment from filling stations.

There seems to be an ironic law governing the contents of toolboxes which dictates that any expensive, well-made and indispensable tool will get lost or 'disappear' quite quickly, but during your all too short ownership of it, it will never break, slip or damage fasteners. Conversely, a cheap, ill-fitting and badly sized tool will be with you for life, even when you thought that you had thrown it away. It will never quite fit properly, and will probably drive you to distraction.

Whilst this may be something of a sweeping generalisation, it does seem to be what happens in the real world. I am sure that there are some very methodical and organised people out there who unfailingly clean and check each tool after use, before clipping it in place on the shadowboard on the workshop wall or placing it in its special compartment in the tool chest. Whilst there are few who practice this disciplined method of tool usage, there is no denying that this is the correct and professional approach to the matter.

There are also those to whom the concept of using the correct tool is entirely alien, and who will cheerfully tackle the most complex overhaul aided only by a handful of cheap open-ended spanners of entirely the wrong type, a single screwdriver with a worn tip, a large hammer and an adjustable spanner. This approach is undeniably wrong, barbaric and to be scorned, but whilst it often results in horribly damaged fasteners, it has to be admitted that for much of the time they seem to get away with it.

It is best to aim for a compromise between these two extremes, and like most mechanics, to cultivate a vision of the perfect workshop that is tempered somewhat by economic considerations. It is this which inevitably leads to a mixed selection of tools, and which seems to end up as the controlling factor in most workshops. The purpose of this Chapter is to give some idea in which areas top quality tools are essential, and also those where economies can safely be made.

2 Spanners and wrenches

When setting out to accumulate a useful range of spanners, it is comforting to consider that, with very few exceptions, it would be almost impossible to purchase tools of inferior quality to those originally supplied in the machine's toolkit. Having said this, all tools can be purchased in varying qualities, and with spanners or wrenches it is in some instances more important to insist on high quality tools. Your spanner set will be your most often-used workshop equipment, and as such demands careful consideration.

Before tackling the matter of selecting spanners, however, it is best to dwell briefly on the systems used in spanner sizing. A quick check through the contents of the toolkit supplied by the manufacturer will give a good indication of the most commonly-needed sizes, though there may well be significant omissions. Just about every machine produced in recent years will use metric fasteners, and in general the spanner jaw size, in millimetres (mm), will be denoted by a number stamped near each end of the tool.

In the case of metric tools, the number denotes the size of the fastener hexagon, measured across the flats, rather than the diameter of the fastener thread. As an example, a 6 mm bolt commonly used on Japanese machines will almost invariably have a 10 mm hexagon, and this will be the size of spanner required to tighten or loosen it. At the risk of confusing the issue, it should be mentioned that the relationship between thread diameter and hexagon size does not always hold true; in some applications, an unusually small hexagon may be used, either for reasons of limited space around the head, or to discourage over-tightening. Conversely, in specialised areas, fasteners with disproportionately large hexagons may sometimes be encountered.

On older British machines, a range of fastener types may be found. Early examples usually used Whitworth or BSF sizings, though the actual thread form varied according to application. With these, the size marked on the tool refers to the thread diameter, not to the hexagon size, and thus a tool marked 1/2 in WHIT 9/16 BSF would be found to have a hexagon almost an inch across the flats. During the 1960s there was a gradual change to Unified fasteners. These used spanners marked with the across-the-flats measurements, in fractions of an inch, eg 1/2 in AF, 5/16 in AF. The situation is complicated by the gradual transition between the two systems, with most manufacturers cheerfully mixing the two systems for some years.

To minimise confusion, the assumption has been made that most owners of British machines will have come to terms with this situation quite early on, so rather than over-complicate the matter, metric fasteners and tools only will be dealt with from now on. To those new to British machines apologies are offered for a rather brief mention of the intricacy of the Imperial sizing systems. Most fellow enthusiasts will be happy to explain in greater depth than is possible here, and you will certainly need their help in locating a supplier of Whitworth spanners.

Choosing tools of optimum quality

You may have noticed that the heading above says optimum quality and not top quality. This was intentional, because you will not always need top quality tools where they are used only infrequently. Conversely, where a tool is used often, it pays to get the best you can afford.

Spanners (wrenches) tend to look rather similar, and it can be difficult to know just by looking at them how well they are made. As with most other purchases, there are bargains to be had, just as there are over-priced tools sold on a well-known brand name. On the other hand, you may well buy what looks like a reasonable value set of spanners only to find that they fit badly or are made from a poor quality material.

With a little experience, it is possible to judge the quality of a tool by looking at it. Often, you may have come across the name before and will have a good idea of the quality of tools sold under that brand name. Close examination of the tool can often reveal some hints as to its quality. Prestige tools are usually polished and chromium plated over their entire surface, with the working faces ground to size. The polished finish is largely cosmetic, but it does make them easy to keep clean. Ground jaws normally indicate that the tool will fit well on the fastener.

A side-by-side comparison of a high quality spanner with a cheap equivalent will often prove illuminating. The better tool will be made from a good quality material, often a chrome-vanadium steel alloy. This, together with careful design allows the tool to be kept as small and compact as possible. If, by comparison, the cheap tool is thicker and heavier, especially around the working ends, it is usually because the extra material is needed to compensate for its indifferent quality. Given that the tool still fits properly, this is not necessarily a bad thing – it is after all cheaper – but on odd occasions where it is necessary to work in a confined area, the cheaper tool may be too bulky to fit.

The safest place to buy hand tools is from a specialist tool shop. You may not find cheap tools, but you should get a large range to choose from and also expert advice. Think carefully about your requirements before purchasing any item and explain what you want to the salesperson.

If you are unsure about how much use a tool will get, the following approach may help. If, for example, you need a set of combination spanners but are unsure of which sizes you will need, buy a cheap-to-medium priced set, making sure that the jaws at least fit the fastener sizes marked on them. After some use, examine each tool in the set to assess its condition. If all the tools fit well and are undamaged, you need not bother buying a better set. If one or two have become worn, replace them with individual high-quality items. In this way you will end up with top quality tools where they are most needed, the cheaper versions being sufficient for occasional use. On rare occasions it will become apparent that the whole set is of poor quality: in which case swallow your pride, buy a better set if necessary, and remember never to use that brand again.

Sources to steer clear of, at least until you get to know a little more about judging quality, are car accessory shops, mail order suppliers and street or market traders. Many of these sources offer good value for money, but a few will be selling unbranded or pattern tools of dubious quality. Tools, like any other consumer product, are often the target of counterfeiting in the Far East. The resulting tools could be quite acceptable, or on the other hand they might be unusable. Unfortunately, it can be hard to judge this by looking at them.

Finally, consider buying secondhand tools from garage sales or similar sources. You may have little choice of sizes, but you can usually gauge from the condition of the tools if they are worth buying. You will end up with a number of unwanted or duplicated tools this way, but it is a cheap way of getting a basic toolkit together, and you can always sell off any surplus tools later.

Open-ended spanners (crescent wrenches)

The open-ended spanner, or crescent wrench, is the most commonly encountered type of spanner, chiefly due to its general versatility. It normally comprises two open jaws connected by a flat handle section. The jaw sizes normally differ by one size, with an overlap of sizes between consecutive spanners within a set. This allows one spanner to be used to hold a bolt head while a similarly-sized nut is released. A typical metric spanner set might thus run in the following jaw sizes: 8 – 9 mm, 9 – 10 mm, 10 – 11 mm, 11 – 12 mm, 12 – 13 mm and so on.

Typically, the jaw end is set at an angle to the handle of the tool, and it is this feature which makes them particularly useful in a confined space; by turning the nut as far as the obstruction allows, then removing it and turning it over so that the jaw faces in the other direction, it is possible to move the fastener by a fraction of a turn at a time. The handle section length is generally determined by the size of the jaw, and is calculated to allow a nut to be tightened sufficiently by hand with minimal risk of breakage or thread damage (though this does not apply to soft materials like brass or aluminium alloy).

The common open-ended spanner of this type is usually sold in sets, and it is rarely worth buying these tools individually unless it is to replace a lost or broken tool from a previous set. Tools invariably cost more bought singly, so the best course of action is to check the hexagon sizes that you are most likely to need regularly, and to buy the best set of spanners that you can afford in that range of sizes. Where money is limited, remember that you will rely on your open-ended spanners more than any other type. It is preferable to make a point of buying a good set and making economies elsewhere.

You are unlikely to have much choice when selecting specialised open-ended spanners. These have the jaws set at unusual angles and are produced to allow the tool to be used in confined spaces, and are rarely sold in sets. You will find that only the better manufacturers produce these tools, so expect to pay quite high prices for individual tools. It is uncommon to need many specialist spanners on most current motorcycles, though special sockets and other tools may be required from time to time.

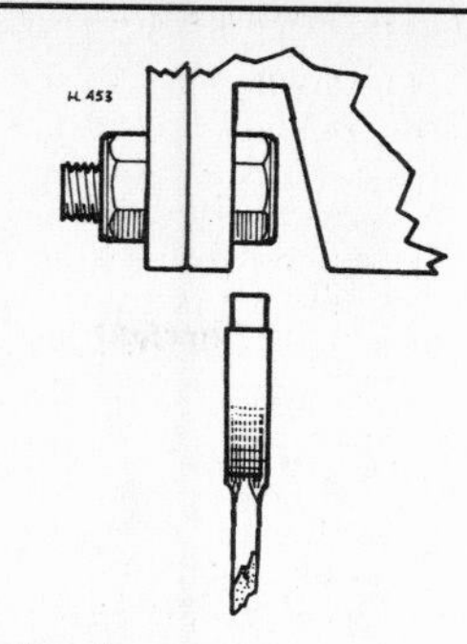

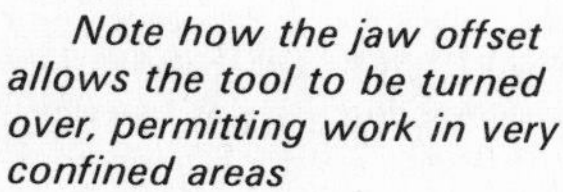

Note how the jaw offset allows the tool to be turned over, permitting work in very confined areas

Where heads are awkwardly recessed like this, only an open-ended spanner will fit

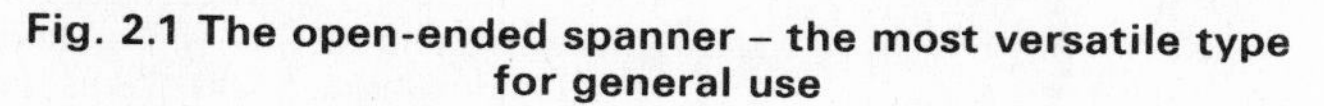

Fig. 2.1 The open-ended spanner – the most versatile type for general use

Ring spanners

The ring spanner comprises a ring-shaped working end inside which is formed a bi-hexagon. This allows the tool to fit onto the fastener hexagon at 30° intervals, rather than at 60° intervals as would be the case were a simple hexagon end used. Normally, each tool has two ring ends of differing sizes, allowing an overlapping range of sizes in a set, as described for the open-ended spanners.

Although available as simple, flat pattern, tools, most ring spanner sets are of the offset pattern. With these tools, the handle is cranked downwards at each end to allow it to clear obstructions near the fastener. Normally, this is an advantage, though on occasions the flat pattern may be needed. In addition to normal length ring spanners, it is also possible to buy long pattern types. These have elongated handles to permit more leverage, and are very useful when trying to shift rusted or seized nuts. It is, however, easy to shear nuts if care is not taken with these tools, and sometimes the extra length impairs access.

Specialised ring spanners are also available, and these can sometimes be invaluable when trying to reach awkwardly-positioned fasteners. Again, these tend to be available singly from specialist tool shops, and are quite expensive because they are made in limited quantities.

As with open-ended spanners, ring spanners are available in varying quality, and again this is often indicated by finish and the amount of metal around the ring end. Whilst the same judgements should be applied when selecting a set of ring spanners, if your budget is limited it is best to go for better quality open-ended spanners and a slightly cheaper set of rings.

Combination spanners

These tools combine a ring and open jaw of the same size in one tool, and offer many of the advantages of both. Like open-ended and cranked ring spanners, they are widely available in sets, and as such are probably a better choice than a set of flat ring spanners. They are generally compact, short-handled tools, and are well suited for motorcycle work where space is often restricted. The compact shape also makes them ideal as a supplementary or replacement spanner set for the motorcycle's tool roll.

Box spanners

Box spanners are essentially lengths of tubing with a hexagon formed on one or both ends. The tubing is drilled at one or two points across its diameter to allow a round bar, generally called a tommy bar, to be inserted to turn the spanner. This type of spanner was once very popular, but has largely been replaced by the socket set. You are most likely to encounter a box spanner in a machine's toolkit supplied as a spark plug spanner.

The main advantage of a box spanner over all other types is its ability to fit closely around a hexagon where it is deeply recessed. It does, however, need plenty of clearance above the fastener, and sufficient room to turn the tommy bar. Another important asset of the box spanner is its cheapness. Although no longer an essential purchase if you have a socket set, it is occasionally useful to have some in the workshop, and they can be an inexpensive alternative where a single, very large socket would otherwise have to be purchased for a one-off job.

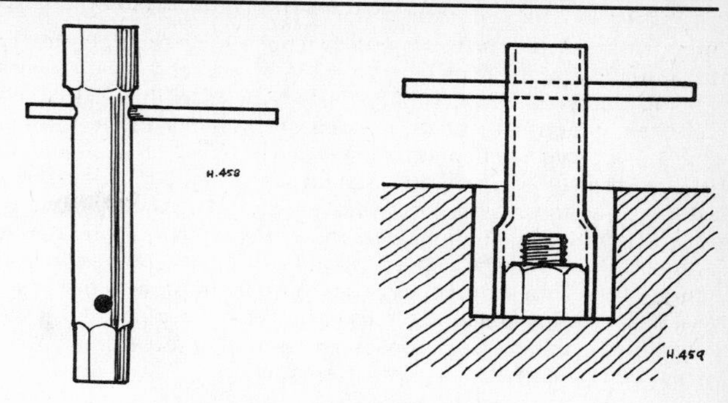

Fig. 2.2 The simple box spanner

An inexpensive alternative to sockets where money is limited. Very useful for unusual or large sizes where the cost of a socket might be prohibitive

Adjustable spanners

These tools come in innumerable shapes and sizes, and with various methods of operation. The principle is the same in each case; a single tool which can cope with fasteners of various sizes. It cannot be argued that any type of adjustable spanner is as good as a specially designed, single-size tool, and it is easy to wreak havoc on unsuspecting fasteners with them. That said, they are an invaluable addition to any toolkit, provided that they are used with discretion. Adjustable spanners are in their true element in a heavier duty world than that provided by the motorcycle. The smaller the fastener and the more restricted its location, the less likely it is that the adjustable spanner will fit. On odd occasions, though, the adjustable spanner is worth its weight in gold.

The most commonly-known type of adjustable spanner is the open-ended type in one of its numerous guises. Common to each is a set of parallel or curved jaws which can be set to fit across the flats of a range of hexagons. Most are controlled by a threaded spindle, though there are various cam and spring-loaded versions available. For motorcycle use, do not be tempted by large tools of this type; you will rarely be able to find enough clearance to use them.

A development of the adjustable spanner is the self-locking wrench, commonly known by its most popular brand names; *Mole Wrench* in the UK and *Vise-Grip* in the USA. Although this tool falls somewhere between being an adjustable spanner, a pair of pliers and a portable vice, it is in its fastener-shifting guise that we will consider it here.

The jaw opening is set by turning a knurled knob at the end of one handle. Placed over the head of the fastener, the handles are squeezed together until the lower jaw linkage goes over centre, locking the tool

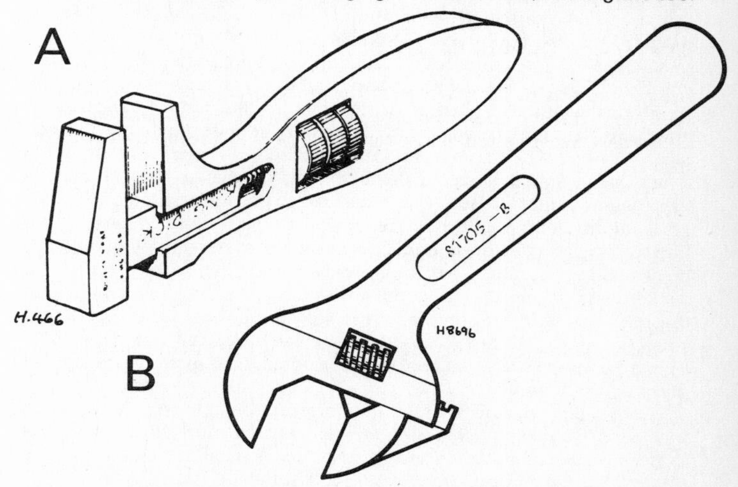

Fig. 2.3 Examples of adjustable spanners

The 'King Dick' type (A) is traditionally favoured by motorcyclists because of its robust and compact design. The more modern 'parrot-jaw' type (B) is more easily found in tool shops

onto the fastener. The design of the tool allows considerable pressure to be applied at the jaws, and a variety of jaw designs enables the tool to grip firmly even on damaged heads. It is in the role of removing fasteners which have been rounded off by badly-fitting spanners that the self-locking wrench excels.

A further variation on the adjustable spanner concept is the variable sized ring spanner. This comprises a semi-hexagonal ring end which is able to pivot on the handle. The end of the handle is formed into a jaw and protrudes into the ring opening. Placed over a hexagon within the range of the tool, turning force on the handle presses the jaw against the base of the hexagon, holding it firmly in place. Like the other adjustables, this type of tool is limited by its bulk and imprecise fit when compared with a single-size equivalent.

Socket sets

A more refined version of the box spanner, and owing much to the ring spanner, the socket consists of a forged steel alloy tube section with a bi-hexagon or hexagon formed inside one end. The other end is formed into the square drive recess which engages over the corresponding square end of the various driving tools.

Sockets are available in 1/4 in, 3/8 in, 1/2 in and 3/4 in drive sizes, the 1/2 in type being the most common. Of these, a 3/8 in drive system is most suitable for motorcycle use generally, though 1/4 in or 1/2 in drive sockets and accessories may occasionally be needed.

The most economical way to buy sockets is in a set, and as always, quality will govern the cost of the tools. Once again, the 'buy the best' approach is usually advised when selecting sockets. Whilst this is sound in as much as the end result is a set of quality tools which should last a lifetime, the cost is so high that it is difficult to justify the expense for home use. Over the years the author has bought three socket sets, all relatively inexpensive, medium quality brands. The three sets still exist, apart from a few sockets which have been lost or damaged due to misuse. A 3/8 in drive ratchet handle was once broken by over-stressing it while attempting to shift a seized bolt. When the cost of a new ratchet handle was checked, it was within a whisker of the cost of the complete combined 1/4 in and 3/8 in drive socket set which was bought instead!

Some of the author's most frequently used (and abused) sockets have worn out over the years. Despite the high cost, these were replaced with quality items by *Britool* or *Snap-on*. There is a noticeable difference in quality between these and the cheaper equivalents; they fit better and are finished to a high standard. On the other hand, they were so much more expensive that it would be difficult to justify the cost of the large range of sizes to match those of their older and cheaper equivalents.

Go shopping for a socket set and you will be confronted with a vast selection, so try to work out what you need before you set out. If you only ever work on metric fasteners, there is no point spending money on combined metric/AF sets, which many of them are. Decide what is the smallest hexagon fastener on your motorcycle (usually 6 mm or bigger) and also the largest (probably around 24 mm) and make sure that the set you buy covers this range comprehensively. Count anything bigger or smaller as a bonus, but don't let it sway your choice.

From the point of view of accessories, in the form of various bars and handles, you will need a ratchet handle, at least one extension bar, and a T-handle or tommy bar. Other desirable, though less essential items, are a speed brace, a flexible joint, extension bars of various lengths and adaptors from one drive size to another. Some of the sets you may find combine drive sizes; these are well worth having if you find the right set at a good price, but avoid being dazzled by quantity of pieces. Some combined-size sets may have a range of 1/4 in drive sockets and a screwdriver handle; very handy for running nuts onto their threads before tightening. It is also worth having an overlap in socket sizes. A 10 mm, 1/4 in drive socket is a good deal less bulky than a 10 mm, 3/8 in drive socket. Occasionally the extra clearance this gives you will be vital.

Above all, be sure that you ignore completely any label which proclaims '86 Piece Socket Set'; this refers to the number of pieces, not to the number of sockets (and in some cases even the metal box and plastic insert seem to be counted in this total!).

Apart from well-known and respected names, you will have to take a chance on the quality of the set you buy. If you know someone who has a set which has lasted well, try to find a set of that brand, if possible. Take a pocketful of nuts and bolts with you and check the fit of a random sample of the sockets. Check the operation of ratchet

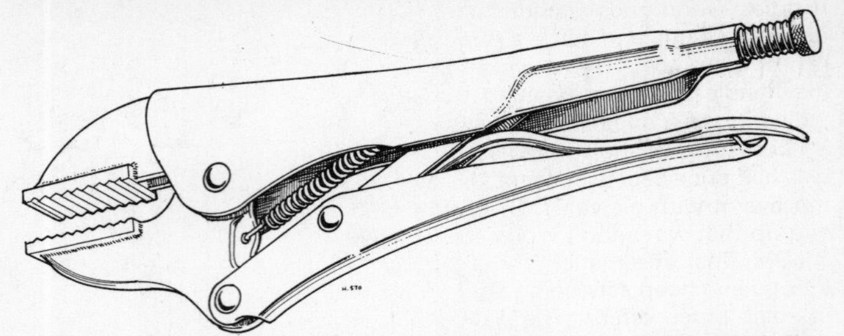

Fig. 2.4 Self-locking pliers ('Mole wrench' or 'Vise-grip')

Indispensible items, allowing fasteners or other parts to be held with the tool locked in place. Various shapes and sizes are available, a small set makes a worthwhile addition to the toolkit carried on the machine

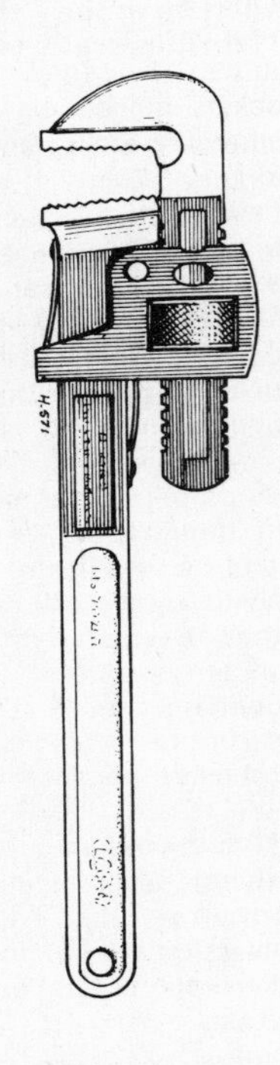

Fig. 2.5 Stillson type wrench

These can be invaluable as a last resort when trying to free a seized nut, but are likely to cause damage if misused. In many applications on motorcycles there will be insufficient room to use this type of tool

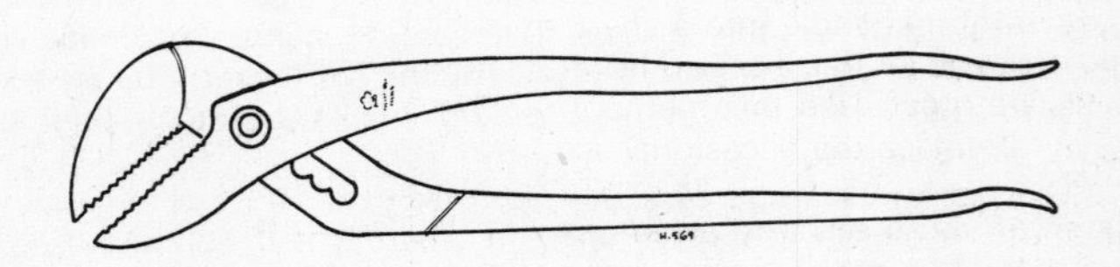

Fig. 2.6 Slip joint pliers, or pipe grips

These are used for general holding jobs. The linked pivot holes allow the tool to cope with a wide range of jaw openings

handles. Good ones ratchet smoothly and easily in small steps, cheap ones are coarse and stiff – a good basis for guessing the quality of the rest of the pieces. Try to avoid ratchets which are reversed by turning the handle over and pushing the driving square through – another indication of a cheap, low-quality set and irritating to use.

Look out for shop soiled sets; the last set purchased by the author was half price because the metal case was dented (the shop owner had run over it with his van!). Given that even the prettiest case will soon end up that way, don't worry too much about this aspect.

One final observation about quality. Some years ago there were a lot of very cheap sets being sold on market stalls and at auctions, and in mail order offers. The sets were impressively large, but used non-standard hexagonal drives and were of staggeringly poor quality. Having not seen any of these for sale for some time, it is to be hoped that they are no longer in circulation. If you should find one, keep well away from it.

Socket set accessories

One of the best things about a socket set is the facility for expansion. Once you have a basic set, you can purchase extra sockets where necessary, and you can also renew worn or damaged sockets. There are specialised sockets having extra deep sections to reach recessed fasteners, or to allow the socket to fit over a projecting thread. You can also buy screwdriver, Allen and Torx bits which will fit the various drive tools, and these can be handy for some applications. Odd pieces can often be picked up at little cost at autojumbles or garage sales, and these are worth considering if you remember to check that they are not worn out. Most socket sets include a special deep socket for 14 mm spark plugs. These have rubber inserts which protect the insulator, and which also hold the plug in the socket to avoid burnt fingers. Other spark plug hexagon sizes are available if needed.

Torque wrenches

Torque wrenches can be considered as complimentary to the socket set, since they require the addition of a socket so that a fastener can be tightened accurately to a specified figure. Modern machines are built to finer tolerances than used to be the case, and to attempt an engine overhaul without a torque wrench is to invite disaster in the form of oil leaks, distortion, damaged or stripped threads, or worse.

The cheapest form of torque wrench consists of a long handle which is designed to bend as pressure increases. At the drive end is fixed a long pointer, and this reads off against a scale near the handle. This type of torque wrench is simple and usually accurate enough for most jobs. A far more satisfying version is the pre-set type, in which the torque figure required is set up on a scale before use. The tool gives a positive indication, usually a loud click, when the desired setting is reached. Needless to say, the pre-set type is far more expensive than the simple beam type, and you alone can decide which you need. For very occasional use, it is probably best to go for the cheaper beam type.

Whichever type is chosen, think about your requirements before buying. Torque wrenches are available in a variety of ranges, to suit particular applications. For motorcycle use, the range of settings required is likely to be lower than those normally found on cars; on a typical machine these might run from around 1.0 kgf m (7.2 lbf ft) for small bolts and casing screws, up to about 10.0 kgf m (72.0 lbf ft) in the case of items like wheel spindles. Check the torque wrench settings applicable to your machine, and select a torque wrench which covers that range.

Impact drivers

The impact driver properly belongs amongst the screwdrivers, but is mentioned here since it can also accept sockets. As is explained elsewhere, the impact driver works by converting a hammer blow on the end of its handle into a sharp twisting movement. Whilst this is a great way to jar free a seized fastener, the loads imposed on the socket used are great. Use this method only with discretion, and expect to have to replace the occasional split or rounded socket.

Special spanners and wrenches

In addition to the various types of standard spanners and wrenches described in the earlier part of this section, there are also innumerable specialised tools designed to perform a particular job. Some of these are of limited general use, being useful on a range of different machines, whilst others are model-specific.

C-spanners, for example, will be needed on almost every machine. The C-spanner consists of a flat strip handle, the end of which is

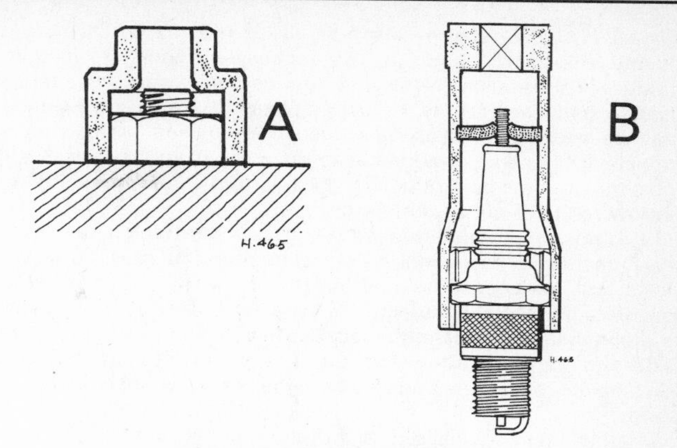

Fig. 2.7 Extended sockets

(A) A cross-sectioned view through a typical socket – note the limited clearance allowed for projecting threads.

(B) shows a special spark plug socket in cross-section. It is also possible to obtain similar deep sockets in other sizes, and these may be needed where a fastener is recessed or where a long projecting thread rules out a conventional socket

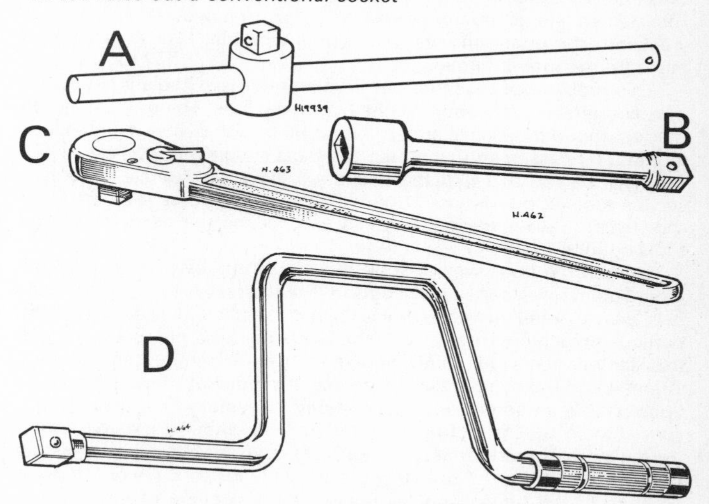

Fig. 2.8 A range of socket accessories

(A) shows a lever bar, the driving piece sliding along the lever bar. Extensions (B) are available in various lengths, and allow awkwardly placed fasteners to be reached. A ratchet handle (C) permits quick fitting and removal of fasteners. The type shown, with its small lever to reverse the ratchet, is preferable to cheaper types where the driving square must be pushed through and the tool turned over to reverse the drive. The brace (D) is less essential than the other accessories, but allows nuts to be 'run up' quickly

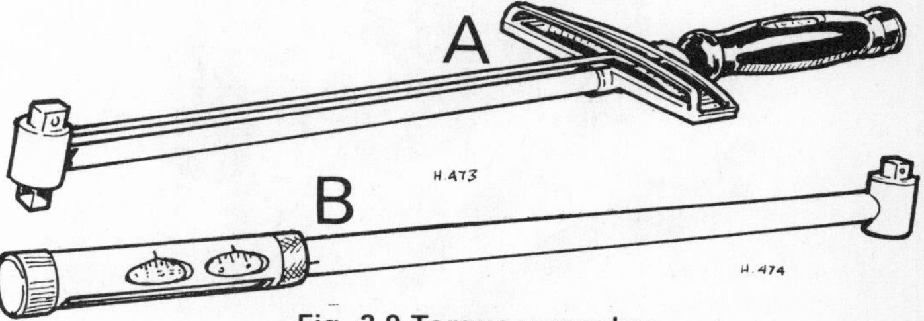

Fig. 2.9 Torque wrenches

Two types of torque wrench. The beam, or load indicating type (A) has a thin pointer which reads off against a scale as the main beam bends in use. The preset type (B) is set to the required torque and indicates when this point has been reached with a sharp click. This type of tool is much more expensive than the beam type, but easier to use in some situations

formed into a 'C' shape with a small hook at its end. The tool fits around a special slotted nut, the hooked end engaging in one of a number of slots machined around its edge. These nuts will normally be used to adjust and secure the steering head bearings, and a similar tool may be needed to adjust rear suspension spring preload on some models. It is not an easy matter to do without a C-spanner for these jobs, the problem being that any given size of C-spanner will fit only a narrow range of nut diameters.

It is possible to obtain adjustable C-spanners which have a joint in the C section to accommodate a greater range of sizes, but several of these will be needed to cope with all the nut sizes likely to be encountered. Most manufacturers list a set of service tools applicable to a particular model, and the local dealer for that make will be able to order the appropriate tool for you, if required. Alternatively, measure the diameter of the nut, and purchase that size of C-spanner from a tool shop.

Rather more specialised is the pin spanner, another flat-handled tool with pins at each end of crescent-shaped jaws. These tools are needed on a few machines where threaded retainers are used to secure wheel bearings. A variation of this design requires a special pin socket to deal with recessed slotted nuts. These are found on the clutch centres of many single-cylinder four-stroke Honda engines, for example. With these tools, it is doubtful if it is worth buying them, since it is unlikely that the average owner will need them more than once. On the other hand, it is almost impossible to do without them. Alternatives are to try to borrow the tool from your dealer over a weekend (if you are on very good terms with him!), or to attempt to make up the tool at home.

This problem of essential but infrequently-used tools is a difficult one, and affects just about every modern machine. The problem lies in the fact that motorcycles are no longer built with owner maintenance in mind. It is assumed that all but the most simple jobs will be carried out by a dealer, and that the dealer will have the full set of factory service tools at his disposal. Some machines are worse than others in this respect, but with very few can you escape the need for special tools entirely.

Later in the book we will look at ways of improvising some of the more commonly-needed service tools. For the really specialised pieces it is best to refer to the Haynes Owners Workshop Manual for the particular machine. Regular users of these manuals may have noticed that factory tools are rarely shown in use. When motorcycles are stripped and reassembled in the course of producing these manuals, a considerable amount of effort is directed to avoiding the need to buy special tools. Sets of factory tools are not kept in the Haynes workshop for the simple reason that it necessitates devising alternatives, just as the owner will have to do. Only occasionally is it necessary to advise purchasing the official tool, normally where damage or injury could result from improvised methods.

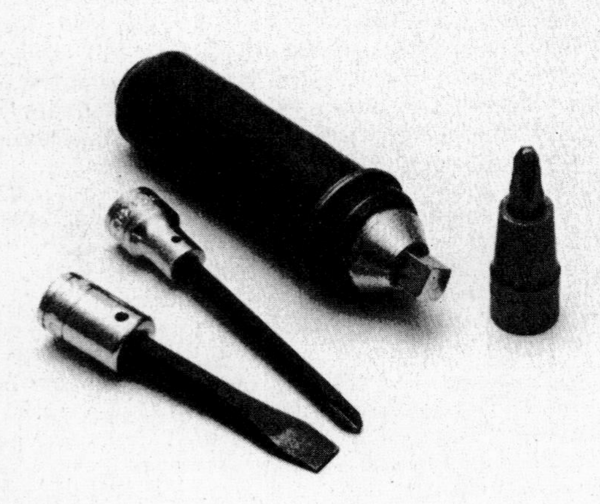

2.1 An impact driver set is indispensable when dealing with casing screws. This version shows extended bits needed to reach recessed fasteners. The square drive will also accept sockets

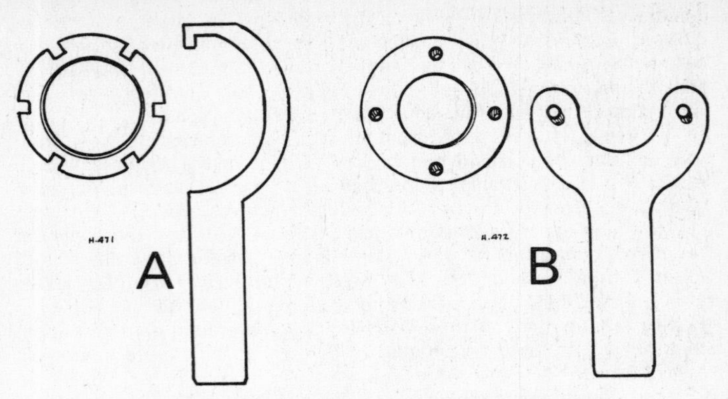

Fig. 2.10 C-spanner and pin spanner

(A) A C-spanner and the slotted nut on which it is used. These will be found on most steering head adjusters, and also on some suspension unit adjusters. (B) shows a pin spanner and a typical bearing retainer. If you come across this type of retainer, a suitable tool can be fabricated at home if required (see Fig. 5.3)

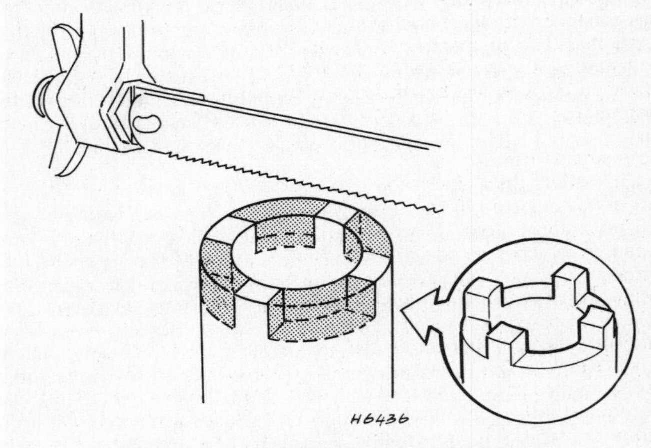

Fig. 2.11 A fabricated peg spanner

A simple peg spanner can be made up in the workshop using steel tubing of the required size. The shaded areas are sawn and filed out to leave the projecting tangs (inset)

2.2 C-spanners are needed to turn slotted adjusters such as those used on steering head bearings and suspension unit preload adjusters. This example is one of a set of adjustable tools; the hinged jaw allows each tool to fit a small range of sizes

2.3 Threaded retainers require a pin spanner to remove them. A tool can easily be fabricated as shown

2.4 This roughly-finished spanner was made up by sawing and filing tangs at one end of a length of tubing

2.5 Despite appearances, simple tools of this type can often replace the official version

3 Using spanners and wrenches

In the last section we looked at some of the vast range of spanners and wrenches available, with a few suggestions about building up a tool collection at reasonable cost. Here we are more concerned about using the tools in practice. Although you may feel that this is self explanatory, it is worth giving the whole matter a little thought; after all, when did you last see instructions for use supplied with a set of spanners?

How was it made?

First of all, think about the machine you will be working on. The way it was built will affect the way you should approach maintenance and repair operations. Most likely, the bike will have been built in Japan, in which case one of the governing factors in its design is the method by which it was assembled. You will probably have read of the highly mechanised factories in which these machines roll off production lines. This means that the bike has been designed to suit this method of production, as distinct from hand assembly. One result of this is that there are few fasteners on recent models which cannot be reached with a socket; fasteners which are positioned in such a way as to require individual hand assembly with spanners means that torque-tightening is not possible, and that production times are increased.

With modern European machines, this automated approach is also evident, but to a lesser extent. Here, major assemblies may have been built up on production lines, but the final assembly of the machine will have been carried out on an individual or batch basis. This means that many, but not all of the fasteners are accessible with sockets.

By way of contrast, British machines were produced using methods far removed from modern assembly techniques. Even the later models produced owed their original ancestry to a machine designed decades earlier, in an era when it was necessary to be able to overhaul the machine at home using simple hand tools. The machines were more or less hand assembled using similar tools, and as a result many of the fasteners are positioned such that it is impossible to get near them with a socket or a torque wrench.

The point of the above is to emphasise the need to give a little consideration to the way the motorcycle in question was originally put together. For most practical purposes, working on a modern motorcycle requires the owner to emulate production line assembly tools as far as possible. In some respects this may make odd components seem awkward to reach or difficult by virtue of the need for a special tool, but in most respects it results in a uniform approach to most tasks.

Which spanner?

Before you start tearing your motorcycle apart, it is a good idea to form a clear idea of the best tool for the job; in this instance, the best spanner or wrench for a hexagon-headed fastener. It is instructive to sit down with a few nuts and bolts and your tools, and to look at how the various tools fit the heads.

A golden rule is to choose the tool which contacts the maximum amount of the hexagon. This distributes the load as evenly as possible, and lessens the risk of damage. The shape most closely resembling that of the bolt head or nut is another hexagon. It follows that a well-fitting hexagonal socket (or box spanner) is the best choice of all. Generally though, sockets, like ring spanners, have bi-hexagonal working surfaces. If you fit a ring spanner over a nut, look at how and where the two are in contact. The corners of the nut engage in alternate corners of spanner head. When the spanner is turned, pressure is applied evenly on each of the six corners. This is fine unless the fastener head was rounded off on a previous occasion. If this has happened, the corners will be damaged and missing, and the spanner will slip. So if you encounter a damaged head, try to use a hexagon head spanner or socket if possible. Failing this, choose the spanner which fits most securely and proceed with care.

If you fit an open ended spanner over a hexagonal head you will see that the tool is in contact on two faces only. This is acceptable provided that the tool and the head are in good condition. It is the necessity of a good fit between the spanner and hexagon which explains the recommendation to buy good quality open ended spanners, even if other tools are of a lower quality. If either the spanner jaws, the bolt head or both are damaged, the spanner will probably slip, rounding and distorting the head. In some applications, the open ended spanner is the only choice due to limited access, but always be careful to check the fit of the spanner on the head before attempting to shift it; if it is hard to get at with a spanner, think how hard it will be to remove it after the head is damaged.

The last resort of all is the adjustable spanner or self-locking wrench. Use these tools only where all else has failed. In some cases, a self-locking wrench may be able to grip a damaged head that no spanner could deal with, but be careful not to make matters worse by damaging it further.

Using spanners, wrenches and sockets

Bearing in mind the remarks above about the correct choice of tool in the first place, there are several points worth noting about the actual use of the tool. First, check that the hexagon head is clean and undamaged. If the fastener has rusted or is coated in paint, the spanner will not fit correctly. Clean off the head, and if rusted, apply some penetrating oil or a proprietary releasing fluid like *Plus Gas,* leaving it to soak in for a while before attempting removal.

It may seem an obvious thing to say, but have a close look at the fastener to be removed before wielding a spanner. On most mass-produced machines, one end of the fastening is fixed or captive. This speeds up initial assembly, and *usually* makes subsequent removal simpler. Where a nut is fitted to a stud, or a bolt screws into a captive nut or tapped hole, you have only one moving component to contend with. If, on the other hand, you have a separate nut and bolt, you must make provision to hold the bolt head whilst the nut is released. In some areas this can be quite difficult, particularly where engine mountings and the like are involved. In this sort of situation you will need an assistant to hold the bolt head with, say, a ring spanner, while you remove the nut from the other side. If this is not possible you will have to try to fit a ring spanner so that it lodges against the frame or engine to prevent it from turning.

Be on the lookout for left-hand threads. These are not common, but are sometimes used on the ends of rotating shafts to make sure that the nut cannot come loose in use. If you can see the shaft end, the thread direction is easily checked by sight. If you are unsure, place your thumbnail in the thread and work out which way you need to turn your hand so that your nail 'unscrews' from the shaft. (It is preferable that you do this without an audience, who might begin to question your sanity if they witness strange gyrations coupled with a puzzled expression!) If you have to turn your hand anticlockwise (counterclockwise) it is a conventional right-hand thread.

Beware of the upside-down fastener syndrome. If you are working on a bolt on the bottom of the frame, for example, it is easy to get confused about which way to turn it. What seems like anticlockwise to you can easily be clockwise, from the bolt's point of view. However long you work with nuts and bolts, this can still catch you out once in a while.

Watch out for semi-captive nuts or nut plates. These are sometimes used on engine mounting points, and you should start to suspect them when a half-removed bolt can be pushed in and out. Check what it is screwed into, and take care to catch it if it drops free. Take even greater care if it happens to be inside the spine frame of a step-thru moped or similar machine, or it may drop down into a dark and inaccessible corner when it comes free.

In the majority of cases, the fastener can be removed simply by placing the spanner on the head and turning it. Occasionally, though, the condition or location of the fastener may make things more difficult. Check that the spanner is fitted squarely over the head. You may need to reposition the tool or try another type to achieve a good fit. Make sure that the assembly you are working on is secure and unlikely to move when you turn the spanner. If necessary, get someone to help steady the workpiece for you. Position yourself so that you can get maximum leverage on the spanner, wherever possible.

If you can, arrange the spanner so that you pull the end towards you. If the situation makes this impossible and you have to push the spanner end towards the item being worked on, remember that it may slip, or the fastener may move suddenly. For this reason do not curl the fingers around the spanner handle or you may crush them when the fastener moves; keep the hand flat, pushing on the spanner with the heel of the thumb. If the tool digs into your hand, place some rag between it and your hand, or better still, wear a heavy glove.

If you fail to move the fastener with normal hand pressure, it is essential that you reappraise the problem before the fastener, spanner or you get damaged. The art of releasing a seized or damaged fastener demands more discussion than is appropriate here, and is dealt with in greater depth in Chapter 4.

Using sockets to remove hexagon headed fasteners is generally straightforward, and if used correctly, is less likely to result in damage than is the case with spanners. Check that the socket fits snugly over the fastener head, then fit an extension bar if needed, and the lever bar (tommy bar). By rights, a ratchet handle should not be used for freeing a fastener or for final tightening because the ratchet mechanism may be overloaded and could slip. In some instances, the location of the fastener may mean that you have no choice but to resort to the ratchet handle, in which case take care to avoid injury in the event of the ratchet failing.

Never use extension bars where they are not needed, and whether or not an extension is used, always support the driving end of the lever bar, **not** the socket, with one hand whilst turning it with the other. Once the fastener is loosened, the ratchet handle can be fitted to speed up removal.

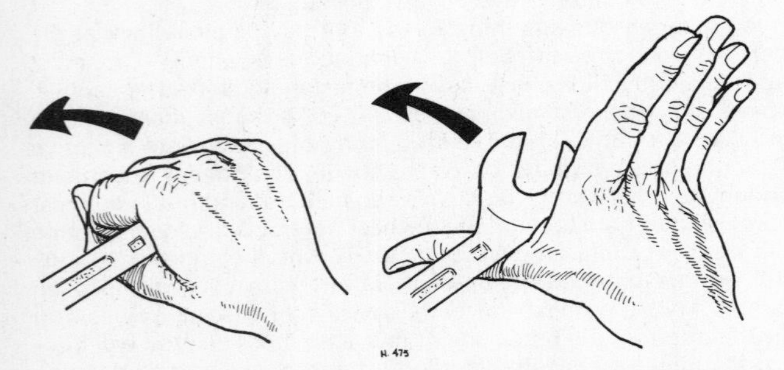

Fig. 2.12 Correct method of spanner usage

The right and wrong way to use a spanner. In the first example, knuckles will get damaged if the tool slips or if the nut loosens off suddenly. Using the tool as shown on the right avoids this risk

4 Screwdrivers

Screwdrivers come in innumerable shapes and sizes to suit the various head designs in use. However well made they may once have been, the screwdrivers in most toolboxes rarely fit their intended screw heads well. This is attributable not only to general wear, but also to misuse. Screwdrivers make very tempting and convenient tommy bars, chisels and drifts, a practice which in turn makes them very bad screwdrivers. The exception to this is the universal screwdriver handle with the reversible blade supplied in the toolkit with most new machines. This will have been a bad screwdriver to start with, and using it on casing screws will almost invariably ruin them.

The screwdriver comprises a steel blade or shank with the driving tip formed at one end. To the other end is attached the handle. Traditionally these were made from wood and secured to the shank by a metal ferrule. More usually, a plastic handle is moulded to the shank, which has raised tangs to prevent it from turning in the handle.

The design and size of both the handle and blade vary considerably. The oval section cabinet handle types are really intended for carpentry work, whilst for motorcycle maintenance, a cylindrical engineer's handle design is normally used. In practice, either handle type will suffice, though a bulky handle may make access difficult on occasions. The shank diameter, tip size and overall length vary too. As a general rule it is advantageous to use the longest screwdriver possible, this allowing greater leverage to be obtained.

Where access is seriously restricted there are a number of special screwdriver types designed to fit into confined spaces. The short dumpy screwdriver has a specially shortened handle and blade. Alternatively, there are cranked screwdrivers, or special screwdriver bits which attach to a socket ratchet handle.

Various types of ratchet screwdriver can be bought, these being designed to speed up the fitting and removal of screws. Whether or not you like these tools is a matter of personal preference, but bear in mind that they are inclined to be somewhat bulky when compared with equivalent conventional versions.

Flat blade screwdrivers

These are used to remove and fit conventional slotted screws, and are available in a wide range of sizes. You will encounter only a few of these screws on most Japanese machines, cross-head screws being preferred for most fastening applications. On European models, a mixture of slotted, cross-head and Allen headed screws will be found.

It is important to have a selection of screwdrivers so that screws of various sizes can be dealt with without damage. The blade end must be of the same width and thickness as the slot to work properly, and without slipping. When selecting flat blade screwdrivers, choose those of good quality, preferably with chrome molybdenum shanks. The tip of the shank should be ground to a parallel, flat profile (hollow ground) and not to a taper or wedge shape which will twist out of the slot when pressure is applied.

All screwdrivers wear in use, but the flat blade type has the advantage that it can be re-ground to shape a number of times, unlike cross-head types which must be renewed when worn. When doing so, start by grinding the end of the tip dead flat at right angles to the shank. Make sure that the reprofiled tip fits snugly in the slot of a screw of the appropriate size, and take care to keep the sides of the tip parallel. Remove only a small amount of metal at a time to avoid heating the tip and destroying the temper of the steel.

Misuse of a screwdriver – the blade shown is both too narrow and too thin and will probably slip or break off

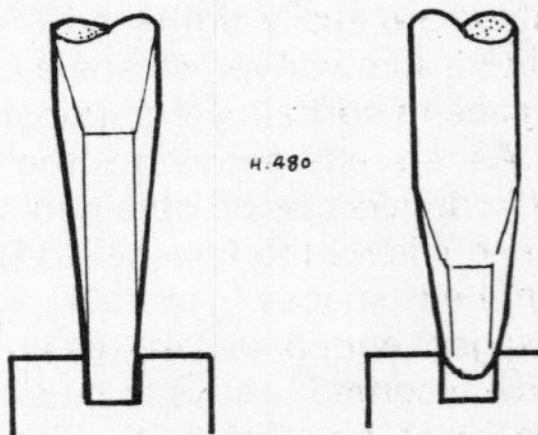

The left-hand example shows a snug-fitting tip. The right-hand drawing shows a damaged tip which will twist out of the slot when pressure is applied

Fig. 2.13 Use of the flat-bladed screwdriver

Cross-head screwdrivers

Most Japanese machines make extensive use of cross-head screws on casings and covers. These are installed during assembly with air tools, and are always next to impossible to remove later, without ruining the heads, particularly where the wrong size of screwdriver tip is used. It is essential that only Phillips type screwdrivers are used on these screws; other cross-head patterns being of a different design will succeed only in damaging the head and the screwdriver tip. In the UK, an improved design known as Posidriv is now well established as a more reliable head recess form. Posidriv pattern screws can be identified by small lines at 45° to the driving slots, and will not be found in motorcycle applications. Be careful to check that any screwdriver you buy is Phillips, rather than Posidrive pattern.

The safest way to ensure that the screwdrivers you buy really do fit the screws you need them for, is to take a couple of screws with you on screwdriver-buying trips to make sure that the fit between the screwdriver and screw is snug. If the fit is good, you should be able to angle the blade almost vertically downward without the screw dropping off the tip. I cannot stress too highly the need to use **only** those screwdrivers which fit exactly. Anything else is guaranteed to chew out the head instantly.

The idea of all cross-head screw designs is to make the screw and screwdriver blade self-aligning. Provided that you aim the blade at the centre of the screw head it will engage correctly, unlike conventional slotted screws which need careful alignment. This makes the screws suitable for machine fitting. The drawback with these screws is that the driving tangs of the screwdriver blade are very small, and thus must fit very precisely in the screw head. If this is not the case, the huge loads imposed on small parts of the screw slot simply tear the metal away, at which point the screw ceases to be removable by civilised methods. The problem is made worse by the strangely soft material chosen for the screws.

To deal with these screws on a regular basis, you will need to equip yourself with high quality screwdrivers of exactly the right profile. It is possible to obtain T-handled screwdrivers, and these are very helpful in this respect. Using the T-handle, pressure can be applied to the screw head to discourage it from throwing the blade out of the head recess.

Particularly good cross-head screwdrivers have full-length hardened steel shanks which emerge at the top of the handle in a mushroom head. By applying a little pressure to the handle, twisting it, and simultaneously striking the end with a hammer, the screw can usually be shocked loose without damage to the head. This approach should only be used with screwdrivers designed for this type of use – if you strike the handle of an ordinary screwdriver, in which the shank extends only partly through the handle, you will shatter it.

Socket and Torx screw keys and screwdrivers

Hexagonal socket screws, commonly referred to by the trade name Allen screws, are widely used on European machines and on a few Japanese models. These fasteners have a cylindrical head in which is formed a hexagonal socket. The design of the screw is inherently strong, and unless made of very poor quality material it is almost impossible to damage the head or shear the screw off through over tightening. The screws are normally turned using one of a set of hexagonal steel keys, and these should break before the screw does.

Each key is made in a size which suits a particular diameter of screw, and this combined with the length of the key effectively limits the amount of torque which can be applied by hand. Even so, socket screws are easily tightened to greater torque settings than is possible using a screwdriver, so some care must be taken to avoid stripping the threads in soft alloy components.

As an alternative to the simple keys, sets of hexagon-ended screwdrivers can be obtained. Some types even have a sort of ball end, which allows the fastener to be turned from a slight angle – useful in confined spaces. Another alternative is to get a set of special hexagon-ended sockets, as an add-on to the normal socket set. These allow recessed screws to be reached easily, and will also allow a torque wrench to be used on the screws if necessary.

Whether these expensive alternatives to a set of keys is warranted depends on the frequency with which you are likely to need them. For occasional use, perhaps to reach a recessed screw head, you might consider cutting off a straight length from a normal key, using this held in a normal socket to allow access.

If you own an Italian machine you will find that socket screws are used almost exclusively, in which case a set of socket screwdrivers may be a worthwhile purchase, but keep a set of simple keys in the toolkit on the machine; they do the same job and take up far less space.

The Torx screw (another trade name) is a development of the hexagon socket screw, probably best described as a stellated hexagon screw. The unique design of the heads of these screws means that if you encounter them, a set of special keys will be needed. Fortunately, although widely used on cars, they are quite uncommon on motorcycles, though BMW have recently taken to using them quite widely.

In an emergency, you can often adapt a conventional hexagon key to suit a Torx head by filing grooves in the flat faces as shown in the accompanying illustration. It must be stressed that this will not give as good a fit as the correct key, and if you own a machine which uses a number of these fasteners, a set of the correct Torx keys should be considered essential.

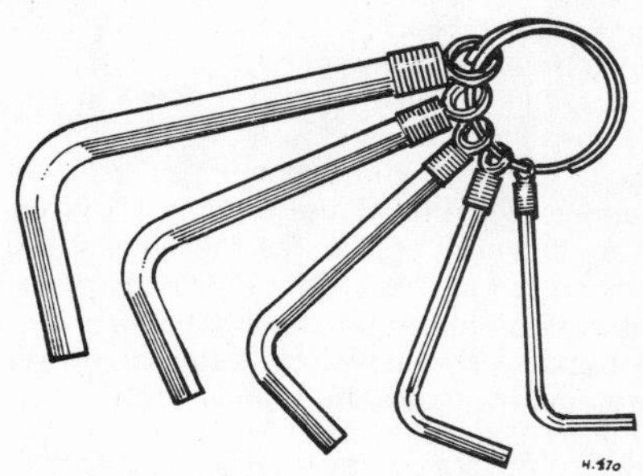

Fig. 2.14 A typical hexagon wrench (Allen key) set

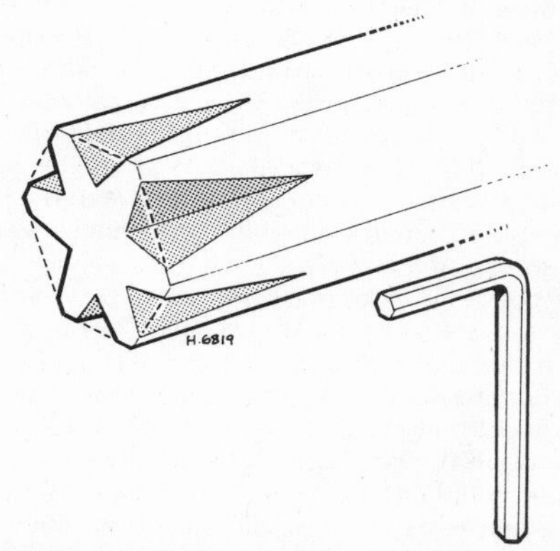

Fig. 2.15 In an emergency it is possible to modify a hexagon wrench to fit a Torx screw

4.1 Hexagon socket (Allen) screws are turned using a set of hexagon keys (left). It is also possible to buy special hexagon wrench sockets (right) and these are worth buying in frequently-used sizes

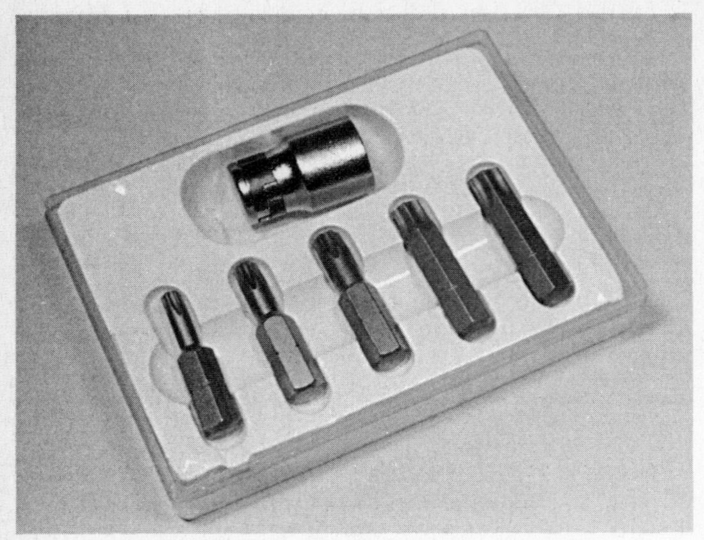

4.2 Torx screws may be found on some models, and require a set of special keys or sockets. This set will fit either an impact driver, or using the adaptor supplied, a normal socket set

5 Hacksaws

A hacksaw comprises a handle to which is attached a frame supporting the flexible steel blade under tension. Blades are available in various lengths, and most hacksaws can be adjusted to accommodate the different sizes. The most common blade length is 12 inches.

Most hacksaw frames are suitable for use in the workshop, and being simple tools, there is little to choose between rival makes. Try to pick one which is rigid when assembled, and which allows blades to be changed or repositioned easily.

The grade of the blade to be used, described by the number of teeth per inch (TPI), is determined by the material to be cut. As a rule of thumb, always make sure that at least three teeth are in contact with the metal being cut at any one time. In practice this dictates a fine blade for cutting thin sheet materials, whilst a coarser blade can be used for faster cutting through thicker items such as bolts or bar materials. It is worth noting that when cutting thin materials it is useful to angle the saw so that the blade cuts at a shallow angle. In this way more teeth are in contact and there is less chance of the blade snagging and breaking or teeth being broken off the blade. This approach can also be used when a sufficiently fine blade is not available; the shallower the angle, the more teeth are presented to the workpiece.

When buying blades it is best to choose a reputable brand. Cheap, unbranded blades may be perfectly acceptable, but you cannot tell this by looking at them. Poor quality blades will be insufficiently hardened on the teeth edges, and will cut very badly. Most reputable brands will be marked 'Flexible High Speed Steel' or something similar to this, giving some indication of the material used in its manufacture. It is possible to buy 'unbreakable' blades in which only the teeth are hardened, leaving the rest of the blade less brittle.

In some applications a full-sized hacksaw will be too big to allow access. Sometimes this can be overcome by turning the blade at 90° to the workpiece, and most saws allow this to be done. Occasionally you may have to position the saw around an obstacle and then fit the blade on the other side of it. Where space is really restricted, you may have to use a junior hacksaw, which as its name suggests is a scaled-down version of the normal hacksaw which uses miniature blades. A further useful tool is a pad handle which allows a hacksaw blade to be clamped into it at one end. This allows access to awkward corners, and has another advantage in that you can make use of broken off hacksaw blades instead of throwing them away. Note that because only one end of the blade is supported, and that it is not held under tension, it is difficult to control and less efficient at cutting than a proper saw.

Before using a hacksaw, check that the blade is of a suitable grade for the proposed job, and that it is fitted correctly in the frame. Note that the saw cuts on the forward stroke, and so the points of the teeth must face away from the handle end. This might seem obvious, but it is easy to fit the blade the wrong way round by mistake. Check also that the blade is held under tension in the frame; if you omit to tighten it correctly it will whip and chatter in the cut, and will probably break. Use safety glasses or goggles while sawing; the blades are usually very brittle and can shatter unexpectedly, and there is always a risk of metal filings entering the eyes.

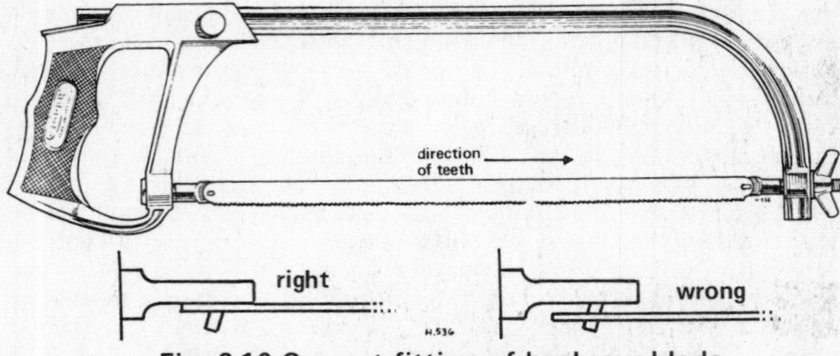

Fig. 2.16 Correct fitting of hacksaw blade

Blades must be fitted in the hacksaw frame with the teeth facing forwards. Check that the side of the blade buts against the locating lugs

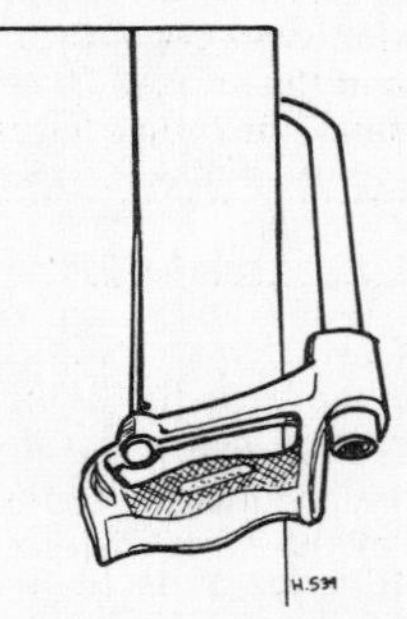

Fig. 2.17 For long cuts, turn the blade through 90° as shown

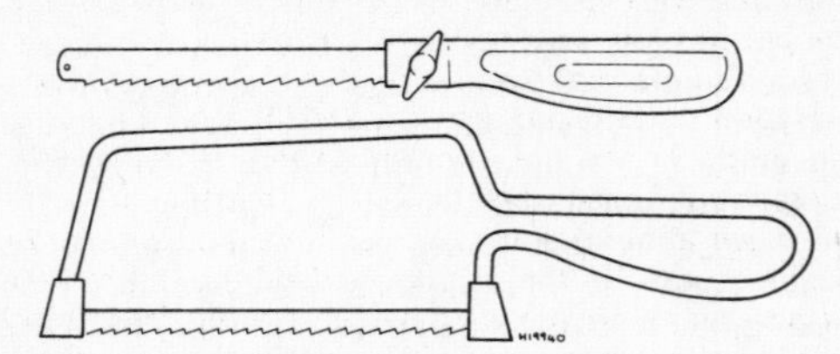

Fig. 2.18 In confined spaces, use a broken-off hacksaw blade clamped in a handle (top) or if space permits, a junior hacksaw (bottom)

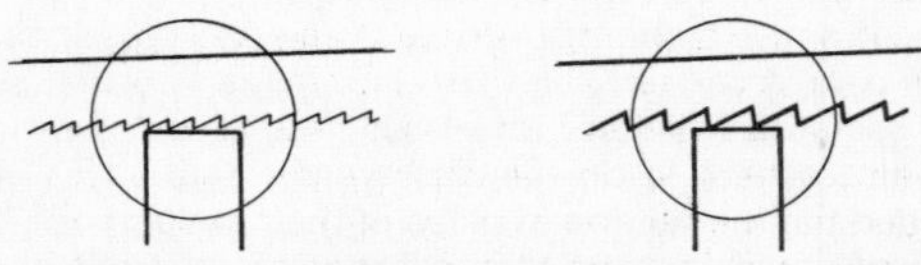

When cutting thin materials, check that at least three teeth are in contact with the workpiece at any time. Too coarse a blade will result in a poor cut and may break the blade. If you do not have the correct blade, cut at a shallow angle to the material

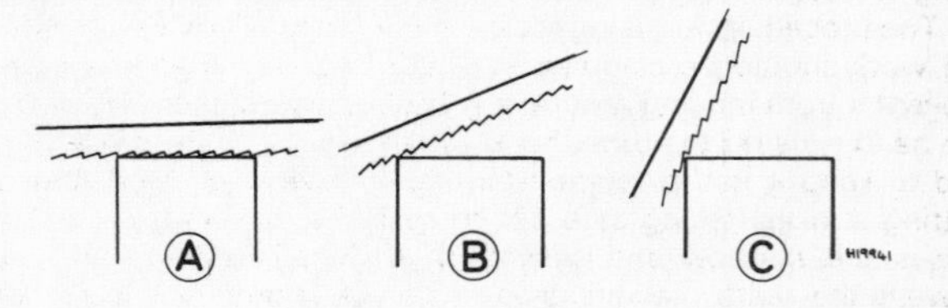

The correct cutting angle is important. If it is too shallow (A) the blade will wander. The angle shown at (B) is correct when starting the cut, and may be reduced slightly once under way. In (C) the angle is too steep and the blade will be inclined to jump out of the cut

Fig. 2.19 Correct cutting procedure

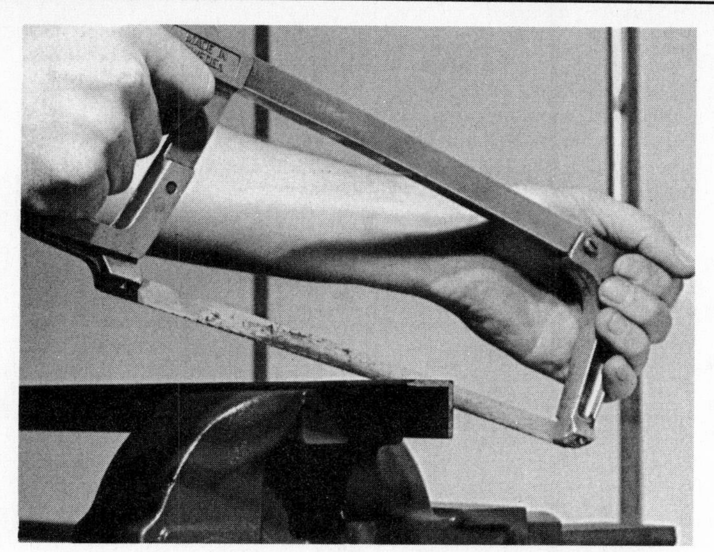

5.1 Note angle of blade to workpiece, 90°. One 'leg' of the angle has already been cut, and the second is just being started. This method prevents the metal pinching the blade

6 Files

Files are available in a wide variety of sizes and types to suit specific jobs, but are all used for the basic function of removing small amounts of metal. For most motorcycle applications this will amount to removing rough edges where a piece of metal has been cut or drilled (deburring) or final finishing of a made-up part to size.

Files come in a wide range of sections, including flat, half-round, round, square and triangular. Each of these is available in a range of sizes and grades. The file blade is covered by rows of diagonally-cut edges which form the cutting teeth. These can be aligned in one direction only (single cut) or in two directions (cross-cut) to form a diamond shaped cutting pattern. The depth and spacing of the teeth control the grade of the file, ranging from coarse to smooth in four basic grades; rough, bastard, second and smooth.

You will need to build up a range of files, and this can best be done by purchasing tools of the required shape and grade as they are needed. A good starting point would be one each of flat, half-round, round and triangular, in bastard or second cut grades. In addition, you will need to buy one or more file handles. Files are usually sold without handles, these being purchased separately and pushed over the tapered tang of the file when in use. You may need to buy more than one size of handle to suit the various files in your toolbox, but never attempt to economise by doing without them. A file tang is fairly sharp, and you will almost certainly end up stabbing it into the palm of your hand if you use a bare file and it catches in the workpiece during use.

Exceptions to the rule on handles are the fine swiss files, which have a rounded handle section in place of the normal tang. These small files are usually sold in sets of a number of different sections and shapes. Originally intended for clockmaking and similar fine work, they can be very useful for detail finishing or use in inaccessible corners. Swiss files are normally the best choice where piston ring ends require filing to obtain the correct end gap in a bore.

The procedure for using files is fairly straightforward. As with saws, the work should be clamped securely in a vice wherever possible to prevent it from moving around while being worked on. Hold the file by the handle, using the other hand at the far end of the blade to guide it and to keep it flat in relation to the surface to be filed. Use smooth cutting strokes, taking care not to rock the file in an arc as it passes across the surface, and sliding it diagonally along the surface to prevent the teeth making grooves in the workpiece. Think carefully about the job you are doing. If, for example, you are trying to form a square hole in a bracket, care must be taken not to undercut adjacent sides when working on another. This is best done by using a flat file with smooth edges.

Files do not require maintenance in the accepted sense, but they should be kept clean and free of swarf and filings. Steel is a reasonably easy material to work with, but softer metals like aluminium tend to clog the file teeth very quickly. This can be avoided by rubbing the file surface with a stick of ordinary blackboard chalk before commencing work. General cleaning is carried out using a file card, or failing that, a fine wire brush. Kept clean, files will last a good while, but when they do eventually become blunt they must be renewed; there is no satisfactory way of sharpening a worn file.

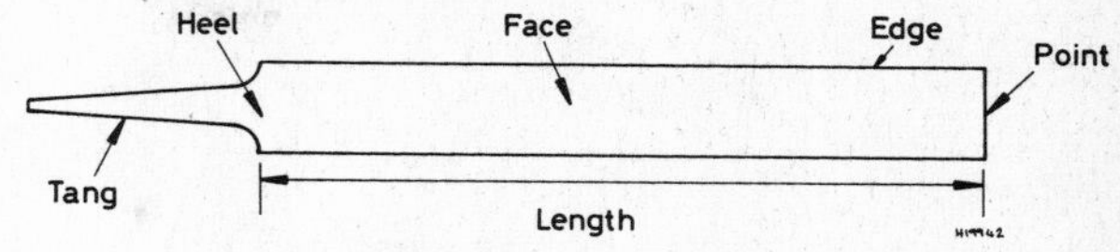

Fig. 2.20 A typical file

Always use a file handle – serious injuries can occur if a bare file jams in use

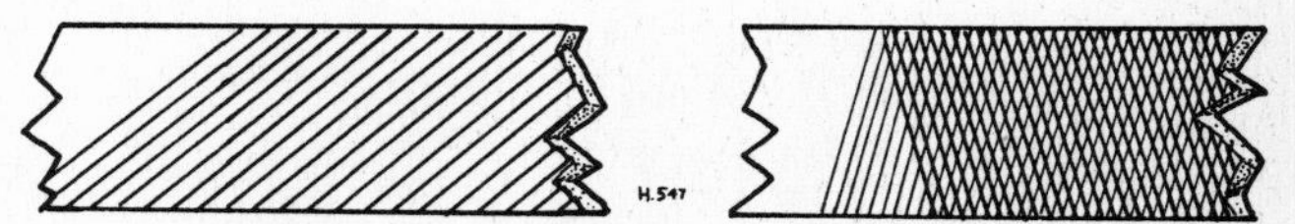

Fig. 2.21 Note the difference between single-cut and cross-cut file types

6.1 Files are available in numerous grades, sections and sizes. The flat file shown has cross-cut teeth

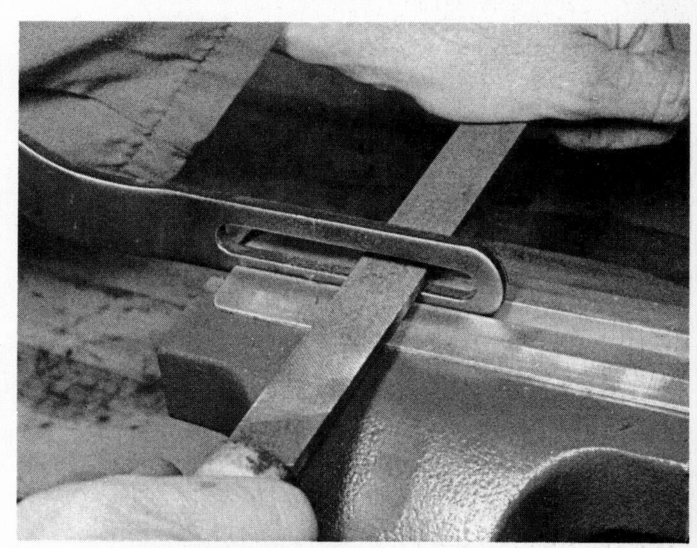

6.2 This flat file is being used to finish a slot drilled and sawn into a fabricated bracket. Note how the file is being controlled with two hands to ensure a flat edge

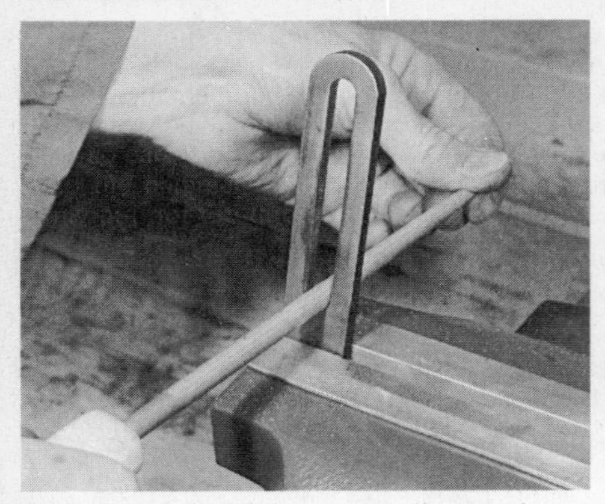

6.3 A round file of the correct diameter is used to smooth the radius at each end of the slot

6.4 A round file is also useful for enlarging or shaping drilled holes

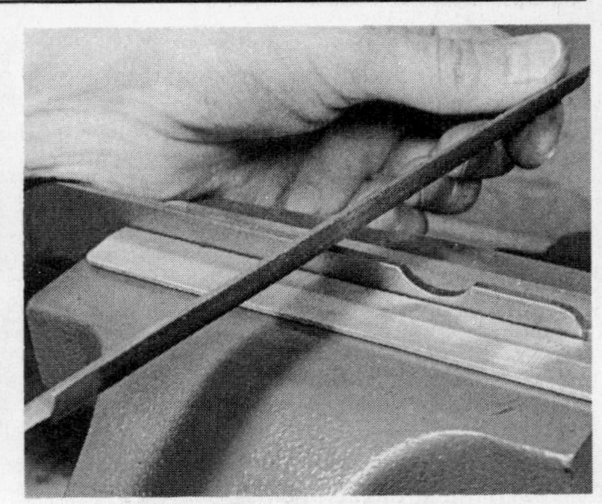

6.5 Half-round files are generally used for irregular shapes. Here, a curved recess is being formed after the shape has been rough sawn

6.6 Rubbing a stick of chalk over the file teeth ...

6.7 ... prevents them clogging when filing aluminium

7 Hammers and mallets

You will need at least one engineer's ball pein hammer in the workshop. This type of hammer has a head with a conventional cylindrical face at one end and a rounded ball end at the other, and is a general-purpose tool in any workshop. Always choose a fairly large hammer in preference to a small one; although it is possible to get small versions, you will not need them often, and it is much easier to control the blows from a more substantial head accurately. As a rule, a single 1lb ball pein hammer will suffice for most jobs, though occasionally larger or smaller types may be useful.

Mallets are used where a hammer would be likely to cause damage to the component or tools being used. A typical application on motorcycles would be to tap around a cover to assist in freeing it from its gasket. A steel hammer head might easily crack a light alloy casting, whereas a rubber or hide faced mallet can be used with relative safety. It is possible to buy mallets with several interchangeable heads, and these represent good value for home use. If finances are really limited, you can usually get by without a mallet by placing a small hardwood block between the workpiece and the hammer head to prevent damage or marking the softer alloy components.

The use of both hammers and mallets is similar; the head should strike the desired object squarely and with the correct amount of force. For many jobs, little muscular effort will be needed, it being sufficient to allow the weight of the head do the work, the force being controlled by the length of swing. With practice, a hammer can be used with surprising finesse, but this will take a while to achieve. Initial mistakes include striking the object at an angle, in which case the hammer head may glance off to one side, or hitting the edge of the object. Either error can result in damage to the workpiece or to your thumb, if it gets in the way, so care should be taken to strike positively to minimise this risk. Hold the hammer handle near the end, not near the head, and grip firmly but not too tightly.

Care should be taken to check the condition of hammers on a regular basis. The danger of a loose head coming off is self-explanatory, but check the head for signs of chipping or cracking too. If damaged in this way, buy a new hammer. The head may chip in use and the resulting fragments can be extremely dangerous. It goes without saying that eye protection is essential whenever a hammer is in use.

8 Drifts, punches and chisels

These tools are used along with a hammer for various purposes in the workshop. Drifts are often simply a length of steel bar, purpose made or improvised, used to drive a component out of a bore in the engine or frame. A typical use would be in removing or fitting a bearing. A drift of the same diameter as the bearing outer race is placed against the bearing, and is then tapped with the hammer to knock it in or out of its bore. Many manufacturers offer specialised drifts for the various bearings on a particular model, and these can be ordered through dealers. Whilst these are useful for the busy dealer service department, they are prohibitively expensive for the owner, who may only need to use them once. In such cases it is better to improvise. For bearing removal and fitting it is usually possible to employ a socket of the appropriate size to tap the bearing in or out; an unorthodox use for a socket, but one which will save time and expense.

Smaller diameter drifts can be bought from tool shops or made up from offcuts of steel bar. In some cases you will need to drive out items like corroded engine mounting bolts. Here it is essential that you avoid damaging the threaded end of the bolt, in which case the drift must be of a softer material than the bolt. Brass or copper is the usual choice for such jobs; the drift may become damaged in use, but the thread will be protected.

Punches are available in various shapes and sizes, and it is useful to obtain a set of assorted types. Fundamental amongst these is the centre punch, a small cylindrical punch with the end ground to a point. This will be needed whenever a hole is to be drilled. The centre of the hole is first carefully measured and marked out, and the centre punch is then used to place a small indentation at the intended hole centre. This acts as a guide for the drill bit, ensuring that the hole ends up in the right place. Without a punch mark the drill bit will wander, and you will find it impossible to drill with any real accuracy. You can also buy automatic centre punches. These are spring loaded and need only be pressed against the surface to be marked. An internal spring mechanism fires the point at the surface, marking it without the need to use a hammer.

Parallel pin punches are intended as tools for removing devices like roll pins from their holes. Roll pins (semi-hard hollow pins formed from steel sheet, and a fairly tight fit in their holes) are relatively uncommon on motorcycles, though you may encounter them on fuel filler cap and seat hinges occasionally. Parallel pin punches have other uses, however. You may on occasion have to remove rivets or bolts by cutting off the heads and driving out the remains, in which case a close-fitting punch is necessary. They are also very handy for aligning holes in components while retaining bolts or screws are inserted.

Of the various sizes and types of chisel available, the simple cold chisel can be considered essential in the motorcyclist's workshop. This is normally about six inches long, with a blade about 1/2 in wide. The blade end is ground to a rather blunt cutting edge (usually around 80°), the rest of the tip being ground to a shallower angle. The primary use of the cold chisel is rough metal cutting, and this can be anything from sheet metal work (uncommon on motorcycles) to cutting off the heads of seized or rusted bolts.

With all of the tools described in this section, it is worth buying good quality items. They are not particularly expensive, so it is not really worth trying to make economies here. More significantly, there is a risk that with cheaper tools, fragments may break off in use; a potentially dangerous situation.

Even with good quality steels, the heads and the working ends of these tools will inevitably become worn or damaged in time, so it is a good idea to have a maintenance session on all such tools from time to time. Using a file or a bench grinder, remove all burring, or mushrooming, from around the head area. This is an important task because the build-up of material around the head can fly off in the form of splinters, and is potentially dangerous. At the working end, make sure that the tool retains its original profile, again filing or grinding away any burrs. In the case of cold chisels, the cutting edge will have to be re-ground quite often because the material in the tool is of necessity not much harder than the material being cut. Check that the edge is reasonably sharp, but do not make the tip more acutely angled than it was originally; it will just wear down faster.

The techniques of using these tools vary according to the job to be done, and is something best learnt from experience. Common to all is the fact that they are normally struck with a hammer. It follows that eye protection will be needed. Always make sure that the working end of the tool is in contact with the item to be punched, drifted or cut. If this is not the case the tool will bounce off the surface and damage may be caused.

9 Twist drills and drilling machines

Drilling operations in the workshop are carried out using twist drills, either in a hand brace or in an electric drilling machine. Twist drills (or drill bits, as they are often known) consist of a cylindrical shank, along about two-thirds of which is formed spiral flutes which clear the swarf produced while drilling, keep the drill true in the hole and finish the sides of the hole.

The remaining length of the drill is left plain and it is this which is used to hold the drill in the chuck. For our purposes, we will consider only normal parallel shank drills. There is another standard type in which the plain end is formed into a specially sized taper designed to fit directly into a corresponding socket in a professional bench or pillar drilling machine. These drills are known as Morse Taper drills, and will be found in most machine shops and engineering works.

At the working end of the drill, two cutting faces are ground, forming a roughly conical point. These are generally angled at about 60° from the drill axis, but they can be re-ground to other angles for specific applications. For general use, though, the standard angle is correct, and this is how the drills are supplied.

When buying drills, it is probably best to start with a set of good quality items. Make sure that the drills are marked 'High Speed Steel' or 'HHS'. This indicates that the drills will be sufficiently hard to withstand continual use in metal; many cheaper, unmarked drills will be less hard, and suitable only for use with wood or other soft materials. Buying a set eliminates the bewildering choice of sizes in which individual drills are available.

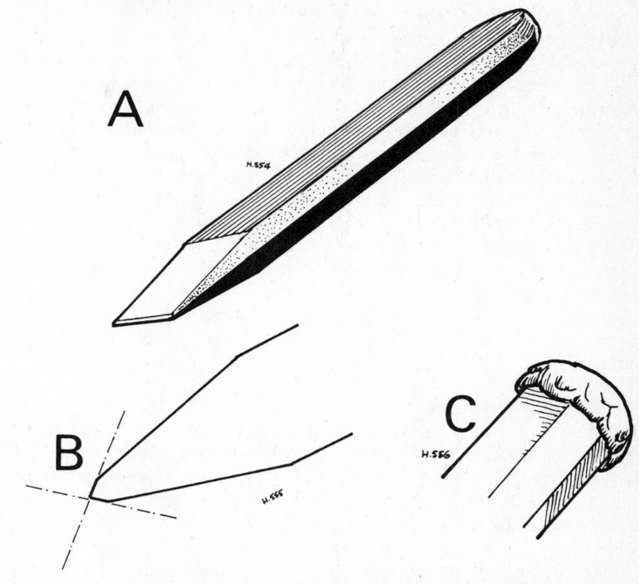

Fig. 2.22 The general-purpose cold chisel

(A) A typical general-purpose cold chisel. (B) Note the angle of the cutting edge – this should be checked regularly and resharpened as required. (C) A common cold chisel malady – the mushroomed head. This is potentially dangerous and should be filed back to the shape shown at (A)

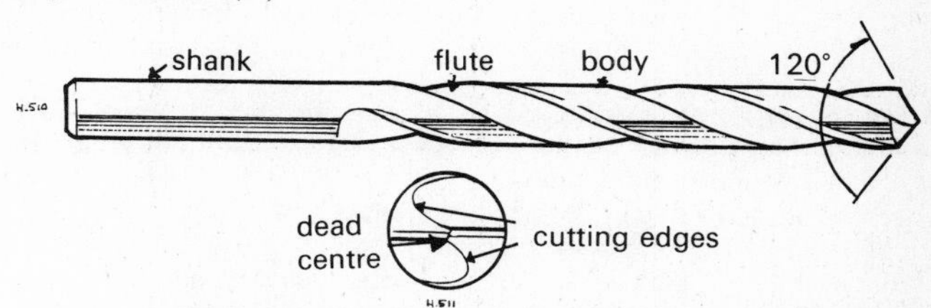

The main parts of the twist drill. Note chisel-like tip

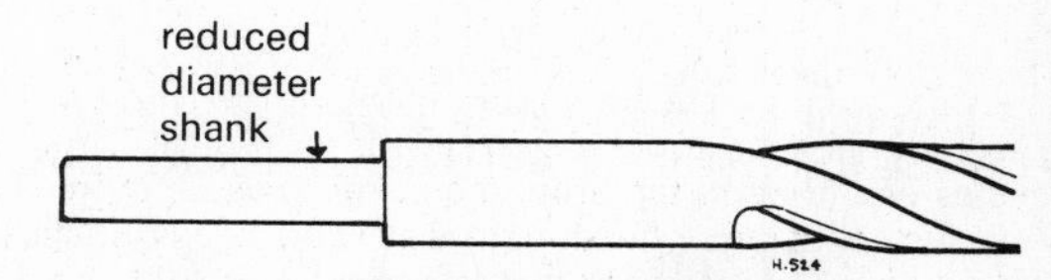

Twist drills in larger sizes can be obtained with a reduced-diameter shank to fit small chucks

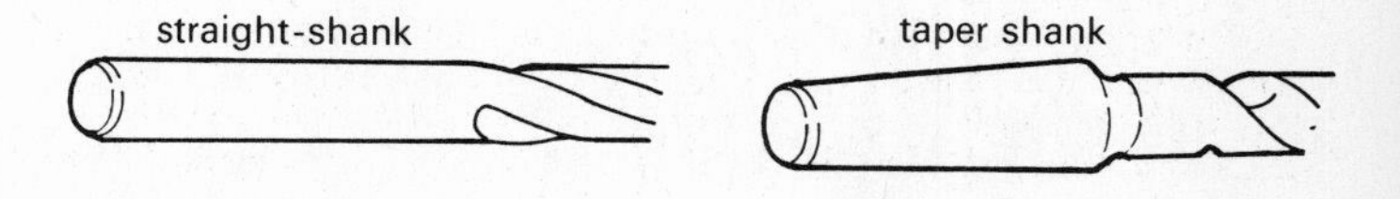

Straight shank twist drills are the most commonly-used type for the home workshop. If you use a bench or pillar drill, Morse taper twist drills give more accurate location and are quickly changed

Fig. 2.23 The twist drill

Twist drill sizes

As has been hinted, twist drills can be obtained in a vast range of sizes, most of which you will never need. There are three basic systems of drill sizing. Starting with fractional inch sizes, these start at 1/64 in, and continue upwards in increments of 1/64 in.

The corresponding metric range runs from 0.2 mm in rather irregular increments; for example, the sequence from 0.9 mm to 1.0 mm runs 0.90, 0.92, 0.95, 0.98, 1.00. Above 1.00 mm, the interval between sizes becomes a more regular 0.05 mm. At 3.00 mm and above, it becomes 0.10 mm intervals, whilst from 14.00 mm upwards, the interval between sizes drops to 0.25 mm. This may seem somewhat confusing, though in practice it means that the smallest increments in size relate to the smaller drill sizes, where this is most necessary.

Finally there are the gauge and letter sizes. These run in descending numerical order from 80 (0.35 mm) to 1 (5.80 mm). At this point the sequence continues in ascending alphabetic sequence from A (5.85 mm) to Z (10.50 mm).

This bewildering range of sizes means that it is possible to drill an accurate hole to almost any size within reason. In practice you will be limited by the size of chuck used, normally 3/8 in or 1/2 in. In addition, no tool shop will stock the entire range of possible sizes, so you will have to shop around for the nearest available size to the one you require.

Sharpening twist drills

Like any tool with a cutting edge, twist drills will eventually become blunt. The frequency with which sharpening becomes necessary is dependent to some extent on whether the tools are used correctly, but a blunt twist drill will soon make itself obvious. A good indication of the condition of the cutting edges is to watch the swarf emerging from the hole. If the tip is in good condition, two even spirals of swarf will be produced; if this fails to happen or the tip becomes hot, it is safe to assume that sharpening is required.

With the smaller sizes, below about 1/8 in or 3.0 mm, it is easier and more economical to throw the worn drill away and use another, and a stock of smaller sizes should be kept for this reason. With larger and more expensive sizes sharpening is a better bet. When sharpening twist drills it is essential that the included angle of the cutting edges is maintained at the original 120°, and that the small chisel edge at the tip is retained. With some practice, this can be achieved freehand, using a bench grinder, but it should be noted that it is very easy to make mistakes.

For most home mechanics, a sharpening jig or machine is preferable. The simplest of these consists of a special jig in which the twist drill is clamped at the correct angle. The jig is then pushed to and fro across abrasive paper, the eccentric wheels on the jig forming the required profile to the cutting edges. In the UK, a tool of this type is produced by the Eclipse company. Also available are specialised electric grinders in which twist drills are quickly sharpened correctly. Both types can be found in tool shops and hardware stores.

Drilling machines

These range from the simple hand brace, to expensive pillar drills, and all have their uses. Ideally, all drilling should be carried out using a bench or pillar drill, with the workpiece clamped solidly in a drilling vice. In practice, these machines are expensive and take up a good deal of bench or floor space, and so are out of the question for all but the most dedicated owner. An additional problem lies in the fact that many of the drilling jobs you may need to do will be on the machine itself, in which case the drilling machine has to be taken to the work.

This leaves us the choice of a manual or powered hand drilling machine. The cheapest and simplest of these is the hand brace. This is operated by a crank handle, which turns the drilling spindle and chuck through bevel gears. Prior to the advent of power tools, many sizes and types of these machines were available, some with a choice of gearing ratios to allow different materials to be drilled efficiently. Most machines available now are simple single-ratio devices. The big advantage of a hand brace is that it requires no external power source other than the user, and it is relatively compact and inexpensive.

Probably the most popular device is the power drilling machine. Known generally in the UK as an electric drill and in the US as a drill motor, these machines offer most of the advantages of the manual equivalent, without the hard work. If you are thinking of purchasing one for the first time, look for a well-known and reputable brand name, a 1/2 in chuck capacity, and variable speed as minimum requirements. A small capacity single-speed machine will suffice, but it is worth paying a little more for the extra facilities mentioned.

An alternative to the above types, and worth considering if electrical power is not readily available, is the cordless drilling machine. These devices are of limited power when compared with a mains-operated version, but their small size and rechargeable power packs make them ideal for occasional use in the workshop. They are also electrically safer than a mains machine if exposed to wet conditions in use.

All drilling machines likely to be used at home use a chuck to hold the twist drill in position. On very small hand braces the chuck may be tightened by hand, but most chucks use a toothed key to tighten the jaws down onto the drill shank. Whenever removing or fitting a twist drill from a power tool, make sure that the power supply is switched off to avoid accidents. Tighten the chuck by hand, checking that the drill is aligned correctly. This is especially important when using small drill sizes which can get caught between the jaws. Once hand tight, use the key to tighten the chuck fully, *remembering to remove the key afterwards.*

10 Drilling and finishing holes

Preparation for drilling

If at all possible, make sure that the part you intend drilling is securely clamped in a vice. If possible, use a bench or pillar drilling machine. Failing this, try to use a portable drilling machine in a bench stand as an equivalent to the bench type machine. If it is impossible to get the work to a vice, make sure it is stable and secure. Twist drills often dig in during drilling, and this can be dangerous if the work suddenly starts spinning on the end of the twist drill. Obviously, there is little chance of a complete motorcycle doing this, but you should make sure that it is supported firmly and stably before work commences.

Start by measuring and marking out the position of the intended hole centre, using a scribing tool or a fine spirit-based marker. Once the centre is established, centre punch it to provide a location point for the

9.1 If at all possible, a bench or pillar drilling machine should be used for work on all small parts

9.2 Using a bench drill fitted with a vice to ensure accurately aligned holes

9.3 Twin spirals of swarf from the cutting edges of the twist drill; an indication of a correctly angled and sharpened tip

bit. It is essential that the punch mark is positioned with absolute accuracy, since the final hole position is largely determined by this.

Unless you are drilling only a small hole (below about 5 mm) you should drill a pilot hole first. As the name suggests, this will guide the larger drill bit through on the correct line, and minimises the risk of the tip jumping out of the hole during drilling. Before any drilling commences, check that the area immediately behind the emerging drill is clear of anything which you would prefer unperforated. (It is easy to drill through a frame bracket to find that you have hit an electrical component – the author once bought a car and found that the previous owner had inadvertently drilled through the bodywork and into the battery when fitting an accessory switch, a mistake which remained undiscovered until he tried to remove the battery and found it firmly screwed to the car!)

If you are working with a bench machine, or with a portable drilling machine fixed in a bench stand, it is quite easy to set up the stops so that the proposed hole does not go deeper than intended. When drilling by hand this is less easy, and you will have to improvise a depth gauge by winding PVC tape around the end of the twist drill at the necessary position.

Drilling

When drilling steel, especially with smaller sizes of twist drill, no lubrication is needed. If using larger drills, it can be advantageous to use a little cutting lubricant to ensure a clean cut and to limit heating of the drill tip. With aluminium, which tends to smear around the cutting edges and clog the flutes, use paraffin (kerosene) as a lubricant.

Wearing safety glasses or goggles, position yourself so that you are stable, and able to control the pressure on the drill accurately. Hold the drill tip in contact with the punch mark, and make sure that if you are drilling by hand, the drill is perpendicular to the surface of the workpiece. Start drilling without applying significant pressure, until you are sure that the drill is positioned accurately. If the hole starts off centre, it can be very difficult to correct this. You can try angling the machine slightly so that the hole centre moves in the opposite direction, but this must be done before the flutes of the twist drill have entered the hole. It is at the starting stage that a variable-speed machine is invaluable; the low speed allows fine adjustments to be made before it is too late. Continue drilling until the desired hole depth is reached or until the drill tip emerges at the other side of the work piece.

Cutting speed and pressure are important during drilling operations. As a general rule, the larger the diameter of the twist drill, the slower the drilling speed should be. With a single-speed machine there is little that can be done to control this, but two-speed or variable speed machines can be controlled with this in mind. If the drilling speed is too high, the cutting edge of the drill will tend to overheat and be damaged. Pressure should be varied during drilling. Start with light pressure until the drill tip has located properly in the work. Gradually increase pressure so that the drill cuts evenly. If the tip is sharp and the pressure correct, two distinct spirals of swarf will emerge from the flutes. If pressure is too light, the drill will not cut properly, whilst excessive pressure will overheat the drill tip.

Care should be taken to decrease pressure as the drill tip breaks through on the far side of the hole. If this is not done the drill may jam in the hole, and if you are working with a portable machine of any size there will be a tendency for the machine to twist out of your hands, especially when using larger sizes of drills.

Once any pilot hole has been made, fit the larger drill into the chuck and drill out to size. The second drill tip will follow the pilot hole through, and there is no need to attempt to steer it; to do so is likely to end in the drill breaking off. It is important, however, to ensure that the drilling machine is held at the correct angle.

After the hole has been drilled to the correct size, remove the burring which will be left around the edges of the hole. This can be done using a small round file, or by putting a small chamfer on the edges of the hole, using a larger twist drill or a countersink bit. Choose a twist drill that is much larger than that used for the hole, and simply twist it around each side of the hole by hand until any rough edges are removed.

Enlarging and re-shaping holes

The biggest practical size for twist drills used freehand is about 1/2 in (13 mm). This is partly governed by the capacity of the chuck (although it is possible to buy larger drills with stepped shanks). The real limit lies in the biggest drill size which can be comfortably

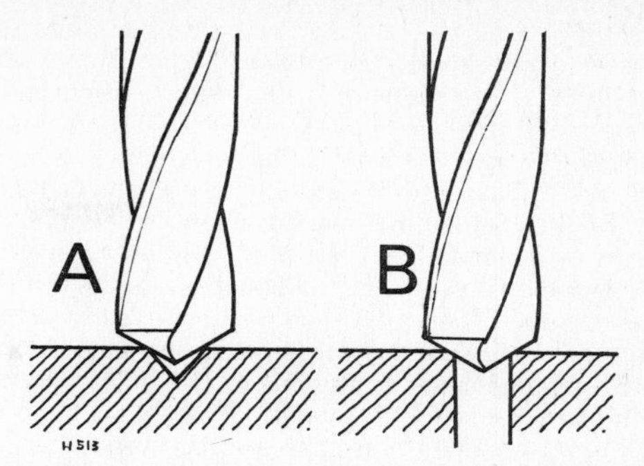

Fig. 2.24 When drilling holes, use either a well-defined punch mark (A) or a pilot hole (B) to guide a larger drill bit

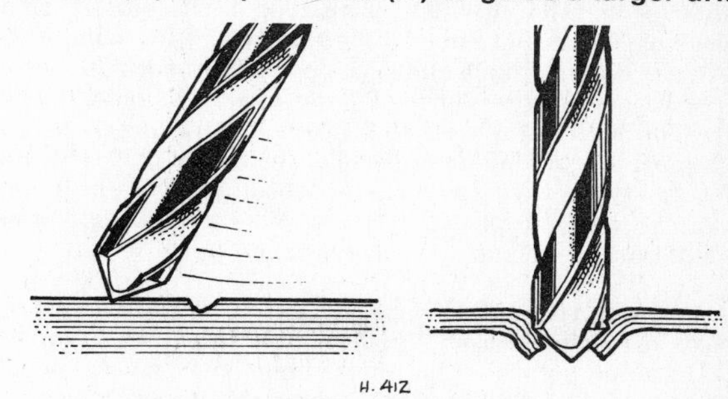

Fig. 2.25 Correct drilling procedure

An indistinct punch mark, or attempting to drill at an angle will cause the drill to slip (left). Be especially careful as the drill tip breaks through on the far side of the hole – it may jam if under too much pressure (right)

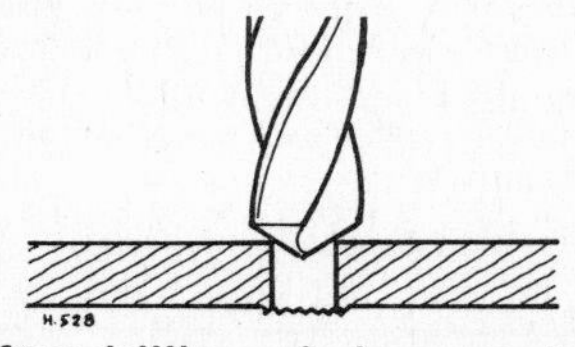

Fig. 2.26 After drilling a hole, use a larger drill tip to remove burring

Fig. 2.27 Rotary files, some of which incorporate a drill tip, are useful ways of enlarging holes

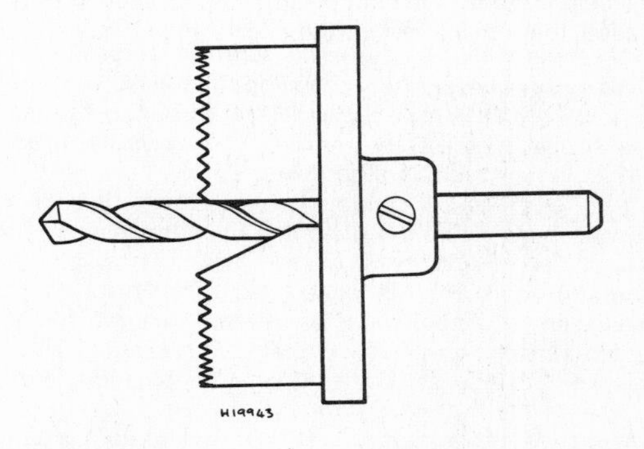

Fig. 2.28 A hole saw

Where large circular holes need to be cut in thin materials, hole saws can be useful. Most of these devices include a series of concentric blades, allowing a range of hole sizes

controlled by hand; drills over the $^{1}/_{2}$ in limit tend to be too boisterous for use in anything other than a bench or pillar drill. If you need to make a much larger hole, or if a shape other than round is required, different techniques are necessary.

If a hole simply needs to be enlarged slightly, a round file is probably the best tool to use. Where the hole needs to be very large, some sort of hole cutter will be needed. The most common type uses what is effectively a circular hacksaw blade supported in a metal boss. At the centre of the boss is a twist drill which is used to drill a pilot hole to locate the hole saw. The tool is then pushed, at a low cutting speed, against the work, where it gradually saws its way through. Other versions of this arrangement employ one or two adjustable cutting tips in place of the saw blade.

Large or irregular shaped holes can also be made by chain drilling. In this instance the finished hole is carefully marked out onto the workpiece using a scriber. The next stage depends upon the size of the twist drill to be used, the idea being to drill a sequence of almost touching holes just inside the finished outline of the hole. If you are using a 4 mm drill size, mark out a line about 2.5 mm inside the finished hole outline, then mark this line at 2.5 mm intervals. Centre punch each position, then drill out the marked holes. A sharp chisel can then be used to knock out the waste material at the centre of the hole, which can then be filed to size. This is a time consuming process, but it is the only practical method for use at home. Note that success is entirely dependent on your accuracy when marketing out.

Reaming holes to size

Finally, remember that although the hole size produced by drilling will be close enough for most applications, you cannot guarantee a precision hole size by drilling alone. Where greater accuracy is needed, drill the hole to within $^{1}/_{64}$ in of the finished size and use a reamer to finish it. Reamers have spiral flutes running along their length, and can be used either hand-held or in a bench drill running at low speed to take the final cut off the walls of the hole. An alternative is to use an expanding reamer. These can be adjusted within a certain size range, and this allows the hole to be enlarged gradually until it reaches the correct size. Reamers, particularly expanding reamers, are expensive tools if used only infrequently, and it is worth finding out if a local tool hire shop can provide one on the few occasions you will need them.

10.1 For really accurate sizing of holes, use a reamer after initial drilling has been completed

10.2 Reamers are also useful for finishing bushes to size after they have been fitted. The adjustable reamer shown allows an accurate selective fit to be attained

11 Taps and dies

Taps

Taps provide a method of forming internal threads, and are also very useful for cleaning up damaged threads. Each tap consists of a fluted shank with a driving square at one end. It is threaded along part of its length, a cutting face being formed where the flutes intersect the threads. The tap is made from hardened steel, and is thus able to cut a thread in most materials softer than itself.

The profile of the threaded portion of the tap varies, giving three profiles described as Taper, Intermediate and Plug. Taper taps, as the name suggests, have a heavily tapered profile in which the thread is almost missing at the tip. This allows the tool to be started in a plain hole where the first few threads serve to locate it and to cut shallow grooves in the hole walls. As the tap is screwed in further, the thread form becomes progressively more developed, and so the final thread form is cut gradually. Only the last few threads are fully formed, and so the tap must pass completely through the workpiece before the finished thread is cut properly along its length.

Whilst the Taper tap is suitable for threading through holes, a problem arises where the hole is blind; the fully threaded section of the tap never reaches the bottom of the hole. To cope with this, Intermediate and finally Plug taps are used, after an initial thread has been formed with the Taper tap. The intermediate tap has a steeper taper near the tip, and thus can reach deeper into a blind hole. The Plug tap has only a slight taper over one or two threads and can effectively thread the hole to the bottom.

The cheapest source of taps is to purchase a combined tap and die set covering the range of sizes which you are most likely to need. Some of these sets can be remarkably inexpensive, but they do have their problems. Leaving aside the quality of the steel used and the hardness of the cutting edges, the biggest drawback is imprecise or badly formed threads; even a small inaccuracy or slight burring can make the tool utterly useless. The alternative is to buy high quality taps as and

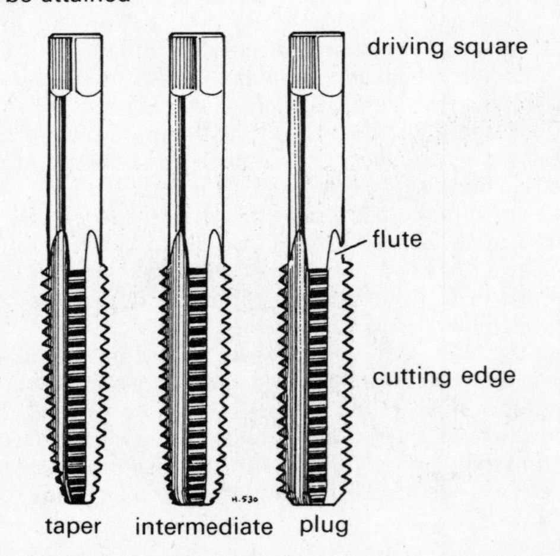

Fig. 2.29 Taper, intermediate and plug taps

Note how the tapered section progressively decreases across the range. Plug taps are normally needed for finishing tapped holes in blind bores

when you need them. Individual high-quality taps are not cheap, especially if you need to buy two or more profiles in a given size. Despite this, it is the best option. You will probably only need taps on odd occasions, and so a full set is rarely necessary.

Taps are normally used by hand (they can be used in machine tools, but this is rather outside the scope of most home workshops). The square driving end of the tap is held in a stock; an adjustable T-handle. For smaller sizes a T-handled chuck is also used. The tapping process starts by drilling a plain hole of the correct diameter. For each tap size there is a corresponding twist drill which will produce a hole of the correct size. This sizing is important, given that too large a hole will leave the finished thread with the tops of each thread missing, producing a weak and unreliable fastening. Conversely, too small a hole will place excessive loads on the hard and brittle shank of the tap, and this can result in it breaking off in the hole. The problems of removing the remains of the tap from an alloy casting can be imagined, and are discussed in Chapter 4.

The correct tapping drill size may be marked on the container in which the tap was supplied. Alternatively, buy an Engineer's Reference Booklet giving details of all commonly used threads, drill sizes and other reference tables. In the UK, an invaluable reference booklet is produced by *Zeus Precision* and is available from tool shops and engineering suppliers. It has plastic laminated pages to resist oil, has all the basic data you are likely to need.

Once the initial hole has been drilled, the tap can be fitted into the stock and threading can commence. It is preferable to use a cutting lubricant on the tap; this will minimise the risk of breakage and will produce a cleanly cut thread. When starting the thread, using a taper tap, make sure that the tool enters the hole squarely. Once three or four threads are into the material the tap will guide itself, but careful alignment when starting the cut is essential. After three or four turns, or whenever the tap is felt to go tight in the hole, back it off by about two turns before continuing. This will break off and clear the swarf which is accumulating around the flutes in the tap. Occasionally, on large diameter or particularly deep threads, it is desirable to remove the tap for thorough cleaning. The hole should be cleaned out too, and fresh cutting lubricant applied before re-starting the threading work.

If working on a blind hole, take care not to force the tap when its tip reaches the bottom of the hole. The tap should be unscrewed and it and the hole cleaned out before repeating the operation using intermediate and finally the plug tap to take the thread to the bottom of the hole.

When using a tap to clean up a damaged thread, check the exact condition of the thread before starting. If the first few threads are damaged it must be remembered that it is essential that the tap engages with the good threads past this point. If you simply screw the tap through the damaged area this is unlikely to happen, and you may well end up destroying the sound threads as a result. You may need to use a drill bit to remove the damaged area so that the tap can locate in the surviving threads accurately. Minor damage and burring can then be cleaned up.

Insert taps

If the thread is completely stripped or seriously damaged, re-tapping may not be successful. In such cases a thread insert will be needed. The most commonly used insert systems work on the principle of forming an oversized thread using a special tap. The resulting new thread is then reduced to the original size by fitting a diamond-section wire insert, allowing the original bolt or stud to be fitted to the repaired thread. This type of repair process is often referred to as a *Helicoil*, after an American manufacturer of that name. In the UK, the *Cross Manufacturing Company* produce similar insert systems, and the widespread use of their products in both general and aircraft engineering applications is indicative of the reliability of this type of repair. It is worth noting that some manufacturers fit these inserts as a matter of course to give greater strength and reliability than would be possible by tapping a thread directly into aluminium alloy.

The thread inserts look rather like springs prior to fitting, and have a small driving tang across the lower end. An inserting tool is fitted inside the coils of the insert, the tang of which engages in a slot in the end of the tool. The insert is then screwed home until the upper end is flush with or fractionally below the surface of the tapped hole. Once in position, the driving tang is broken off, leaving the insert firmly in position.

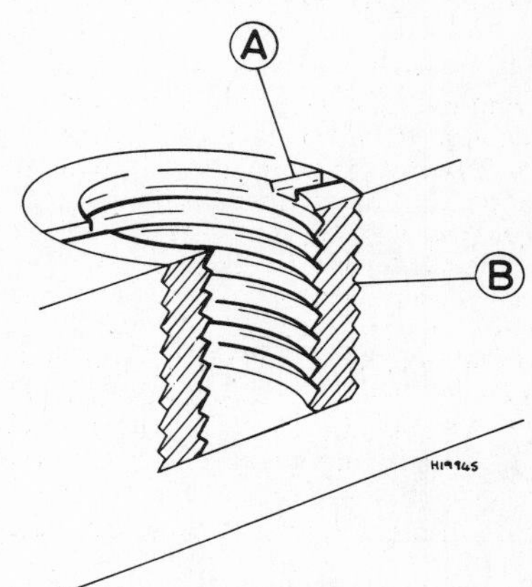

Fig. 2.30 Section through a solid thread insert

The damaged hole is tapped oversize, and the insert (B) is screwed home using the screwdriver slots (A) or a special inserting tool. Thread locking compound is used to secure the insert

11.1 A tap clamped in place in a stock (tap wrench)

11.2 The tapping operation in progress, showing the curls of swarf ahead of the cutting edges

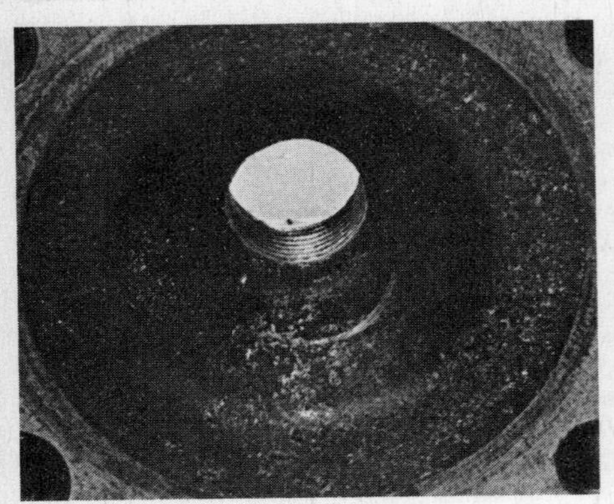

11.3 This stripped spark plug thread is to be restored by fitting a wire thread insert

11.4 The first stage is to cut a new oversized thread using a special tap ...

11.5 ... leaving a sound thread into which the insert will be fitted

11.6 The wire insert has a diamond section to suit the newly-tapped thread and to provide the new internal thread

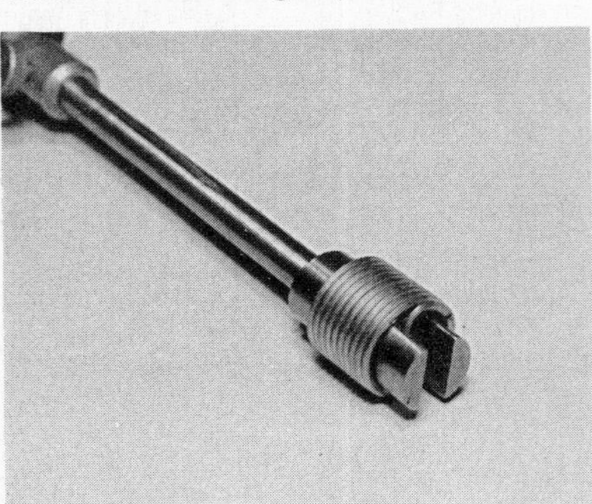

11.7 The thread insert is positioned on the end of the fitting tool with the driving tang located in the slot

11.8 The screwdriver-type fitting tool is used to screw the insert home in the tapped hole

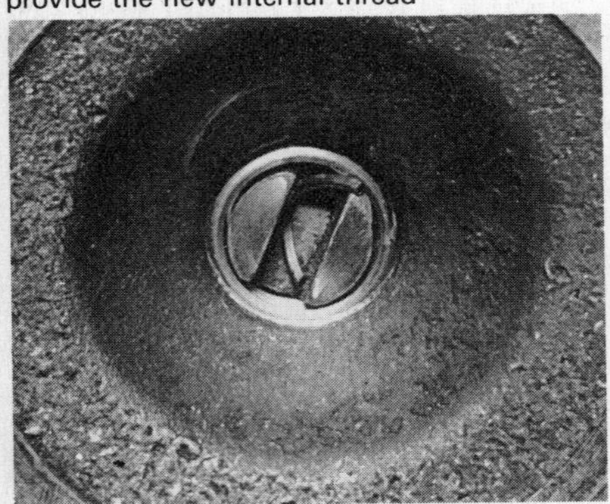

11.9 Once in place, the tool is tapped downwards to break off the driving tang, leaving the restored thread ready for use

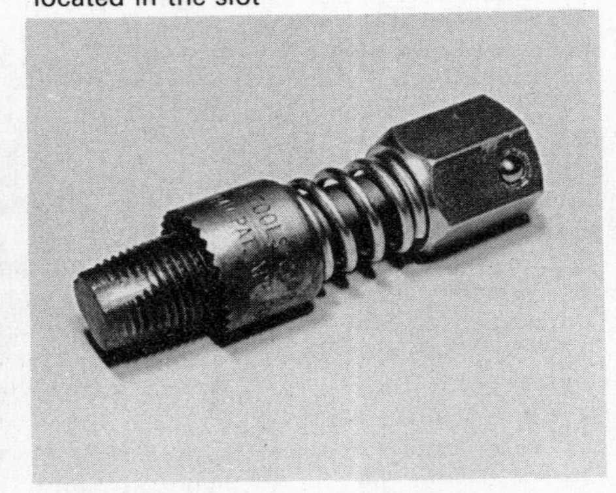

11.10 This special spark plug thread restoring tap is used to clean out carbon deposits and to make good light damage.

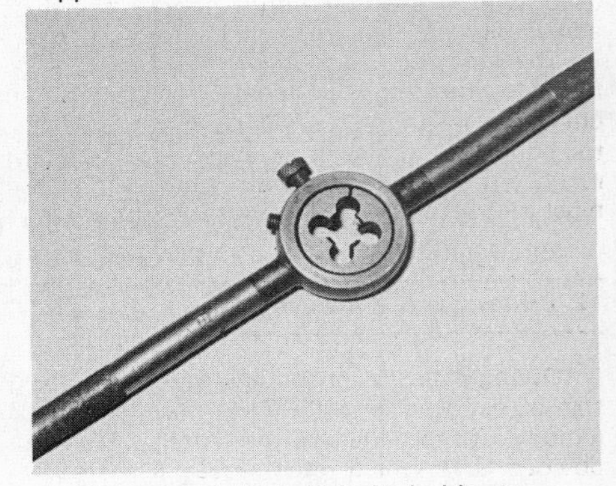

11.11 A typical die held in its holder or stock. The screws locate the die and allow small adjustments to the cutting depth

This type of repair is a popular method of reclaiming badly worn or stripped threads in alloy castings, the new thread formed by the stainless steel insert being permanent and more durable than the original. In most cases, the original hole need not be enlarged prior to re-tapping, making the repair quick and simple to carry out. The only drawback for the home user is the high cost of purchasing a range of taps and inserts. The taps are not of standard sizes, being made specially to suit each insert size. The inserts are not especially expensive individually, but to build up a stock of the most commonly-needed diameters and lengths is probably too expensive for most motorcycle owners.

Most motorcycle dealers, however, stock this type of inserting system, and this is well worth considering where damaged threads are noticed during an overhaul. The most common application is the repair of stripped spark plug threads in alloy cylinder heads, and here the demand is sufficient to have encouraged the manufacturers to produce DIY repair kits. These are available from normal sources of tools, and also from many car accessory shops.

As an alternative to the wire thread insert it is sometimes possible to fit a solid thread insert. After drilling and tapping the hole oversize the insert is screwed into position, having applied a thread locking compound to its external thread.

Dies

Dies are used in a similar fashion to taps, but for cutting external threads. Most dies take the form of a cylindrical block of hardened steel, the central thread being interrupted by several circular cutouts. These equate to the flutes on taps and allow swarf to escape during the threading process. The die is located in a T-handle holder, or stock, in which it is clamped by screws. The die is normally split at one point, and this allows it to be adjusted to some extent, giving fine control of thread clearances.

The use of dies is less common than taps, for the simple reason that it is normally cheaper and easier to buy and fit a new bolt than to attempt to make one. Nevertheless, it can be useful to be able to extend the threaded area of a standard bolt, or to clean up damaged threads

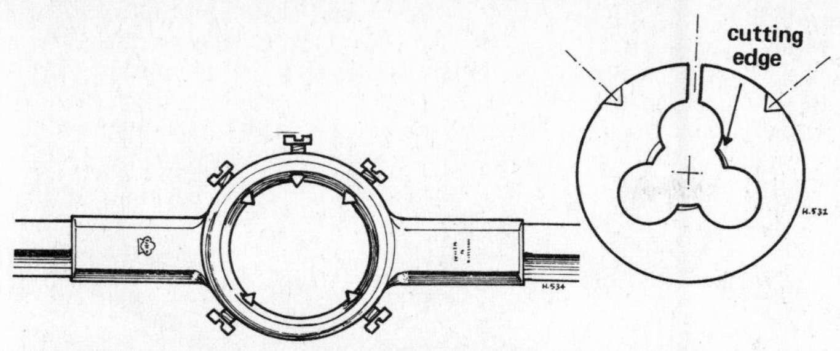

Fig. 2.31 A die used for external threading, and below it, a die holder or stock

The small screws both locate the die, and allow minor adjustment of its size. This enables the thread to be cut accurately

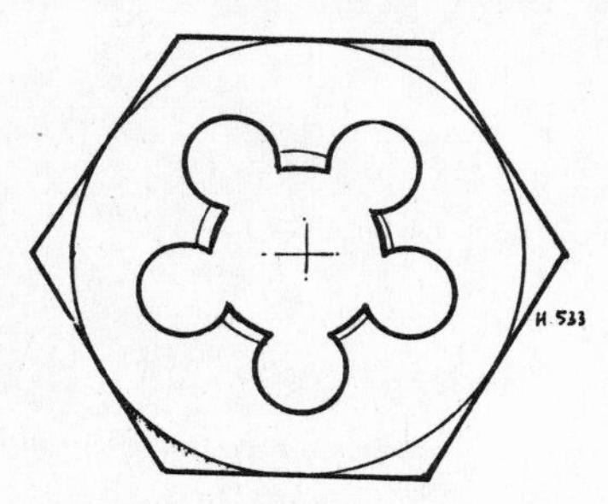

Fig. 2.32 Die-nut

Die-nuts can be turned with a conventional spanner. Lacking the adjustability and accuracy of a normal die, they are usually used to clean up and restore existing threads

using a die. This last function can be carried out using die-nuts, a simplified type of die having an external hexagon which allows it to be used with a spanner rather than a die stock. Die-nuts are non-adjustable, and thus less suitable for cutting new threads.

The procedure for cutting threads with a die is broadly similar to that described above for taps. When using an adjustable die, the initial cut is made with the die fully opened, the adjustment screw being used to reduce the diameter of successive cuts until the finished size is reached. As with taps, a cutting lubricant should be used, and the die must be backed off every few turns to clear swarf from the cutouts.

12 Pullers and extractors

During almost every engine overhaul there will be a need for some sort of extractor or puller, the most common application being the removal of the generator rotor from its tapered shaft end. Other less obvious tasks which often require some sort of extractor include the removal of bushes and bearings, and the gudgeon (wrist) pin from pistons. Common to all of these jobs is the need to exert pressure on the part to be removed whilst placing minimal strain on the surrounding area, and the best way of achieving this is to use an extractor specifically designed for the purpose.

Specialised extractors

A good example of the need for a specialised extractor is demonstrated by the method of removing the generator rotor. On most current machines, the alternator rotor (or flywheel generator rotor) fits over a tapered shaft end, where it is secured by a retaining nut and located by a Woodruff key. Even after the nut has been removed, it will still be necessary to exert considerable pressure to draw the rotor boss off the shaft. This is because the securing nut will have drawn the tapered faces tightly together during assembly.

The method normally used to remove the rotor is to fit a specially designed extractor which screws into a thread provided for this purpose in the rotor boss. An extractor bolt at the centre of the tool is then tightened down on the shaft end until the taper joint is broken and the rotor comes away. Occasionally it will also be necessary to tap the head of the extractor bolt to 'shock' the joint apart, and this general procedure must always be applied in preference to using excessive force with the extractor. Tighten the extractor bolt firmly then tap the bolt head smartly to jar the taper joint apart. Some discretion must be applied when doing this, though; it is possible on some machines with pressed-up crankshafts to drive the mainshaft out of line if it is struck too hard.

As you may appreciate, this method of removal means that the pressure from the extractor is applied where it is most needed; at the centre of the rotor boss, rather than at the edge where it would be more likely to distort the body of the rotor than to draw the boss off the shaft end. In practice, it will also mean that you will need the appropriate extractor for your machine, and this will have to be ordered through an official dealer. In the case of many lightweight two-stroke models, a standard extractor thread is used, and with machines so equipped an inexpensive pattern extractor is available, and these are often kept in stock by dealers. On other machines you may have to place an order for the official tool. On a few machines it is possible to get around the need for a puller, and where this can be done without risk of damage an improvised method is described in the relevant Haynes Owners Workshop Manual.

General-purpose extractors and pullers

You are likely to need some sort of general-purpose extractor at some stage in most overhauls, often where parts are seized or corroded, or where bushes or bearings need to be renewed. Universal two- and three-legged pullers (sometimes described as sprocket pullers) are widely available in numerous designs and sizes. Whilst these can be invaluable, it should be noted that they are of less use on current motorcycles than was once the case, mostly due to the different design and manufacturing techniques employed. The usual problem is that there is just not enough clearance around the part to be removed to allow this type of tool to be used. Nevertheless, the universal puller makes a good basis for an improvised tool.

The general design of these tools is based upon a central boss, to which is attached two or more arms. The ends of the arms terminate in hooked jaws which locate on the part to be drawn off. Normally, the arms can be reversed to allow the tool to fit internal openings if needed. As with specialised extractors, the boss is threaded to accept a central extractor bolt, and it is this which does the removal work. It is also possible to obtain hydraulic versions of these tools. This allows much greater pressure to be applied, but the bulk and cost of the tool makes them something of a luxury in the home workshop. As supplied, the tool is useful for removing gear pinions and sprockets, provided that clearance problems do not prevent this.

It is possible to adapt these tools by making up specialised arms for specific jobs, but this will all take time and may not be successful. If you do decide to try this approach, remember that the extraction force should be concentrated as near to the centre of the component as possible to avoid placing undue strain on it. In use, the tool should be assembled and a careful check made to ensure that it does not foul the adjacent casing, and that the loadings on the part to be removed are even. Where you are dealing with a part retained on a shaft by a nut, slacken the nut, but do not remove it entirely. This will support the shaft end and reduce the risk of it belling out under pressure, and will also stop the assembly from flying off the shaft when it does come free.

All extractors of this type should be tightened down gradually until the assembly is under moderate pressure. **On no account** should the extractor bolt be screwed down excessively hard or damage will be done. Once under pressure, try to jar the component free by striking the extractor bolt head a few times. If this does not work, tighten the bolt a little further and repeat the process. With luck, the component should come off its shaft with a satisfying 'crack'. The puller can then be dismantled, the nut run off the shaft end and the part lifted away.

If you are unsuccessful, it is time to stop and reconsider the problem. It is vitally important to be a little discriminating at this stage; at some point a decision must be made as to whether it is safe to continue applying pressure in this way. If the component is abnormally tight, something is likely to break before it comes off. If you find yourself in this situation, you could try applying a releasing fluid around the taper and leaving it overnight, with the extractor in place and tightened firmly, to work into the joint. With luck, the taper will have separated and the problem will have resolved itself by the next morning.

If you have the necessary equipment, and are skilled in its use, you can try heating the component with a blowlamp or gas welding torch; this is often a good way of releasing stubborn tapers, but is not to be recommended unless you have some experience of these methods. In particular, this approach should **never** be tried on rotors because the heat can easily demagnetise it or cause damage to the coil windings. The heat should be applied to the boss area of the component to be removed, keeping the flame moving to avoid uneven heating and the risk of distortion. Keep pressure from the extractor applied, and make

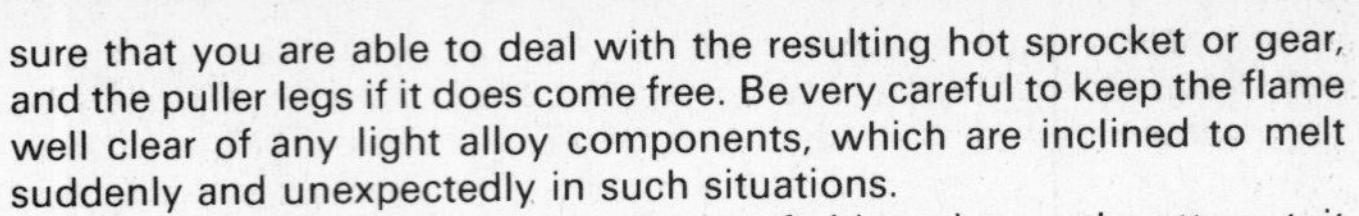

sure that you are able to deal with the resulting hot sprocket or gear, and the puller legs if it does come free. Be very careful to keep the flame well clear of any light alloy components, which are inclined to melt suddenly and unexpectedly in such situations.

If all rational methods fail, never be afraid to give up the attempt; it

12.1 An example of the commonly-used rotor extractor

12.2 In use, the extractor is screwed into the rotor and held while the extractor bolt is tightened

12.3 In some instances it may be possible to use a legged puller in place of the official tool, but this must be done with care and discretion if damage is to be avoided

12.4 In other instances the design of the rotor may allow a bolt to be used to draw it off the shaft. Where this is possible it is described in the relevant Haynes Owners Workshop Manual

12.5 Legged pullers are useful when drawing off bearings. Note the nut on the shaft end; this prevents damage to the shaft

12.6 This more elaborate hydraulic puller may be needed where the component to be removed is an unusually tight fit on its shaft

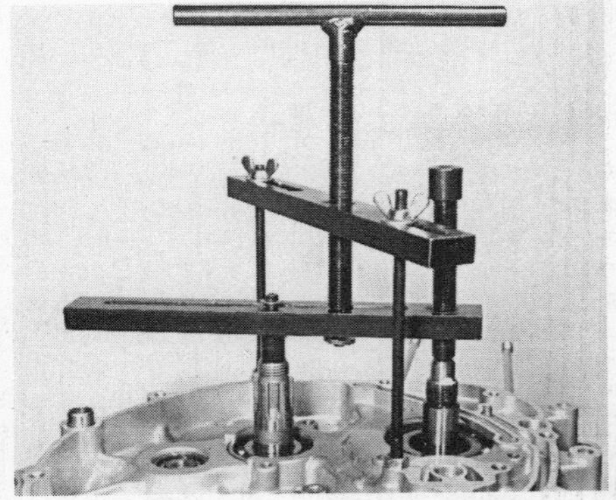

12.7 Another specialised puller, this special tool for separating crankcase halves can be made up at home if required.

12.8 Never overlook the obvious; it is often possible to adapt simple methods of removing small parts like bearings, provided they are not too tight

is cheaper to do this than to repair a badly damaged engine. Either buy or borrow the correct tool, or take the engine to a dealer and ask him to remove the part for you. Alternatively, reassess the problem. It is often possible to leave the offending sprocket or gear in place, continue the dismantling process, and then to make another attempt at removal when access is improved.

Drawbolt extractors

The simple drawbolt extractor is easy to make up and invaluable in every workshop. There is no standard extractor of this type; you simply make up a tool to suit a particular application. You can use a drawbolt extractor to deal with stubborn gudgeon pins, or to remove bearings or bushes.

To make up a drawbolt extractor you will need an assortment of threaded rod (studding) in various sizes, together with nuts to suit. These can be obtained from engineering suppliers. In addition you will need assorted washers, spacers and tubing. These latter items will be found in boxes or dark corners of your workshop, because you will have put them there years ago in case they came in useful at some later date (and you were right, they have!). For those of you not disposed to hoarding junk, or those who have hoarded so much of it that you can never find that one elusive piece of tube offcut, try an engineering supplier, scrap metal dealer or local workshop for suitable offcuts or scrap parts. In the case of the tubing, fairly thick-walled material is best. Don't forget to improvise where you can. Your socket set offers a range of sizes for short parts like bushes, though for things like gudgeon pins you will usually need a longer piece of tube.

Some typical drawbolt arrangements are shown in the accompanying line drawings, and these also give some idea of the order of assembly of the various pieces. The same arrangement, less the tubular spacer section, can normally be used to draw the bush or pin back into place. Using the tool is reasonably obvious, the main point to watch out for is to ensure that you get the bush or pin square to the bore when you attempt to fit it, and to lubricate the outside face to ease fitting, where appropriate.

If you make up a gudgeon pin extractor using this method, it is worthwhile making it well, and also a little on the big side so that it can be kept as a permanent workshop tool and used on other machines. The end of the tubular spacer that contacts the piston must be smooth, and it is a good idea to file a radius to match that of the piston skirt. To

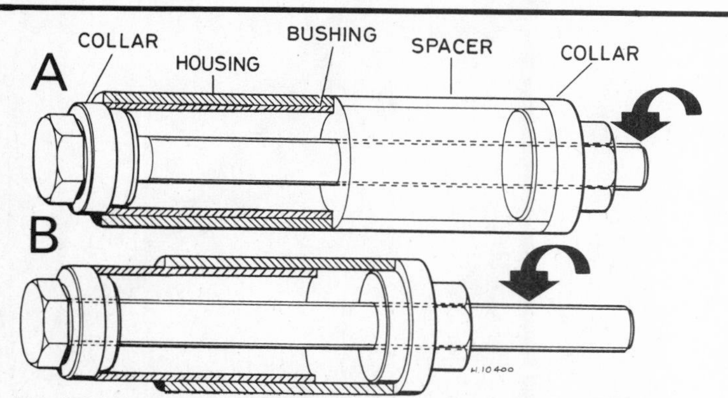

Fig. 2.33 Typical applications for drawbolt arrangements

In (A), the nut is tightened to pull the collar and bush into the large spacer, whilst in (B) the spacer is omitted and the drawbolt rearranged to allow the new bush to be installed

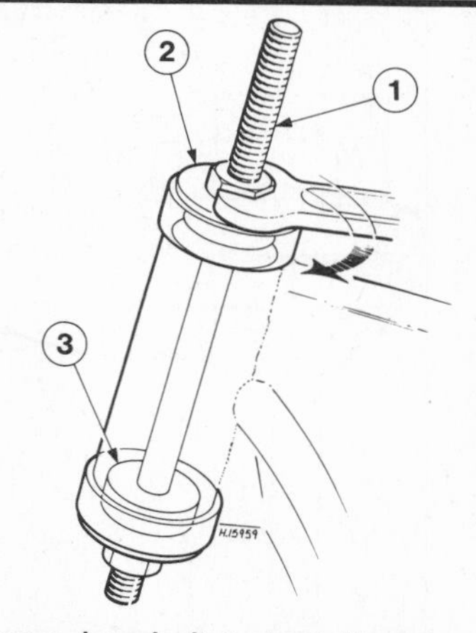

Fig. 2.34 Using a drawbolt to install new steering head bearings

In this example, a length of studding (1) and two nuts are used instead of a bolt. The large washers (2) are used to pull the bearings (3) into their bores

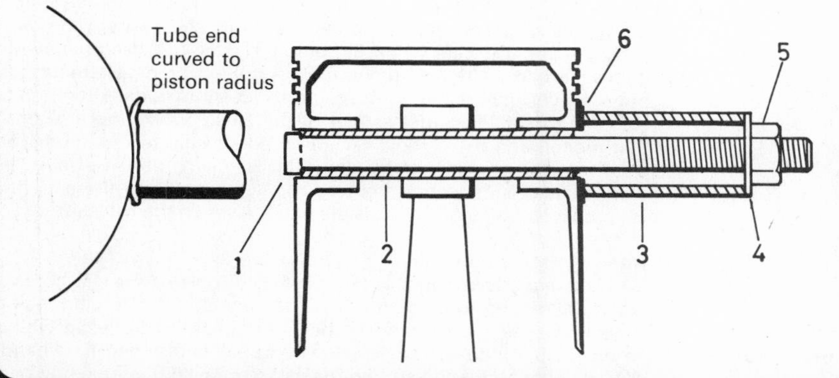

Fig. 2.35 Using a drawbolt to remove a gudgeon pin from a piston

The bolt head (1) has been ground down to fit inside the gudgeon pin bore. A spacer (3) and a washer (4) provide a recess into which the gudgeon pin (2) can be drawn as the nut (5) is tightened. Note the plastic packing (6) to prevent damage to the piston surface.

prevent damage to the soft alloy of the piston, use some thin-walled plastic or neoprene rubber tubing slit lengthways and pushed over the edge of the tubing.

Extractors for use in blind bores

You are likely to encounter a bush or bearing fitted into a blind bore at some stage, and there are types of extractor designed to deal with these situations. It should be noted that, in the case of engine bearings, it is often possible to remove them without an extractor. In some cases it may be possible to heat the casing (evenly in an oven or by submerging it in boiling water) and tap it face downwards on to a clean wooden surface to dislodge the bearing. If using this method take suitable precautions to prevent the risk of scalding when handling the heated components. If an extractor is needed, it will probably take the form of a slide-hammer with appropriate attachments for the working end. These range from universal legged puller arrangements, to specialised bearing extractors. The latter take the form of hardened steel tubes with a flange around the bottom edge. The tube is split at several places, and this allows a wedge arrangement to be used to expand the tool once fitted. The tool fits inside the bearing inner race, and is then tightened so that the flange or lip is locked under the edge of the race.

The slide-hammer consists of a steel shaft with a stop at its upper end. The shaft carries a sliding weight, and this is slid along the shaft until it strikes the stop. This allows the tool holding the bearing to drive it out of the casing. A bearing extractor set is an expensive and infrequently-used piece of equipment for home use. As such it is probably preferable to hire this type of equipment when it is needed.

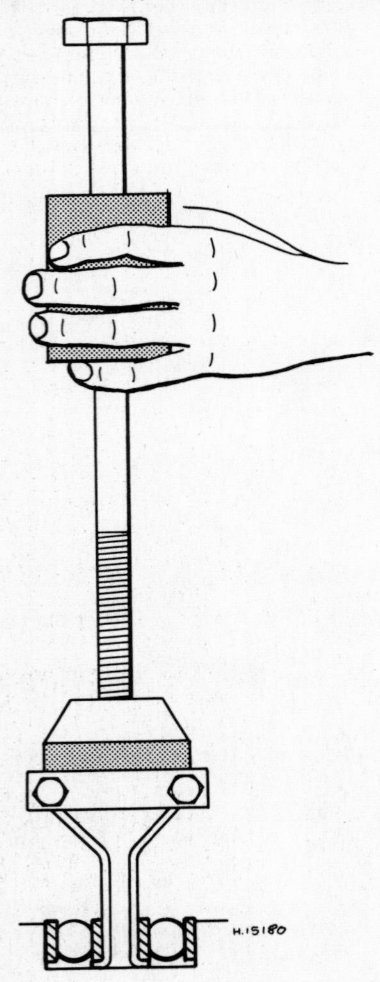

Fig. 2.36 Where a bearing or bush is a tight fit in a blind bore, this type of slide-hammer and bearing extractor arrangement will be needed

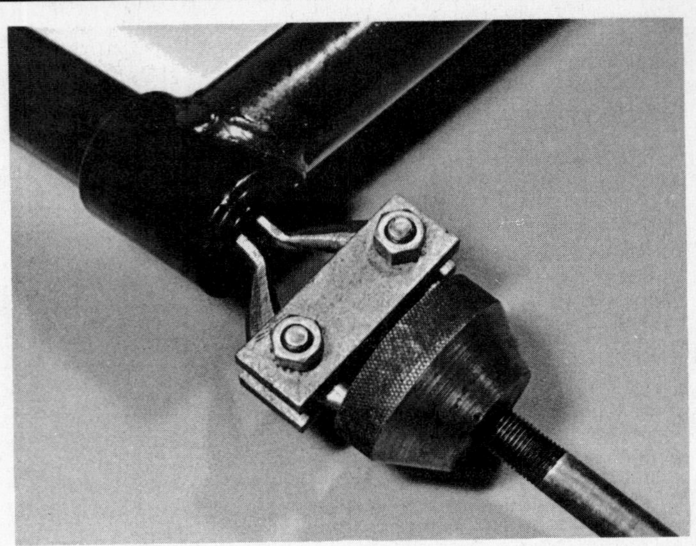

12.9 Where a bush or bearing is housed in a blind bore you will need a slide-hammer extractor to remove it. Either the universal legged puller ...

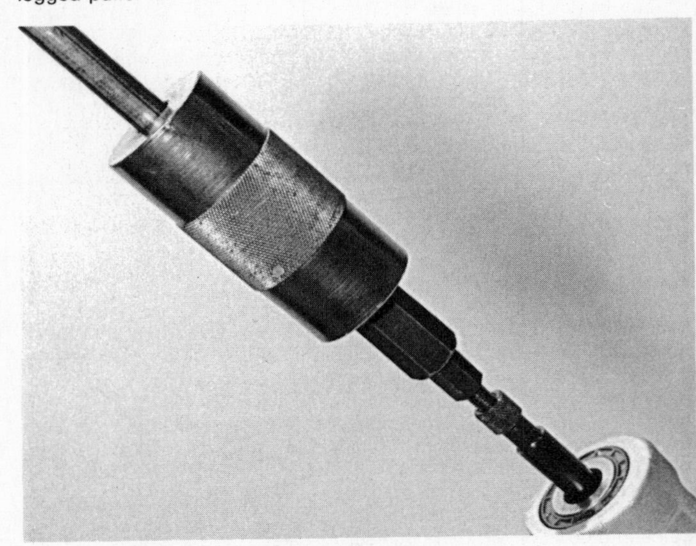

12.10 ... or the wedge arrangement type

13 Holding tools

During engine rebuilds, it is not uncommon to be faced with the problem of holding certain components stationary while a retaining nut is released. This applies to almost all clutches, where the clutch centre is free to rotate when the pressure plate has been removed to gain access to the retaining nut. It also applies to most alternator rotor nuts. As you may have guessed, nearly every manufacturer supplies some sort of holding tool as part of the dealer workshop tool set for that model, and these tools can often be specific to one model or a small range of similar models. Whilst this is fine for the dealer, it is a less than satisfactory situation from our point of view. In this section we will consider some of the possible alternatives to the official tool.

Ready-made tools

There are a few commercially-produced tools which can be useful when attempting to stop an engine component from turning, namely strap wrenches and chain wrenches. In the case of the strap wrench, an adjustable loop of tough nylon webbing is fastened to a handle. With the loop tightened around the part to be held, the loop grips onto the metal quite securely when the handle is pulled against pressure from the spanner. The chain wrench works in exactly the same way, except that a loop of chain is used instead of the webbing. This tends to grip more positively, but the chain links tend to dig into the part being held, and can damage alloy parts quite badly.

Either of the above tools has its place in the workshop, but be warned that the main drawback is finding some way of fitting the tool. As an example, it is almost impossible to use either tool on a clutch centre, because the outer drum masks it. The same problem applies to alternator rotors, though on small machines with flywheel generators, the large flywheel rotor is easily accessible, and thus ideal for holding by this method. It must not be forgotten that any tool not specifically designed for a job may cause damage if not used with great care. It is possible to distort components like flywheel rotors, and where external ignition pickups are fitted, beware of causing damage.

A fabricated clutch centre holding tool

Over the years countless one-off holding tools have been made up when dismantling various machines in the course of originating Haynes Owners Workshop Manuals. These have ranged from pieces of scrap metal cut and filed to lock a component by wedging it, to more generally useful tools. Of these, the most successful and widely used tool has been a scissor-type clutch holding tool. No credit can be claimed for its design – many manufacturers produce a similar item as part of their service tool set for dealers. It is, however, easy to make and works on almost any clutch.

The construction of the tool is most easily seen in the accompanying illustration, using ordinary mild steel strip, obtainable from steel stockists or engineering shops. The tool was made up into the scissor-shape shown with the jaws bent at 90° so that they lined up with the slotted clutch centre. The working edges of the jaws were filed to a shape which matched the profile of the clutch centre slots to reduce the risk of the tool slipping. There is no real need to harden the jaws – they will be gripping on aluminium alloy which is a good deal softer than mild steel. The handles were made quite long so that the jaws would grip the centre firmly without having to apply undue pressure.

A fabricated rotor holding tool

This is another useful tool which will work on the majority of machines. The tool requires that the rotor has slots or plain or tapped holes to accept a holding tool on the outer face. Once again, mild steel strip was used to make up a scissor-shaped device. At each jaw end is a hole through which was fitted a high tensile bolt and a nut to secure it to the jaw. This arrangement will work on most rotors with either plain holes or slots in the outer face. If the rotor has tapped holes, the normal bolts should be removed and replaced with a size which can be screwed directly into the holes in the rotor. It is important to check that the bolt ends do not protrude too far into the rotor, otherwise they may come into contact with, and damage, the coils on the stator assembly. This should be checked, and the effective bolt length adjusted before use.

An improvised chain wrench

If you find that you need a chain wrench when the tool shop is closed, it is relatively easy to make a quick improvised version using an

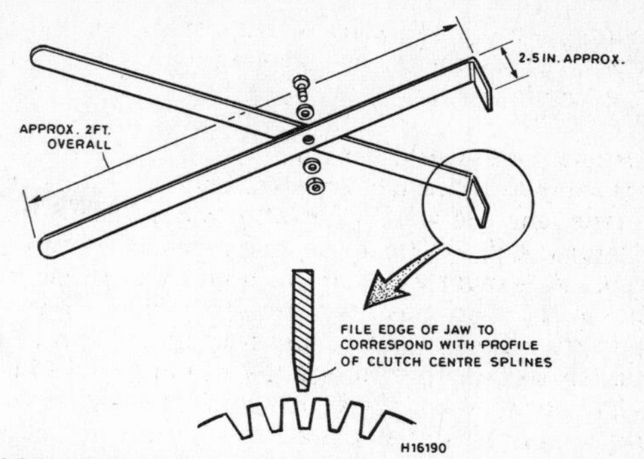

Fig. 2.37 This scissor-shaped holding tool can be adapted to hold almost any clutch centre

It is important to shape the jaw ends to fit the splines if damage is to be avoided

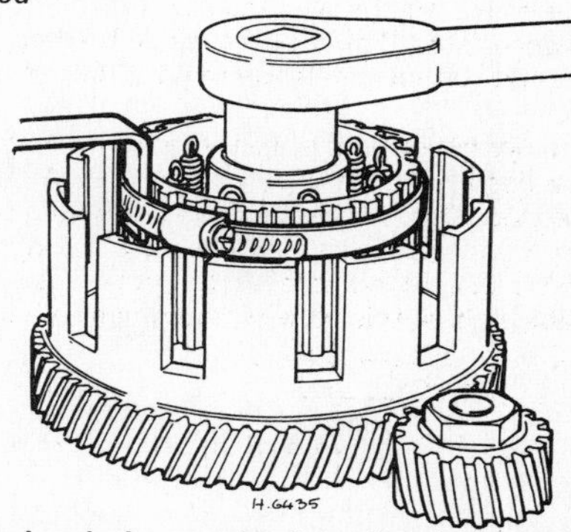

Fig. 2.38 A simple improvised tool to hold a clutch centre while the retaining nut is released

The handle was formed from a piece of scrap steel strip and secured by a large worm-drive hose clip

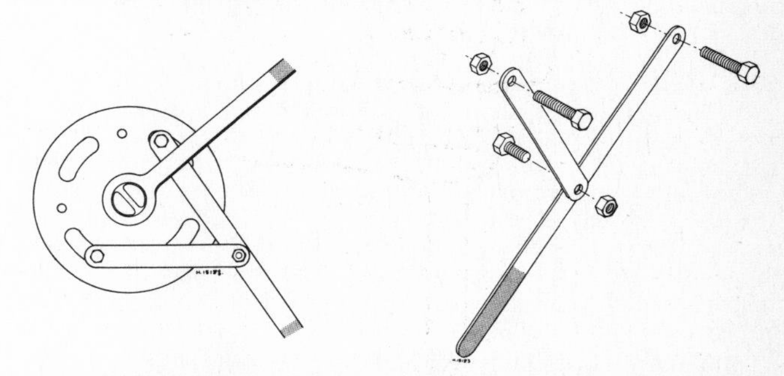

Fig. 2.39 An improvised tool to hold a flywheel generator rotor

The pivoting arm allows a range of spacings to be accommodated

13.1 An improvised chain wrench in use. Note packing to prevent damage to fork seal holder

13.2 A strap wrench in use for slackening oil filter

13.3 The fabricated scissor-type clutch centre holding tool in use

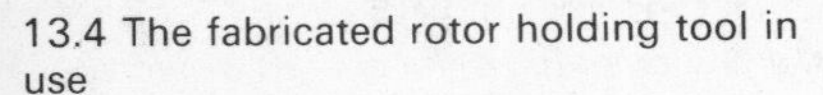

13.4 The fabricated rotor holding tool in use

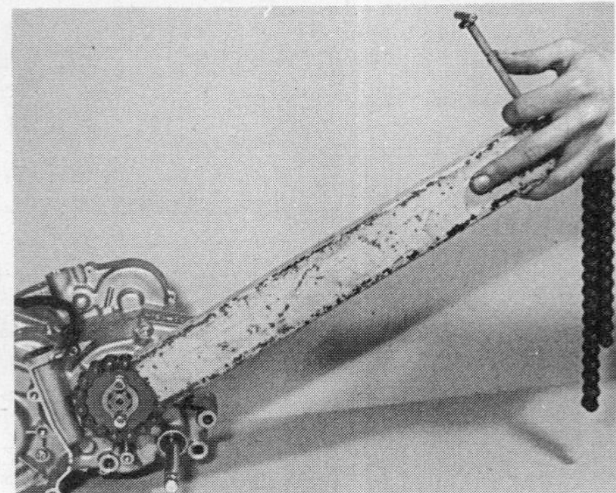

13.5 An improvised chain wrench

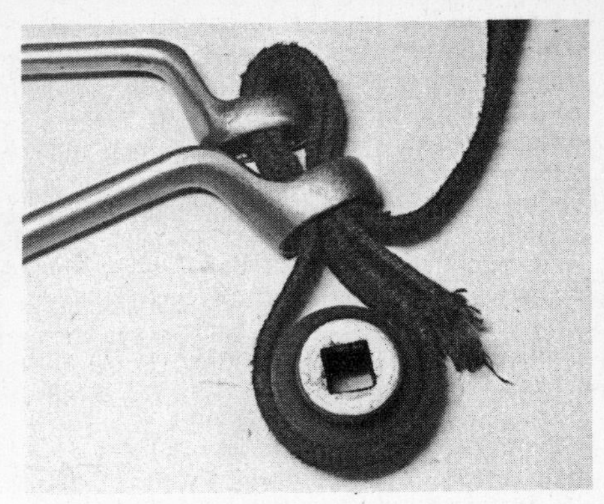

13.6 Use of ring spanners to lock strap wrench in position

old final drive chain and a length of steel tubing. Pass a loop of chain through the centre of the tube and fit this around the item to be held. Pull the handle as tight as possible, then pin it in this position by sliding a small high-tensile bolt or pin through the chain ends. The arrangement described is dependent on the strength of the bolt, which of necessity must be quite small to fit between the chain rollers. Despite this limitation (which could be overcome by a little modification of the basic design) the tool works adequately for most jobs, and is above all quick and simple to make up.

An improvised strap wrench

As with the chain wrench described above, this is quick and easy to make up. Ideally, a length of nylon webbing should be used to form the strap, although in the absence of this a length of rope could be used. The handle could be made up as for the chain wrench described above, passing the loop of webbing through it, and then tightening it by passing a bar through the loop above the handle and twisting it until the webbing tightens around the object to be held. You could also employ a pair of ring spanners as shown in the accompanying photograph. Used in this way the spanners act as a handle and also lock the webbing in position.

Ways of avoiding the need for holding tools

Before we leave the subject of holding tools, it is worth mentioning that as an alternative to the methods described above, it is often possible to avoid the use of holding tools if the matter is considered early enough in the dismantling operation.

If working on a motorcycle with a conventional transmission system note that the entire gear train can be locked via the rear brake if the engine is still in the frame. If top gear is selected and the rear brake applied hard, the engine will be effectively locked in position through the gears and final drive. Since this method requires the engine to be in place in the frame, the clutch centre nut, primary drive nut or rotor nut will have to be slackened **before** engine removal takes place.

Alternatively, once the cylinder head(s) and barrel(s) have been dismantled and the piston(s) removed, rotation is prevented by passing a smooth round bar through the connecting rod small-end eye

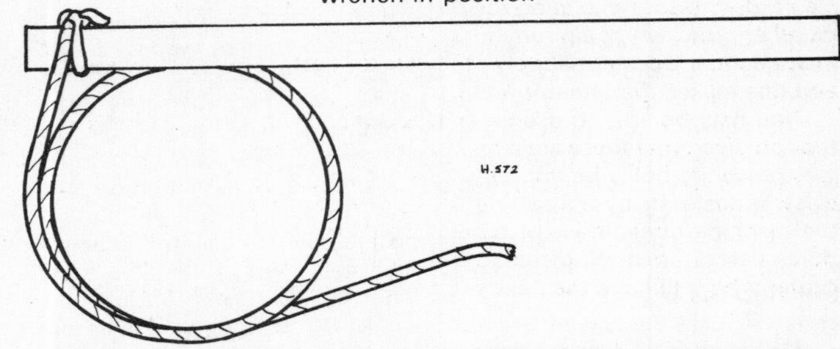

Fig. 2.40 This simple holding tool can be fabricated from a length of steel tubing and some rope

It is quick to make up and can be useful for immobilising rotors

and supporting the ends of the bar on wooden blocks placed across the crankcase mouth. If using this method, make sure that the bar is smooth, or damage may be done to the small-end bush. The wooden packing pieces are essential too, both to prevent the crankshaft from turning past bottom dead centre (BDC), and to spread the load applied to the crankcase and thus avoid damage to its gasket face.

14 Spring compressors

Valve spring compressors

A spring compressor of some description will be required on most machines, the most common requirement being for a valve spring compressor during cylinder head overhauls. As with most other service tools, the manufacturer almost invariably supplies some sort of tool for this job, although in most cases a cheaper 'universal' spring compressor can be used. These are generally designed for use on car engines, and the design of many of them is not suitable for use on motorcycle heads. This is usually due to the additional bulk of the cooling fins on air-cooled engines – a factor not normally considered by the tool manufacturers.

13.7 Where circumstances permit, try bolting a steel strip to the gearbox sprocket as a holding method

13.8 The crankshaft can be immobilised by passing a smooth round bar through the small-end eye. Note use of wooden blocks to protect crankcase surface

Fortunately, it is still possible to obtain tools which will fit motorcycle heads. These usually take the form of a large C-shaped frame, the ends of which are threaded to take adjustable jaw ends. One of these is ground to a blunt taper, while the other normally carries a tubular foot with cutout to permit removal of the valve collets after the spring has been compressed. Note that you are unlikely to find valve spring compressors suitable for motorcycle use on sale in car accessory shops or tool shops; you will probably have to try several large motorcycle dealers, or you could purchase one by mail order from tool suppliers who advertise in the motorcycle press.

Other types of spring compressor

Occasionally, you will need to compress other springs during dismantling operations, suspension springs being the most common candidates. These are not easy to deal with, due to the high spring rates often used, especially on single-shock rear suspension units. Whatever method you use, you should be sure that it is safe; if the arrangement used slips during use you could be badly injured as a result. Before contemplating the removal of this type of spring, find out whether you will gain anything by doing so. All too often the manufacturer does not supply replacement parts, only complete units, and this makes the attempt futile.

You may be able to use a car-type suspension spring compressor, though given that these are made for use on car suspension struts, they may prove too bulky for motorcycle use. It is worth seeking advice from either a dealer or by consulting the appropriate workshop manual for the machine concerned for advice about removal and fitting procedures before attempting this job. If all else fails, take the unit to the dealer who will have the necessary equipment to do the work safely.

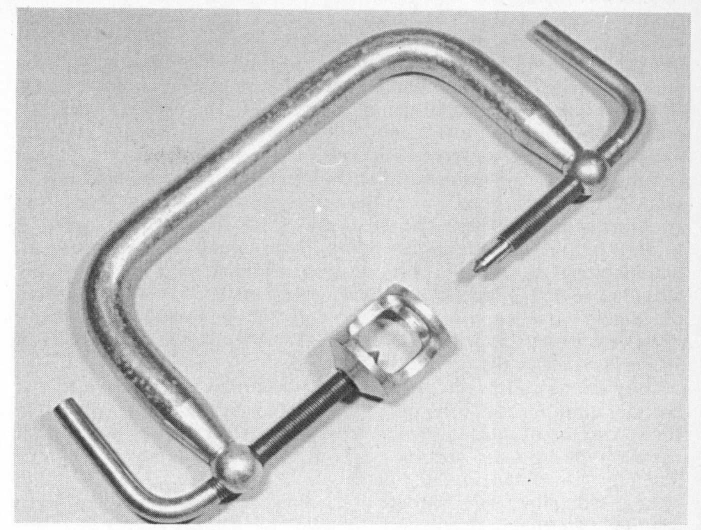

14.1 When purchasing a valve spring compressor, choose a special motorcycle type to ensure that the jaw opening is deep enough; car versions may not fit some heads

14.2 The compressor must be able to span the head as shown. The cutout or notches at the end allow the collet halves to be removed

14.3 In special applications, such as this drive shaft shock absorber assembly, an hydraulic spring compressor may be needed to release the spring safely

15 General test equipment

In addition to the various tools described earlier in this Chapter you are likely to need a certain amount of test equipment. The choice in this area is vast, and effectively limited only by your own budget. Rather than attempt to catalogue all that is available, this section will concentrate on a few items which can be considered either essential or extremely useful when setting up a workshop. Later, you will no doubt develop your own ideas about additional pieces of equipment, and these, like extra tools, can be purchased as the need arises and funds allow.

Compression testers

These devices are basically pressure gauges which connect to the engine via the spark plug hole and allow the pressure developed inside the cylinder to be measured. Most types consist of a clock-type gauge connected by a short rigid or flexible tube to a tapered rubber nozzle. In use, the gauge nozzle is pushed hard into the spark plug hole and the engine is then cranked for a few seconds. When the gauge needle reaches a stable reading it is removed and the reading noted, before the pressure is released to reset it. The exact pressure figure applicable varies widely according to the make and model of machine.

The purpose of the tool is to diagnose loss of pressure in a cylinder without having to dismantle the engine to check for physical wear. Used in the way described above, the gauge will show up bore or ring wear, and this can be extended on multi-cylinder engines to diagnosing compression differences between cylinders. By introducing a little oil into a cylinder showing an abnormally low reading, a worn bore can be temporarily sealed; if the test is repeated, the pressure reading should have improved. Where the reading stays about the same, this points to the leakage being due to a worn or burnt-out valve, rather than in the bore. It follows that this last stage of testing is applicable only to four-stroke engines.

When buying a compression tester, bear in mind the problem of restricted access on motorcycle engines. Like many other tools, compression testers are produced primarily for the car owner, who does not normally have such restricted access in the area above the cylinder head. It follows that a tool with a flexible hose between the gauge and nozzle could be an advantage.

Vacuum gauges

These are used to measure the vacuum effect developed in the inlet tract while the engine is running, and provides an accurate method of setting up and synchronising two or more carburettors. Two basic designs are available, having either clock-type gauge heads, or manometer types relying on mercury filled columns. Either arrangement will work satisfactorily, though with the clock-type instruments it is important that the gauge heads used are accurate. This is less critical with manometer gauges, which cannot easily be inaccurate. Many manufacturers will supply sets of gauges for two, three, or four carburettor installations, and when choosing a set you should consider the maximum number of carburettors you are likely to want to deal with, now and in the future. Check also that a range of adaptors is included with the set.

In use, the synthetic rubber hoses from the gauges are connected, via adaptors, to the inlet port for each cylinder. Most machines are provided with tapped holes which are normally covered by sealing plugs, and these are intended for the connection of vacuum gauges. If no take-off point is provided, you will have to drill and tap the inlet adaptors to allow the gauges to be fitted.

With some gauge sets, you will have to attach each gauge to one cylinder in turn, so that the gauge damper valves can be set up so that each instrument gives the same reading. Other types use preset damping, and do not need setting up. Once connected, the vacuum reading for each cylinder will be shown with the engine idling and at normal operating temperature. The throttle stop screw for each cylinder can then be adjusted to give the same reading for each cylinder. Most manufacturers quote a reading for a particular model, though in practice it is more important to obtain an identical reading on each gauge than to get a specific figure. Note that there are a few exceptions to this general rule; some four cylinder models must be set up so that the inner pair of cylinders show a slightly different reading than the outer two.

Once the idle vacuum settings are correct, the engine should run noticeably smoother at tickover. Indeed, one popular four cylinder model was so sensitive to synchronisation that it could sound as if the clutch or primary drive was in need of urgent attention if the setting was incorrect. Once adjusted, the backlash which was causing the noise was reduced and the noise disappeared.

Attention can now be turned to synchronising the throttles of the various instruments. If the twistgrip is opened up gradually, the gauge readings for each cylinder should move with each gauge staying in synchronisation. If one or more cylinders read differently, it is necessary to adjust the throttle cable or linkage until this is achieved. Once set up, the gauges can be removed, the plugs refitted and the machine tested. You should find that a few minutes work will be well rewarded, with the engine feeling smoother and more flexible. Keeping the carburettors in synchronisation will also tend to improve fuel economy and minimise mechanical wear due to backlash in the primary drive.

Setting up carburettors without vacuum gauges is possible, but far from easy. With twin-cylinder machines it is reasonably easy to do so by ear; a length of plastic tubing can be used as a stethoscope, and the sound of the intake hiss through each carburettor compared. If you can accrue enough experience in this method, you can get fairly good results, but for most people a set of gauges is a better bet. On engines with more than two cylinders, gauges can be considered essential.

Stroboscopes

Stroboscopic test lamps, usually called strobes or timing lights, are used to check dynamically the ignition timing settings. There are two basic types, based on either neon or zenon tubes. The neon versions are cheap, but the weak reddish-coloured light produced is dim and poorly defined. Given that they are hard to see clearly unless it is dark, it is best to avoid them. The zenon tube types give a much brighter white light and are far clearer in use. Zenon lamps normally require a power supply, usually from a 12 volt battery source, but occasionally mains voltage is used. If supplied from a battery it is preferable to use a separate battery – connecting it to the machine's battery can result in spurious signals triggering the lamp. Two other leads are fitted, these connecting to the spark plug and plug cap of the cylinder under test, effectively connecting the lamp into that cylinder's high tension circuit. When the engine is running, the lamp is triggered each time the plug fires, producing a short pulse of light.

The light is aimed at the ignition timing marks, which are usually found near the contact breaker assembly or pickup unit. The rapid pulses of light have the effect of 'freezing' the movement of the rotating pickup or cam unit, and the position of the marks in relation to a fixed index mark can be seen at the moment that the plug fires. If the two do not align, this indicates that the ignition timing setting is incorrect.

In the case of machines with contact breaker ignition systems (though mopeds and some small two-strokes are exceptions here) you can alter the ignition setting by turning the contact breaker baseplate once the retaining screws have been slackened. On most machines with electronic ignition, and in the case of the small two-strokes mentioned above, adjustment is not possible. If you are dealing with electronic ignition, incorrect timing usually means a fault in the ignition amplifier, whilst on fixed contact breaker systems, worn or badly adjusted contact breakers are usually to blame.

On four-cylinder engines, and some twins, a spare-spark ignition system is used. With this arrangement, two plugs are run from a shared ignition coil. This means that both cylinders spark at the same time, even though only one is at compression. In the case of most Japanese fours, the system is arranged so that the cylinders 1 and 4 fire together, as do cylinder 2 and 3, so two timing checks are made. The timing marks normally show which pair of cylinders they relate to, often being marked 1-4 F and 2-3 F.

You will note from the above that a strobe can be an invaluable piece of equipment when used on machines with adjustable ignition timing. On models where the ignition is fixed, they are of more limited use; the strobe can tell you if the ignition setting is incorrect, but it is not normally possible to do much about this, short of fitting new parts. With electronic systems it is debatable whether regular timing checks serve any useful purpose, though it is worthwhile being able to diagnose a suspected fault.

Multimeters

A multimeter is an invaluable piece of equipment if you are contemplating any work on the electrical system. Comprising a multi-purpose electrical test instrument, the multimeter is capable of measuring resistance (ohms or Ω) ac and dc volts, and in some cases current (amps). Most meters have several ranges in each of these catagories. These meters are primarily intended for electronics work, and so the cheaper ones do not have current ranges capable of more than a few milliamps. You can either pay a little more to get this feature, or do without it (you can always buy a separate ammeter if you need

15.1 Small compression testers like this can be obtained but may prove difficult to gain access on multi-cylinder engines

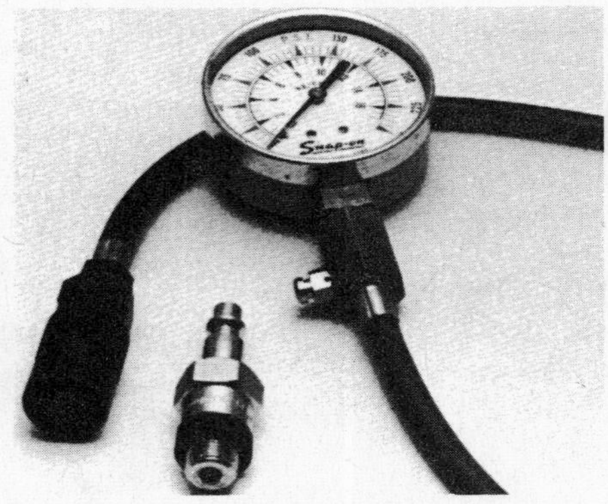

15.2 This type of compression tester with a hose and spark plug adaptor is preferable for motorcycle use

15.3 A stroboscope, preferably the zenon type, is necessary for dynamic timing checks

it). Slightly more expensive versions have an internal resistance which allows current readings up to about 10A.

The single most useful function of the multimeter is its resistance ranges. In addition to its obvious application in checking the internal resistances of electrical components, the resistance range allows the meter to function as a simple continuity tester; invaluable for checking for broken connections or short circuits.

If you are shopping for a multimeter, the best places to find them are electronics hobby shops or suppliers. The cheapest versions are the tiny 'pocket' multimeters. These are adequate for most jobs (except for the lack of current ranges) and are accurate enough for general motorcycle work. The slightly bigger, more expensive models usually have more facilities and are more accurate. Accuracy becomes important if you need to check specific values in some of the electronic assemblies on a motorcycle.

It should be noted that the polarity and internal workings of some meters means that the resistance readings you might get when testing a perfectly sound component could vary from those specified by the manufacturer. The only way to avoid this is to make sure that you use the specified meter for that model. This is normally mentioned in the relevant Owners Workshop Manual for the machine in question, and the approved meter can be obtained from an official dealer, if required. In practice, any reasonable meter will give a good indication of a major fault, even if the values do not agree precisely.

16 Measuring equipment

During any overhaul or repair job, you will need a certain amount of basic measuring equipment in order that wear can be assessed. In addition, some of the more basic devices, such as feeler gauges, are needed for routine servicing work. In this section we will look at the most commonly-needed items, starting with those which should be considered essential, and working up to more specialised and expensive items. Some of these, such as vernier calipers, and certainly micrometers, are specialised pieces of equipment, and might be thought to be somewhat ambitious for home use. You could probably get by for most of the time without them, but be prepared to ask someone else to do any precise measuring for you, should the need arise.

Feeler gauges

These are an essential purchase for work on almost any machine. If it is either four-stroke, or has contact breaker ignition, you will need feeler gauges to set the valve or contact breaker clearances. Feeler gauges normally come in sets. In the smaller sets, different feeler gauges must be combined to make up thicknesses not covered directly. Larger sets have a wider range of sizes which avoids this problem. Feeler gauges are available in either metric or Imperial (inch) sizes; you should buy the range which suits your machine. Normal feeler gauges take the form of thin steel strips, and these are the best choice for most purposes. There are also wire-type feeler gauges which may be preferable in some circumstances.

You will need feeler gauges wherever it is necessary to make an accurate measurement of a small gap, typical applications being the checking of valve clearances on four-stroke engines, endfloat in gearbox shaft assemblies and crankshafts, and for checking clearances in oil pumps. You can also use feeler gauges when assessing distortion of gasket faces; the cover or casting is placed, gasket face down, on a surface plate, and any gap can then be measured directly using the feeler gauges.

To measure a gap using feeler gauges, slide progressively thicker blades into the gap until you find the size which gives a light sliding fit. This can sometimes be awkward, particularly when measuring valve clearances on machines where access is limited. In such cases it is better to open up the adjustment, fit the correct size of blade, and then screw the adjuster down until the correct gap is obtained. On machines where bucket-type cam followers are used with overhead camshafts, this is obviously not practical. In this situation, you will have to make a careful measurement of the existing clearance so that any revised shim sizes can be calculated.

Rulers

A basic steel ruler is another essential workshop item, and can be used both for measuring and marking out, and as a straightedge for checking warpage of gasket faces. Buy the best quality tool you can afford, and keep it well away from the toolbox, or it will soon get bent and damaged.

Dial gauges

The dial gauge, or dial test indicator (DTI) as it is more correctly known, consists of a short probe attached to a clock-type gauge unit capable of measuring small amounts of movement accurately. These test instruments are generally useful for checking runout in shafts or other rotating components. The gauge is fitted into a support bracket and the probe arranged so that it rests on the shaft to be checked. The rotating face is set to zero in relation to the pointer, and then the shaft is turned and the movement of the needle noted.

Another popular adaptation of the dial gauge is in the checking and setting up of ignition timing on many two-strokes. In this guise the gauge is attached to a special adaptor which screws into the spark plug hole. The gauge allows the exact TDC position of the piston to be established, and it is then a simple matter to set the crankshaft so that the piston is at the specified distance before TDC for the points to be just separating. It is possible to purchase dial gauge kits complete with the adaptor for this specific purpose.

Degree discs

The degree disc used to be a popular device used for setting up ignition timing or for checking cam timing settings. In its simplest form it consisted of a printed card disc calibrated in degrees. The disc was fitted to the end of the crankshaft via a fixing hole at its centre, and a wire pointer was arranged close to the edge of the disc, attached to a casing screw. The method can still be used successfully, provided that a degree disc can be purchased or made, and that the required settings are known as a degree figure, rather than the position of the piston before top dead centre (BTDC).

Degree discs were once widely used, and available from most motorcycle dealers. They are generally made from stiff card, and marked in degrees around the outer edge of the disc. The disc centre is marked and could thus be made to fit any size of bolt. Often they were given away free as promotional devices for oil companies, but seem to

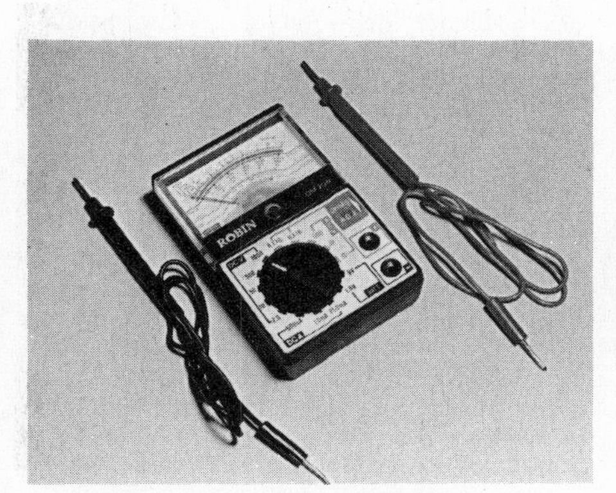

15.4 Cheap 'pocket' multimeters are adequate for simple electrical checks

16.1 A good quality feeler gauge set is essential

16.2 Wire-type feeler gauges for checking spark plug gap. Adjustment by slotted tool

16.3 A typical feeler gauge application is the measurement of gaps or clearances

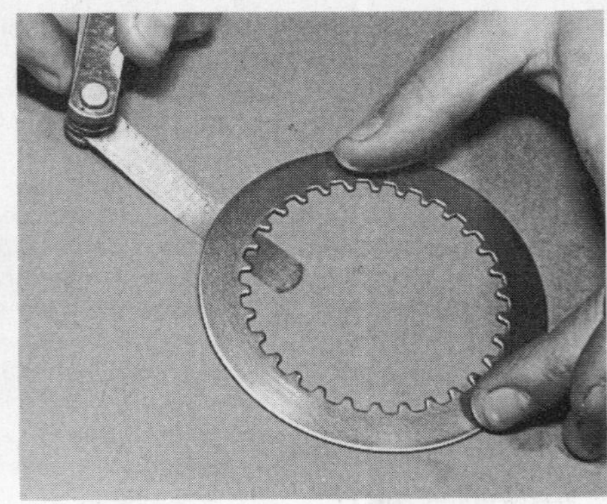

16.4 Feeler gauges can also be used to check distortion – in this case clutch plate warpage

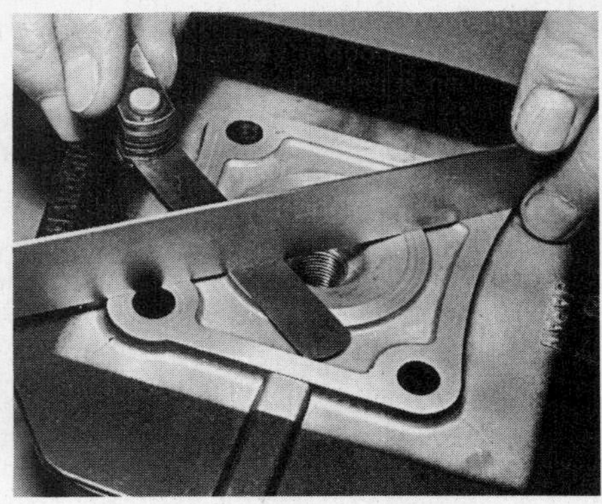

16.5 A ruler can also be used as a straightedge when checking for gasket face distortion

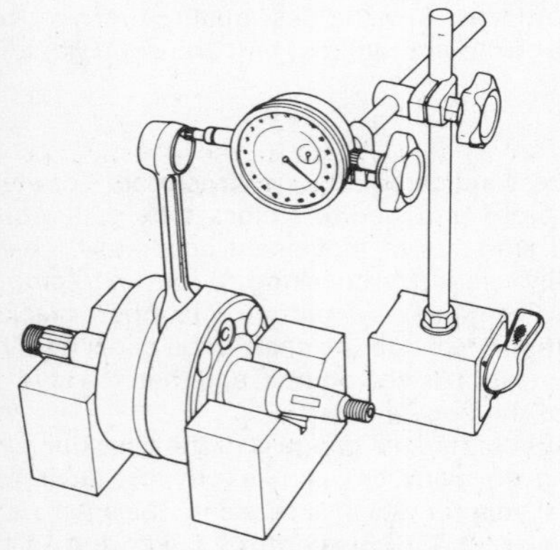

Fig. 2.41 Using a dial gauge set to measure side-to-side deflection of the big-end bearing

16.6 This comprehensive dial gauge set features a wide range of adaptors and probes for numerous measurement tasks

16.7 A degree disc in use. Although mainly used for timing checks its application here is tightening fasteners by degrees of rotation

have largely disappeared in recent years. If you are unable to locate one, purchase a simple 360° protractor from a stationers or office supplies company.

You will now need to attach the disc to one end of the crankshaft, using a bolt or nut, as appropriate, to secure it in place. It may be possible to employ the alternator rotor nut or bolt to secure the disc, possibly using washers or spacers in place of the rotor where its design dictates the need for removal. Try to fix the disc securely so that it cannot slip in use. Fix a simple wire pointer to a nearby casing screw, positioning the end close to the edge of the disc.

The next step is to find the exact position of top dead centre. This is done using a piston stop; a device which will prevent the complete rotation of the crankshaft. A suitable stop can be made up using an old spark plug. Clamp the old plug in a vice, then carefully cut away the spun-over ring holding the insulator to the plug body. The insulator and the centre electrode can then be withdrawn and discarded. Cut off and file flat the earth electrode, leaving the bare metal plug body. A bolt is now passed down through the hollow body and secured with a nut, the resulting bolt end forming a fixed stop protruding from the threaded end of the plug body.

With the stop in position, the crankshaft is turned one way until the piston touches the stop. The disc is then set in relation to the crankshaft so that it reads zero degrees (0°) in relation to the pointer. If the crankshaft is then turned the other way until it again stops, the arc of movement can be read off on the disc.

The exact amount of movement is unimportant – we simply need to divide this figure in half to allow the position of TDC to be established. If, for example, the disc shows 240° of movement between the two stop positions, this means that if the crankshaft is turned 120° from either stop position, the piston will be at the bottom of its stroke, or BDC.

In practice, the crankshaft is turned to one of the stop positions and the disc is then set to read zero (0°) against the pointer. Turn the crankshaft until the halfway position is shown (120° in the above example) to leave the piston at BDC. The disc is now repositioned once again to indicate 180°, denoting bottom dead centre, and the piston stop removed. All that needs to be done now is to turn the crankshaft in the normal direction of rotation until the required position before top dead centre (BTDC) is shown by the disc. Once TDC has been established, it is relatively easy to check ignition or cam settings where these are expressed in degrees.

In more recent times, the need to check cam timing has all but vanished, and most manufacturers now prefer to express ignition timing settings on two-strokes in terms of piston position (see above). As a result the humble and inexpensive degree disc is all but forgotten. It is, however, useful if you are contemplating tuning modifications, especially where different cam timings are being tried.

Vernier calipers

Although not strictly essential for routine work, a vernier caliper is a good investment in any workshop. The tool allows fairly precise internal and external measurements to be made, up to a maximum of about 6 in (150 mm) or so. The object to be measured is positioned inside the external jaws (or outside the internal jaws) and the size read off against a sliding scale. The vernier scale allows this to be further narrowed down, giving an accuracy of about 0.001 in (0.05 mm). The vernier caliper allows reasonably precise measurement of a wide variety of objects, and thus is a versatile piece of equipment. Whilst it lacks the absolute accuracy of a micrometer, it is a great deal cheaper to buy, and much less limited in use.

Being a precision tool, a vernier caliper will not be cheap, but on the other hand, if looked after will last for many years. A cheaper alternative would be a plastic vernier caliper, which would suffice if accuracy was not of paramount importance. Cheaper still would be a pair of ordinary calipers, which can be used for measuring and then read off against a steel rule. Whilst this is certainly an inexpensive alternative, it is by no means as accurate as a good-quality vernier caliper, and as such of limited use if a precise reading is necessary.

Micrometers

The micrometer is the most accurate measuring tool likely to be needed in the home workshop, and indeed you could reasonably argue that its cost balanced against its occasional use makes it an unaffordable luxury. This is particularly so since any one instrument can cover a limited range of sizes, and you would need a set of inside *and* outside micrometers to be fully prepared for any measuring job.

The basic outside micrometer comprises a U-shaped frame covering a limited size range. At one end is a ground stop, called the anvil, whilst at the other is an adjustable stop, called the spindle. This is moved in or out of the frame on a high-precision fine thread by means of a calibrated thimble, usually incorporating a ratchet to prevent damage to the mechanism. In use, the spindle is screwed down carefully until the object to be measured is gripped very lightly between the anvil and spindle. A calibrated datum line on the fixed sleeve below the spindle shows the rough size, whilst the accurate measurement is made by adding to this base figure, an additional clearance shown on the thimble calibrations.

Micrometers are available in a large range of sizes, starting in the range 0-1 inch, 1-2 inch and so on. There are also versions called internal micrometers, which as you will have guessed, are for making internal measurements, such as cylinder bore sizes. Alternatives to the internal micrometer include telescope gauges which allow the reading to be made indirectly, and then read off on a conventional outside micrometer.

All micrometers are precision instruments and are easily damaged if misused. They require regular checking and calibration if their accuracy is to be maintained. Given the fragile nature and high cost of these devices, I would suggest that as a general rule you make do without them until you know that you have a definite need for this degree of accuracy. The only common instance where a micrometer would be useful for routine servicing work is in the case of machines with shim valve clearance adjustment. This can be overcome by marking each shim as it is removed, and taking the shims and a sheet of paper showing the existing clearances to the dealers when ordering new shims. The dealer can then carry out the measurement and calculations for you, and order the shims from this. Whilst you may be charged for this service, it will cost a very small fraction of the price of a micrometer.

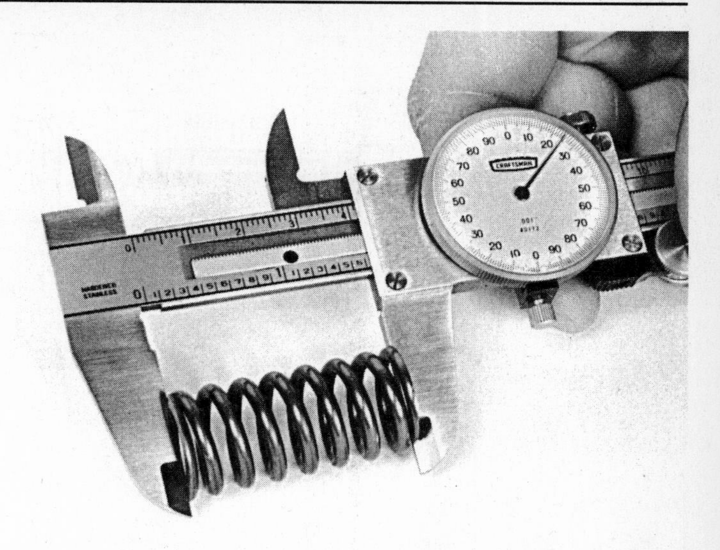

16.8 A vernier caliper is an invaluable general-purpose measuring tool of reasonable accuracy, whether for checking spring free length as shown ...

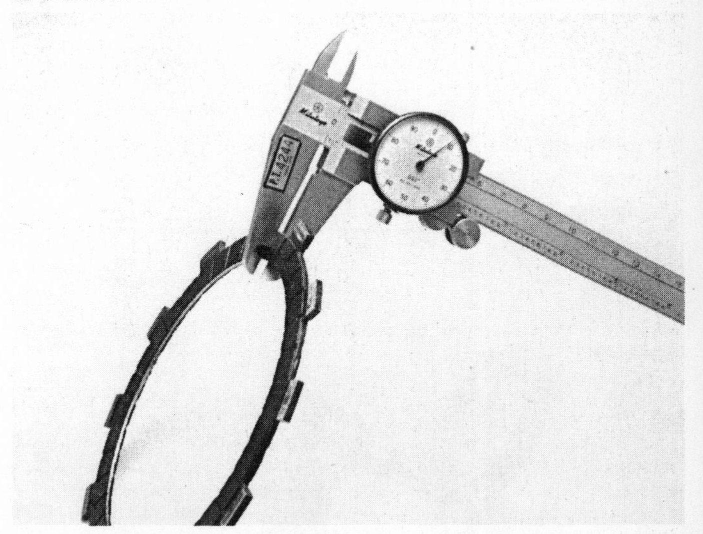

16.9 ... or measuring clutch plate wear. This example features a dial-type scale in place of the normal vernier scale

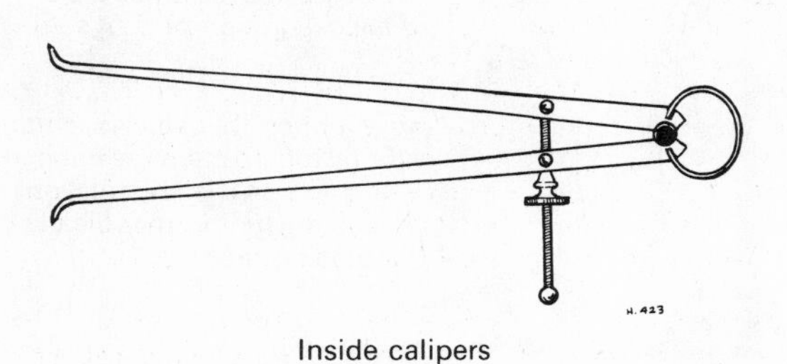

Inside calipers

Outside calipers

Fig. 2.42 Simple measuring can be carried out using inside and outside calipers

The piece is measured, and the distance read off against a steel rule. Note that this method does not give great accuracy

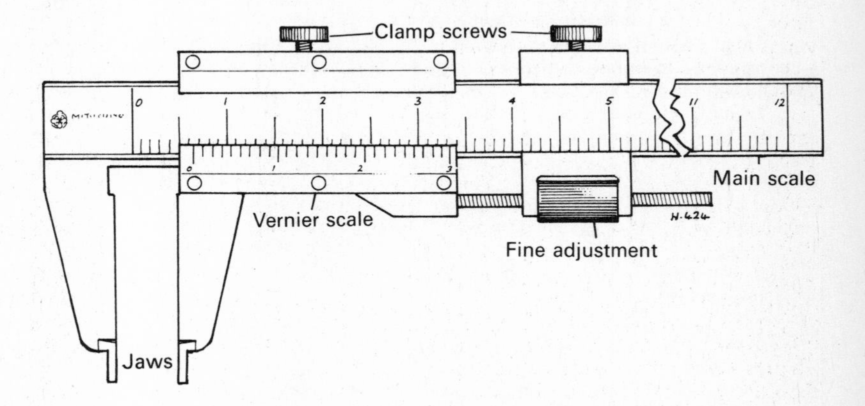

Fig. 2.43 More accurate than calipers is a vernier caliper gauge

The jaw design allows internal and external measurements to be made

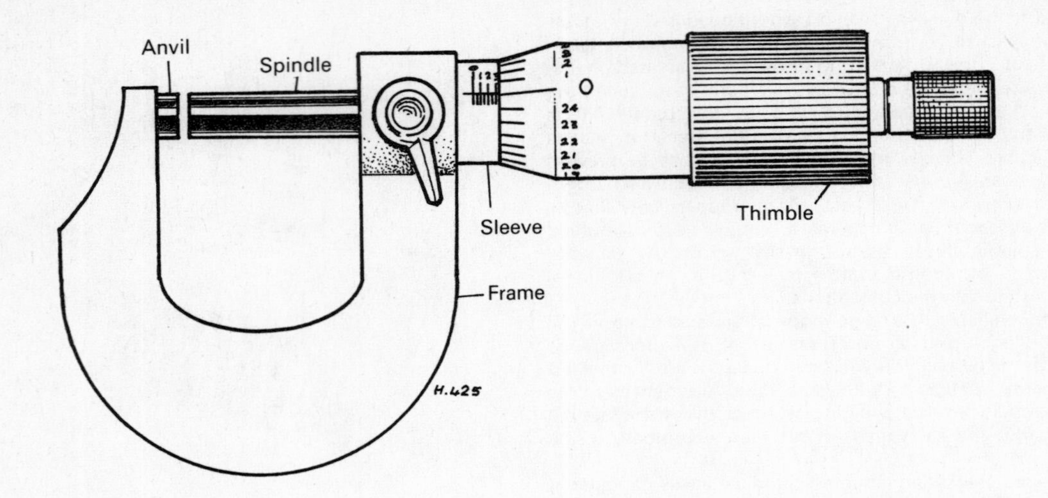

Fig. 2.44 The most accurate workshop measuring tool is the micrometer

Disadvantages include its high cost, the need for several different types to cover most measuring work, and the risk of inaccurate readings if the instrument gets damaged. Even more expensive versions have a battery-powered digital readout in place of the thimble and sleeve scales

16.10 A 0 – 1 inch micrometer is well suited for jobs like shim measurement and similar work

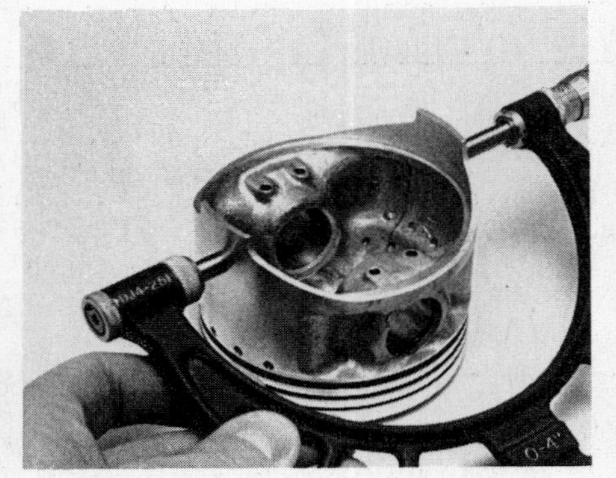

16.11 Bigger components will demand a wider jaw opening

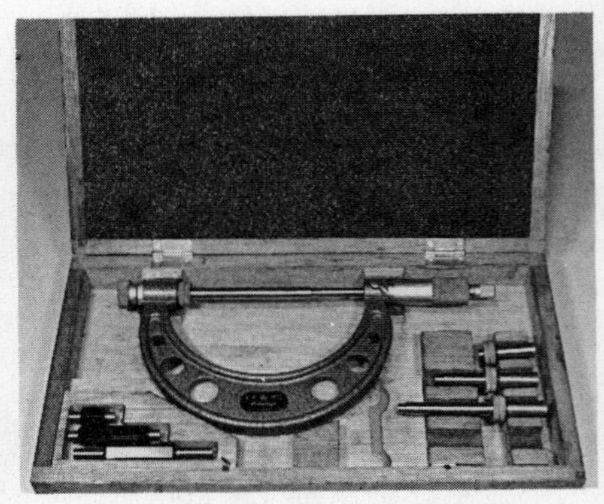

16.12 This set features a large micrometer with interchangeable anvils to allow measurement of smaller parts

Chapter 3 Workshop techniques

Contents

1 Introduction

In this Chapter we will look at some of the general workshop techniques and procedures commonly needed when working on motorcycles. For most of the time, the work is likely to be confined to dismantling, overhaul and reassembly of the motorcycle itself, and to a much lesser degree, repair of worn or damaged parts. For this reason, there will not be great emphasis on fabrication work, although we will consider a few of the simpler working methods and review the uses of more advanced techniques.

Becoming proficient in the skills required to work with various metals and other materials will take time, practice and above all, training. This is true also of other processes such as welding, riveting, lathe work and painting. You will have to make up your own mind whether you have the time or the need to develop these skills, bearing in mind that you may use them very infrequently. Add to this the cost of equipping yourself for these jobs and it soon becomes apparent that it is both quicker and cheaper to have such work carried out for you.

If you decide that you would like to acquire one or more of the more specialised skills, you should consider learning one skill at a time, and find someone to provide the necessary training. To this end, check with your local authority about evening courses in your area. Most colleges run courses of this type, often during evenings, and these are about the best way to learn new skills. It is also possible to obtain books devoted to such crafts, but whilst these will provide a good deal of useful theoretical advice, they should be considered only as reference material to back up actual training, rather than a substitute for it. It should be stressed that it is likely to be ineffective as well as potentially dangerous to attempt to carry out extensive modifications to your motorcycle, unless you first master the necessary skills to do so, as well as acquiring the necessary equipment.

You will not need to develop specialised skills for routine repair and overhaul work, and a basic understanding of the materials you are dealing with and the correct approach to working with them will suffice.

2 Raw materials

At some stage, you are likely to need to indulge in a little simple fabrication work, and on motorcycles this most often means making up brackets and similar fittings. Before you start, consider what will be required of the fitting. A simple bracket to support a silencer or an electrical accessory should present no great difficulty, but if the proposed item is to be in any way structural, great care will have to be taken to make sure it is safe. This means designing the part carefully, and using the correct materials in its construction. Once again the need for experience is underlined. There is no simple rule-of-thumb which can be applied to indicate the correct choice of material and design for any particular job; this is a skill which must be learnt by practice. Until you are sufficiently confident to know what to choose, either avoid any structurally important areas entirely, or get advice from someone with the required training and experience.

The choice of raw materials for a given job is governed by a number of factors, foremost amongst which is ensuring adequate strength. On motorcycles, there is a fairly high level of stresses due to vibration, and the proposed fitting must be able to withstand this. There is also the effects of weather to be considered. On most machines, almost every part is visible, and so the finished appearance must be taken into account. More importantly, you should bear in mind the likelihood of corrosion. In this section, we will look at the more commonly-used materials so that you can develop a general understanding of what types are used for particular jobs, and what properties each of these possess.

Ferrous metals

Iron is used, on its own, or more commonly in an alloy to form steel, for numerous engineering applications. On motorcycles, plain iron may be found as the material used in some castings, such as cylinder barrels, and occasionally for certain applications such as piston rings or valve guides.

Steel is an alloy of iron, together with small amounts of carbon, and in some cases various other metals. These tiny additions to the iron are designed to increase its strength and alter its hardness and flexibility. The proportions of these additional materials are chosen to emphasise a particular characteristic, and in this way various types or grades of steel can be made to suit the application to which the steel is to be put. It is this adaptability, along with the relative abundance and low cost of iron, which has made steel the basic material for most vehicle applications for many years.

In the case of most production motorcycles, steel is most evident in the frame. This is normally assembled from a collection of tubing, pressings and cast lugs, welded together to form the skeleton around which the machine is constructed. Less obvious are the smaller but diverse applications of various grades of steel, from fasteners and bearings through to items like control cables, wheel spokes and the fuel tank.

In some of these applications, the steel used must be ductile and flexible to allow it to be pressed into complex shapes, a good example being the fuel tank which is pressed from mild steel sheet. In other areas, these attributes would be a positive disadvantage, such as in bearing balls and rollers where a hard finish is essential.

Whilst some of the characteristics of a metal are decided at the time the original steel was manufactured, others are imparted at later stages of production. In its original state, the molten steel is poured into moulds to produce cast ingots, and these can be treated in various ways to impart certain properties. In its ingot form, the steel will have a hard, grainy and crystalline structure, similar to that of cast iron. If this material is then 'worked', by forcing it through rollers or by hammering or forging it in a hot state, the structure can be altered to produce a more resilient material with a directional structure to the grain, rather like wood.

Further treatments can impart hardening, either throughout the material, or more usefully, as a hard layer on the outside of a softer but tougher core. This allows items such as crankshafts and camshafts to be made, combining the tensile strength of a ductile steel with a hard working surface to resist wear. The various heat treatment techniques are outside the scope of this book, and indeed many of them employ dangerous chemicals which would be hazardous or illegal to use at home.

Simple case hardening of small steel components can, however, be carried out safely at home at no great expense. These processes require the part to be case hardened to be coated in a carbon-rich powder and then heated. The carbon is absorbed into the surface layer of the metal, altering the molecular structure of the outer layer to give it good resistance to wear. Proprietary chemicals for this process can be obtained from engineering suppliers and will give detailed instuctions for use. More sophisticated processes will demand facilities not easily created in the small workshop, and thus should be left to experts.

In its simplest form, hardening is achieved by heating the steel until a prescribed temperature, usually judged by its surface colour, has been reached. It is then cooled rapidly by quenching it in water or oil. In practice, the process is more closely controlled than this, and other techniques are used to achieve more accurate degrees of hardness. The opposite effect, annealing, is achieved by heating the steel to a cherry-red colour, and then allowing it to cool slowly in air. In each case, the principle of this treatment is to modify the molecular structure of the metal to change its physical properties, usually in the presence of chemicals. For those interested in the subject it is suggested that a more detailed study is made, referring to specialist works on the various techniques used commercially.

The motorcycle manufacturer works from stock materials, in the form of castings or forgings made specifically for a particular model, to tube, bar and sheet stock. The finished parts, be it a frame or a crankshaft, pass through a series of machining and treatment processes to achieve the result specified by the designers. Other parts will have been bought in from outside suppliers who specialise in the manufacture of one type of product, a good example being the numerous fasteners used to hold the machine together.

When working on the machine it is important to distinguish between the various grades of steel. Forged, and in particular, cast steel or iron parts, are fairly hard and brittle, and it is easy to break them if struck hard. Mild steel, as used in sheet, bar and tube form is much less fragile in this respect, and comparatively easy to deal with.

The most commonly-used material for general use, and the type most likely to be needed when making up the odd bracket or spacer is mild steel, and this has numerous advantages. It is reasonably easy to work, being malleable. It can be drilled, tapped, cut and welded where necessary and it is cheap and easily obtained in numerous shapes and sizes. Its main drawback is that it rusts easily, and thus must be painted properly before fitting.

Stainless steel is another extremely useful material for motorcycle use, having a natural bright finish which is corrosion resistant and capable of being polished to a mirror finish, if desired. In most respects it is similar to mild steel to work with, though slightly less easily worked. Most commercially-used grades of stainless steel are alloys of steel and metals such as nickel and chromium, plus traces of other metals. The percentages of nickel and chromium denote the degree of corrosion resistance, the most common grade being 18/8 (18% chromium and 8% nickel). This grade has a high degree of corrosion resistance, and is non-magnetic.

Despite its general usefulness, the marginally greater unit cost incurred when using stainless steel has so far dissuaded most manufacturers from employing this material on production motorcycles, though the two biggest problem areas, fasteners and exhaust systems, are well catered for by after-market suppliers. It is widely assumed by the more cynical that stainless steels are not used because to do so would damage the lucrative replacement parts business. You can draw your own conclusions on this point, with the suggestion that you consider it in greater depth the next time your exhaust system rots through from the inside under the inevitable assault from acidic exhaust deposits.

The uses of stainless steel for the home workshop are almost as great as those for ordinary mild steel, and for making up simple brackets it is unsurpassed. The cost is not much greater than mild steel, and you will not need to spend time and money on painting the finished work, unless it needs to match existing painted parts for cosmetic reasons. If required, the finished part can be buffed and polished to a chromium-like mirror finish, and there is no fear of subsequent rusting, if the correct grade is employed. Any scratches or scrapes made during manufacture, or after fitting can, within reason, be buffed out.

Non-ferrous metals

The most commonly-encountered metal apart from the various steels described above is aluminium. This is a soft, whitish metal, most usually found as the major part of alloys and used extensively on motorcycles, where its light weight and good heat dissipation ability make it an invaluable material. On most machines, the majority of the engine castings will be of aluminium alloy, and so too are many wheels and frame fittings.

Aluminium, in the form of a high-tensile alloy such as Duralumin, can be very useful for making up brackets and similar fittings. It is rather less easy to work than mild steel, being inclined to clog cutting tools, but it will allow presentable-looking items to be made up. It should be noted, however, that welding aluminium alloys is less easy than welding steel. Aluminium is light in weight, and it corrodes less easily than mild steel, forming a thin film of aluminium oxide on its surface which then almost eliminates further corrosion. This allows the fitting to be polished, if required. An exception to this rule is its susceptibility to electrolytic corrosion when it is in contact with a dissimilar metal like steel. This process, which is speeded by the presence of water and road salt, will lead to the rapid erosion and decomposition of the metal. This can be a real problem in some applications, requiring painting to prevent the corrosion from taking place. A secondary disadvantage is that aluminium is not as easy to deal with as steel when a paint finish is required, and you will need to use special aluminium etch primers to prevent the paint from flaking off.

Some of the more intricate castings found on motorcycles are made not from aluminium alloy, but from a zinc-based alloy. This is much heavier than the aluminium types, but is a fine-grained material, and thus suitable for die-casting. In this process, the liquid metal is forced into a complex mould under high pressure to produce a very high quality complex casting requiring little finishing work. It is this which makes it suitable for jobs requiring very intricate casting details to be reproduced accurately. Zinc alloys are soft and very brittle, and very difficult to repair if damaged. The metal will polish to a very bright finish, but will then rapidly tarnish to its more usual dull grey surface finish.

Other non-ferrous metals are used in various applications on most motorcycles. These include copper, which is chosen as the conductor in electrical wiring because of its good electrical conductivity. Copper corrodes easily, however. Thus, the softness of the metal, and its tendency to work harden under the effects of vibration, mean that it is rarely used for other purposes. One odd property of copper which is worth noting, is the method of annealing hardened and embrittled material. This is the exact opposite of ferrous metals; the copper is heated and then quenched in cold water to soften it.

Brass, an alloy of copper, zinc and tin, is also used for electrical applications, and will be found in switches where it is normally used for the contacts. Some carburettor manufacturers still use brass pressings soldered together to form the float, though these are more often made from plastics these days. Brass is a stronger and springier material than copper, and thus suits this sort of application well. Both copper and brass are easy to work, and can be joined without difficulty by soldering.

Plastics

Plastics have been used extensively on motorcycles for many years now, mainly for trim and non load-bearing items. In more recent years, developments in plastics technology has seen the gradual appearance of more sophisticated materials able to withstand greater stresses, and more importantly, heat. These types are often referred to as engineering plastics, and it is likely that these will replace more and more of the metal components in the engine and transmission as this branch of technology is developed still further.

'Plastics' is a very generalised term used to group together a vast range of materials, and it would be futile to attempt to describe the composition and properties of even the major types in this book. With a few exceptions, plastics are not good raw materials for the home workshop. Apparently similar materials often have widely differing properties; eg one may melt at very moderate temperatures whilst a seemingly similar material may be sufficiently heat resistant to be used to form crankcase covers. Many plastics are very prone to attack by solvents, though certain types allow carburettor parts to be made, despite constant immersion in fuel.

As a general rule, your most likely course of action when dealing with a damaged plastic part is to buy a new one. In some cases you may be able to repair cosmetic shrouds and covers by glueing them, though this would be unwise in the case of a more important engine part, like an inspection cover, where subsequent failure could lead to oil loss or worse. Items such as fairings, seat tail sections and side covers can be repaired and repainted, but this is very much a specialised job, requiring heat welding machinery and specialist paints and primers.

Smaller repair jobs like damage to seat covers or small splits in fairings or other moulded panels can be attempted at home, and these are discussed later in this book. You might also need to learn how to use glass reinforced plastic (GRP), more commonly known as fibreglass. As the name suggests, this material consists of a resin which can be hardened by the addition of a catalyst. The liquid resin can be reinforced by adding layers of glass fibres in the form of chopped strand or woven matting. The resulting material is light and strong, and can be 'layed up' in a detailed mould to produce a variety of complex shapes. It is often used to produce small runs of fairings and similar items, and it is a technique which can be copied successfully at home to produce small items. At a more advanced level, GRP can produce structural, load-bearing items like complete car bodies, but this level of expertise is outside the scope of most home workshops.

A development of GRP technology saw the introduction of carbon fibre materials. Initially confined to the aircraft world due to its high cost, the procedure is not dissimilar to working with glass reinforcing fibres. Carbon fibres allow light, flexible and immensely strong structures to be built. To date this material has not seen much use on motorcycles, though several high-specification safety helmets have used carbon fibre reinforcement.

It is convenient to group rubber with plastics, since most of those used are synthetic products. Although the most obvious use of rubber on a motorcycle is in the tyres, you will find innumerable examples of it used to form the hoses, seals and bushes used throughout most machines, where its resilience and flexibility allow it to fit closely around other parts.

3 Measuring and marking out

The first stage in most workshop fabrication jobs is to measure up and mark out the proposed job. How well you are able to do this depends on the type of equipment you are using, and the skill and care with which it is used. You are not likely to be able to achieve high precision using the average home workshop tools, but for the purposes of making up small fittings or improvised tools reasonable accuracy will suffice. This can be accomplished if care is taken at the marking out stage, using a good steel rule, dividers or a vernier caliper, a scriber and a centre punch.

When marking out sheet or plate to fabricate a bracket or a similar item, you can work from a straight edge on the material, assuming it has been cut accurately. You will normally find that materials which have been cut to size on a guillotine will be quite good in this respect. It is easier to treat this straight edge as finished, and this approach will indeed save time, but to ensure accuracy it is better to allow about 1 mm or so and then file back to this after cutting the rough shape out of the metal. Measure and mark out the shape of the work on the metal, using the scriber to trace the outline. To make the scribed lines more visible, the metal can be coated with engineer's blue (you can get this from engineering suppliers or good tool shops) and the lines scribed onto this. A convenient alternative is to use a spirit-based felt marker to darken the surface.

Corners should be marked out by intersecting scribed lines to start with; you may wish to file a radius on them to give a smooth finish, and this can be added later. To save work, and provided the radius size is not important, you will find it quite easy to scribe around coins to obtain a smooth, rounded corner.

Where several holes have to be drilled to locate the bracket, measure up the hole centres very carefully. Even a small error here will mean that the fasteneres will not align, and you will then have to enlarge the holes to compensate for the error. Scribe the hole centres as intersecting lines initially. Check that they are positioned correctly, and then carefully centre punch the intersection to give an accurate starting point for the drill. The holes can be marked out using either two edges of the bracket as reference points, or by dimensioning each hole in relation to the others on the bracket.

When drilling the holes, it is best to use a bench drilling machine with the work clamped in a drill vice; this is the only way to be sure that the hole is drilled at 90° to the surface. Remember that the drill can be at an angle in two places, and this is not easy to check with a hand-held drilling machine. If you do not have access to a bench drilling machine or a stand for your hand-held machine, check carefully before drilling commences that the drill bit is perpendicular to the surface of the metal. Brace yourself against something solid to help keep the drill square. Use a self-locking wrench or a vice to hold the work. Drill a pilot hole at each position, then enlarge this to the required size using successively larger twist drills.

When cutting out the bracket from the sheet, remember to allow for the thickness of the blade. Always err on the large side – this may mean more filing, but it is preferable to having to start again due to a mistake while cutting. As an alternative to hacksawing, you may prefer to use an electric jigsaw. Most of these tools will cope with thin metals, using a special metal cutting blade. The work must be supported securely and clamped to the bench with self-locking wrenches or G-cramps. Note that whichever sawing method is used, tight radiuses will have to be filed to shape, and the edges of the work should be filed smooth to complete the job.

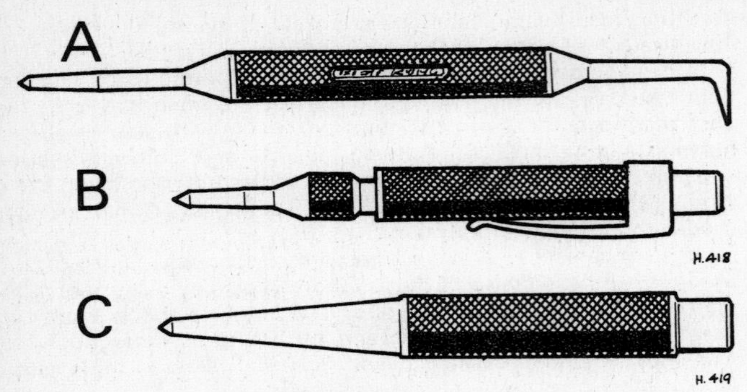

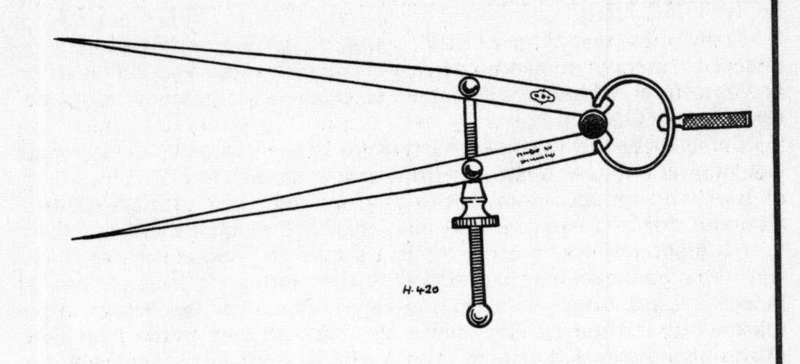

Fig. 3.1 Marking-out tools

(A) A scriber is the traditional tool for marking out work prior to cutting, filing or drilling
(B) An automatic centre punch contains a percussion mechanism which will centre punch the work when the tool is pressed against the surface
(C) A conventional centre punch does the same job when used with a hammer

Fig. 3.2 A pair of dividers is used to trace circles and radii during marking out

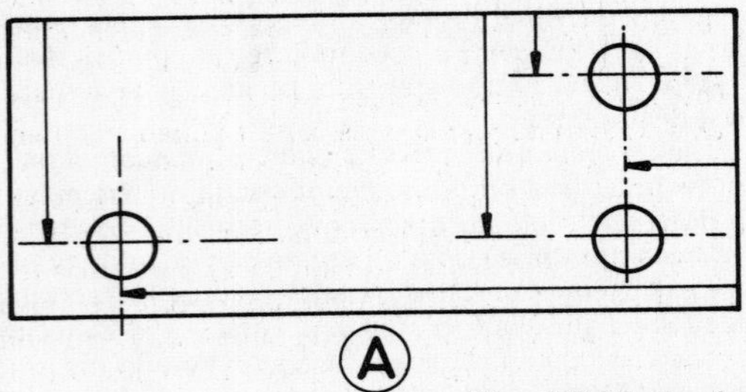

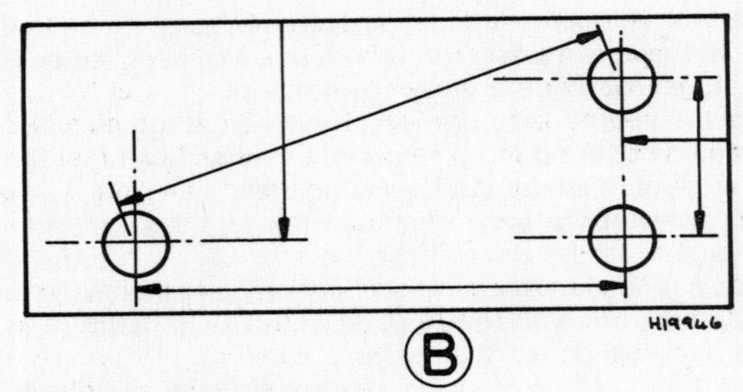

Fig. 3.3 Methods of marking out hole positions

When marking out hole positions, note how in (A) the holes are located in relation to the right-hand and upper edges of the work piece. An alternative method, shown in (B) allows the holes to be located relative to each other – an important point if the hole positions are to be accurately aligned

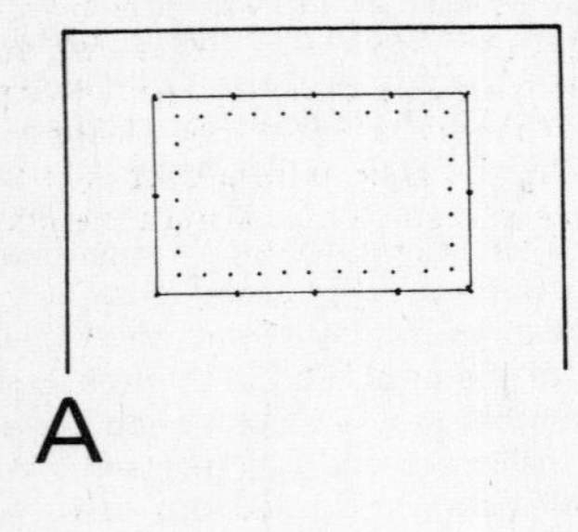

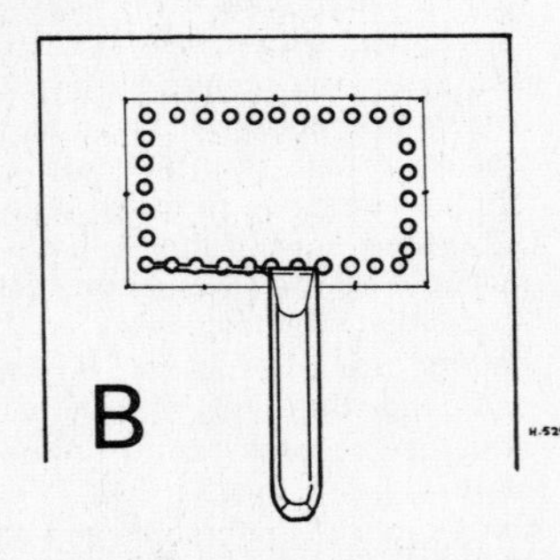

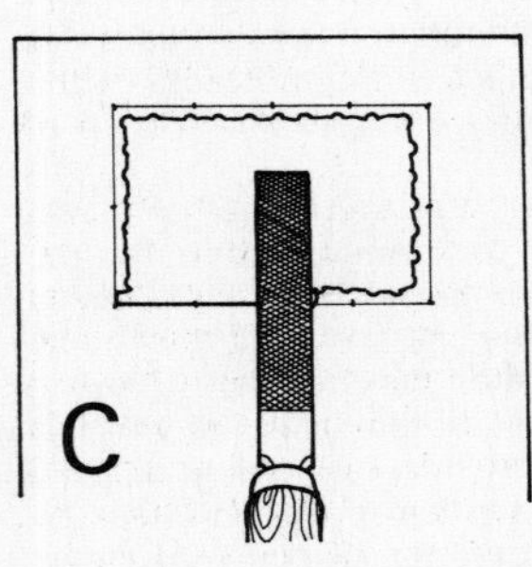

(A) Mark out the hole shape, then scribe a second outline inside the final hole outline. Centre punch at drill-diameter intervals

(B) Drill a series of holes following the punch marks. Cut away the surplus material using a sharp chisel

(C) File the hole to its finished shape to complete the hole

Fig. 3.4 Cutting a square or rectangular hole by chain drilling

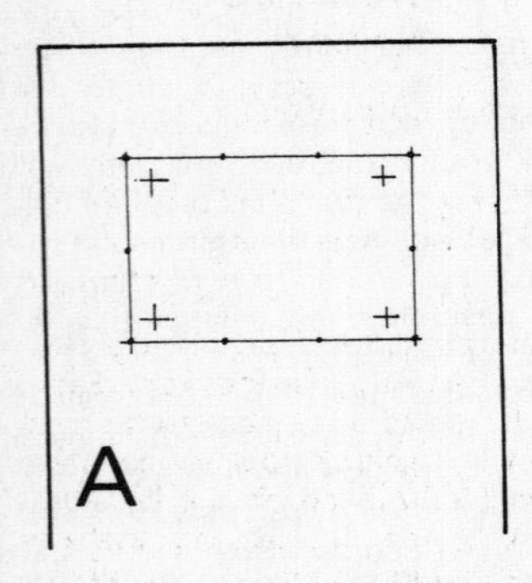

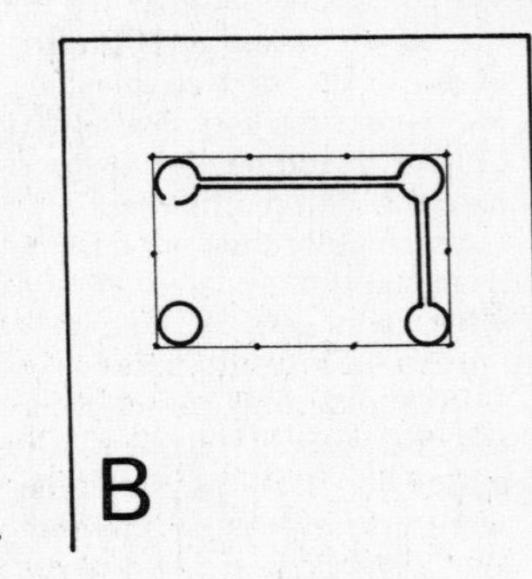

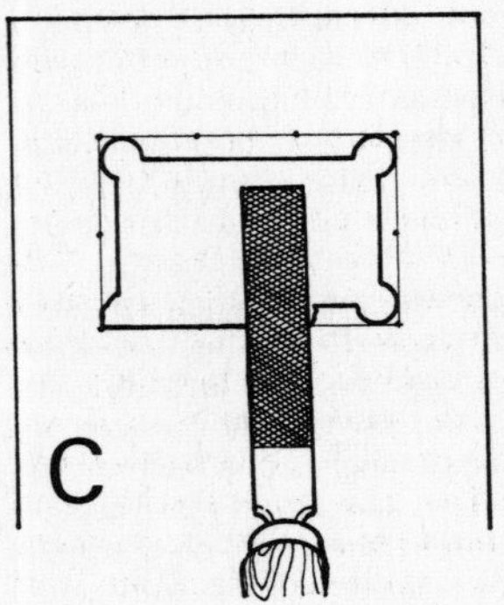

(A) The hole shape is marked out, together with centre punched marks for a pilot hole near each corner

(B) Drill out the four pilot holes, then saw between them to remove waste material

(C) File away remaining metal to marked outline to complete the hole

Fig. 3.5 Cutting a square or rectangular hole by sawing and filing

4 Bending and forming

Many jobs will require that the work is bent in one or more directions, a common example being a bracket where one edge is bent at 90° to form an L-shape. This can often be achieved by cutting the bracket out of steel angle, but in some cases it will have to be made up from steel plate and then bent to size. Whilst it is easier to make up the bracket flat, and then bent it to the required shape, careful calculation of the bend radius is required, and you then face the problem of attaining that radius in practice.

For most purposes it is preferable to make up a template from thick card. This can then be checked in position and any alterations made before starting work on the metal version. Try to use card of a similar thickness to the metal being used for the bracket. In this way the radius of the bend and the position of the fixing holes can be checked. An alternative is to cut a piece of metal rather bigger than the required bracket, form the bend first, and then cut it to shape and drill any fixing holes.

In theory, the bend radius is calculated from the thickness of the material being used, though in practice the material will tend to control the bend radius naturally, using normal workshop hand tools. It is important that the material chosen is fairly malleable, and that you do not attempt to form too acute a bend. If this is not the case, the metal will be stressed across the bend and may fracture.

Thin mild steel sheet can be bent fairly easily by hand pressure across a suitable edge, though the thicker the material the harder this becomes. One useful method is to use lengths of hardwood clamped along the intended line of the bend. The projecting metal can then be folded along the edge of the hardwood, working up and down the line with a hammer. If necessary, you can shape the wood first as a former for the radius.

When using heavier materials such as steel bar or thick strip, you will have to resort to the use of heat and a small anvil to form the bend successfully. The heat will soften the material, allowing it to be bent more easily and with less risk of cracking. Wear stout industrial gloves and eye protection when using this method, and hold the work at one end, clamped securely in the jaws of a self-locking wrench. The heat source required will be dependent on the size of the workpiece, and thus the rate at which it can absorb heat. For small pieces, an ordinary blowlamp will probably prove adequate, whilst larger items will need much more heat. Any solid-fuel fire or stove will prove adequate for most applications (assuming the workpiece will fit!). Heat the metal until it turns bright red, then remove it and start forming the bend quickly, before it loses too much heat.

Lay the metal across the anvil, and start hammering along the intended bend, moving the hammer blows up and down the line, and keeping close to the intersection with the anvil. Avoid moving the hammer too far from the anvil, or an irregular bend will be formed, and in the wrong place. As the metal cools and ceases to be luminous, apply more heat and continue the forming operation once red heat has been restored. When the bend has been formed, allow the workpiece to cool down in air to avoid embrittling it. Remember that it will be hot enough to cause serious burns for some time after it has ceased to glow red hot.

The above technique will work well with all mild steel materials, including the softer stainless steels. Note that the latter will discolour due to the heat, though this can be filed and polished out if required. You can apply the same principle to the straightening of items like footrests (if they are solid) and stands. Do make sure that the repair is safe; if badly bent, straightening will leave serious stresses in the component, which might lead to sudden failure at a later date. If you are unsure, play safe and renew the part rather than risk this.

With aluminium materials, the situation is less easy. Plain aluminium and soft alloys will bend reasonably easily, but are easily fatigued. If you have to use them, bend them only as much as you have to, and with the minimum of working. This will keep to a minimum the risk of work-hardening and fracture. The high tensile aluminium alloys, like Duralumin, are very difficult to bend successfully and usually break before a 90° bend has been achieved. The best advice with such materials is not to attempt to bend them, but to have the bracket made up by an engineering works with the necessary facilities and equipment to do this type of work.

If you are dealing with round or square section tubing, a bending machine and formers are needed to stop the tube walls from distorting and collapsing, though small diameter thin-walled tube can be bent using a bending spring. Angle, U-section and box section are very difficult to bend using normal hand tools, and this sort of specialised work should be left to experts.

5 Riveting

Amongst the methods of permanently joining materials, riveting is one of the easiest to carry out at home, though practice will be needed to form a good joint. This technique is especially suitable where dissimilar metals have to be joined, though this does not remove the risk of electrolytic corrosion taking place between the two. Rivets form a relatively strong and semi-permanent joint, and will have to be drilled out or the head filed off to remove them. In some applications, it may prove easier to use nuts and bolts, though rivets often take up less room.

Cold riveting

For our purposes we will discuss cold riveting only, and in practice this means small rivets which can be 'set' without the need for heat to make them sufficiently workable. Rivets are available in innumerable shapes and sizes, and in various materials. The types normally used have round heads (snap-heads) and are made in steel and aluminium. Other materials, such as brass or copper, are sometimes used for specialised applications, like riveting a soft material to a metal backing. On older machines, brake shoe linings were fitted in this way.

The basic principle requires a clearance hole to be drilled through the two parts to be joined. This is best done by aligning and clamping the two parts to ensure that the hole(s) align correctly. The rivet is passed through the hole and its head supported on an anvil or similar solid support, and this is best done with the help of an assistant who can hold the assembly in place and make sure that the two parts are in close contact. A hammer is then used to form the second head and to spread the rivet shank so that it is a tight fit in the holes.

The technique of riveting requires a certain knack; early attempts will almost invariably produce a rivet with a bent-over shank, instead of an evenly-formed head. The head should be formed gradually until it is about the same shape and thickness as the original head on the opposite side. Needless to say, aluminium rivets are much easier to set than steel ones. Despite this, do not be tempted to use aluminium rivets to join steel items. The close contact between the two metals will lead to rapid failure due to electrolytic corrosion. This condition occurs when two dissimilar metals react electrically, rather like a small battery.

Tubular rivets are easier to set in larger diameters than an equivalent solid rivet. It is best to start forming the head with a drift or a centre punch, and care must be taken to avoid shearing through the edge of the head.

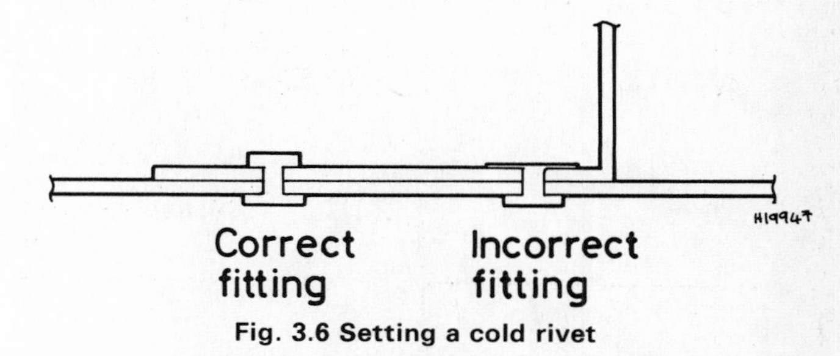

Fig. 3.6 Setting a cold rivet

Take care not to flatten the head excessively – this will weaken the joint

Pop-riveting

Best of all for home use is the ubiquitous 'pop-rivet'. This is a tubular rivet which is set from one side using a special tool. Each rivet has a shank which passes through from the far side of the rivet and emerges through the head. When the setting tool is operated, the shank is pulled into the tool, forming the head on the far side. At a predetermined tension, the shank snaps off, leaving the rivet in position. These rivets are available in various lengths, diameters and materials, and can be bought as a kit together with the setting tool. Apart from the ease with which these rivets can be set, the main advantage is that you need access to one side of the workpiece only, and no form of anvil or support is required.

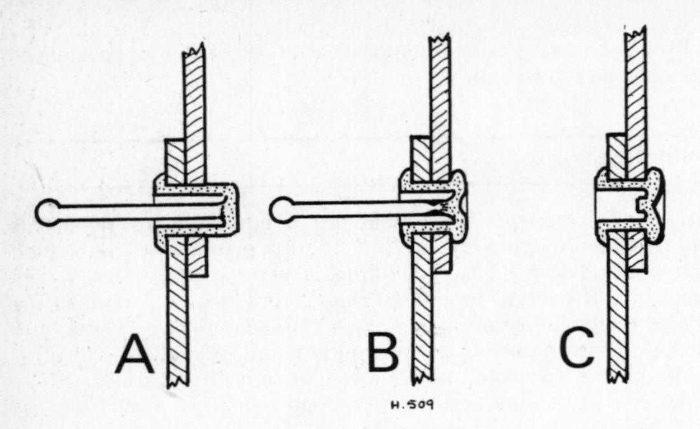

Fig. 3.7 Sectioned view of a pop rivet

(A) shows the rivet positioned through the two items to be joined

(B) depicts the head being formed on the far side of the joint. Note how the pin is beginning to weaken near the head

(C) The pin has now broken off, leaving the two parts secured by the rivet

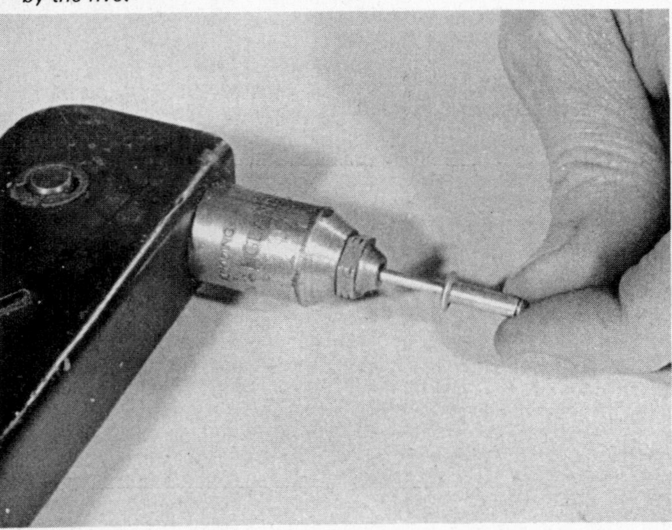

5.1 The pop rivet is held in place in the setting tool by a long steel shank

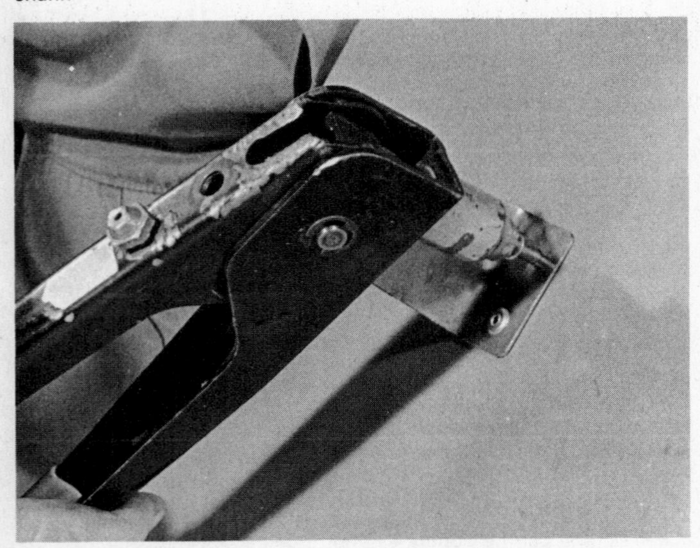

5.2 Once in position, the tool handles are squeezed to form the head of the rivet and eventually the steel shank breaks off leaving the rivet in place

Fig. 3.8 Sectioned view of a tubular rivet

This type of fixing has largely been replaced by the pop rivet

6 Soldering

Soldering is a technique for joining two metals using a third, usually an alloy of lead and tin. It is most often used for joining copper and brass, though there are other more specialised soldering processes. Unlike welding, soldering does not in itself form a strong bond between the two parts. Strength is instead achieved by using the solder as a sort of combined glue and filler, allowing a mechanically strong joint to be made stable and secure.

Soldered joints are used extensively for making electrical connections, where the solder is useful as a good conductor between a wire and its terminal. The other common use is as a method of securing the nipples to control cable inner wires. It is these two operations we will be looking at in detail, but first let us deal with the soldering operation in general.

The process of soldering is entirely reliant upon cleanliness. This does not just mean that the worst of the dirt and grease has to be removed; the metal surfaces to be joined must be absolutely clean and any oxide deposits removed with a file or abrasive paper. If this is not done, the solder will not take to the metal. Once the parts have been prepared, the surfaces to be joined have to be tinned. To do this you will need to heat the parts, and this is usually done with a soldering iron. These are most often electrically operated, but you can get versions which will run off small gas cylinders, or simple irons which are heated in a blowlamp flame. For most applications, a fairly large electric soldering iron is the best choice, provided that a sufficiently heavy-duty model can be obtained. Most irons sold are rated at approximately 25W and are intended for electronics work. We really need something of at least twice this power, though a small soldering iron will be adequate for light work.

Tinning means coating the metal in the area of the joint with a thin film of solder. To assist in this process a flux is used, and this cleans away surface oxides and helps the solder to flow. The most convenient form of flux is carried in solder wire in small cores, solder of this type being called 'multi-core'. Alternatively, plain solder can be used in conjunction with a paste flux which is applied to the metal before tinning. Apply the soldering iron bit to the area to be tinned. It may help to apply a small amount of solder to the bit to assist in heat transfer. Once the metal has heated through, apply the solder to the metal (not to the soldering iron bit). The solder should flow out smoothly over the heated area. If it fails to do so and simply collects in a bead, the metal is not clean enough. Let the metal cool off, then use fine abrasive paper to produce a bright finish and try again. Repeat the tinning operation on the remaining piece to be joined.

Once tinned, the two parts can be sweated together. Arrange the two tinned surfaces so that they are in firm contact, then apply the soldering iron bit. The solder applied during the tinning operation will heat up and melt, flowing together to bond the two pieces of metal. This basic procedure can be applied to all soldering jobs, though you will have to adapt the technique in some instances. For example, if you are soldering a large piece of metal, the heat loss may be so great that the solder will not melt onto the surface. To get around this problem, try pre-heating the metal with a blowlamp, and supporting it on wooden blocks so that the heat loss is slowed. If you need to protect a part from the effects of heat, use a large piece of metal as a heat sink. As an example, a fragile electrical component could be protected by gripping the lead or tag being soldered in pliers or self-locking wrench jaws.

Soldering control cables

For most purposes, it is easier and quicker to buy ready-made control cables than to attempt to make them up at home, but on occasions it may be useful to be able to make up an emergency replacement cable, or to modify an existing one. In addition to solder and a soldering iron you will need a strong pair of side cutters capable of cutting cleanly through the inner and outer cable material. It is also preferable to have a selection of new cable nipples to hand. Whilst it is possible to re-use nipples on some occasions, many are now crimped onto the cable and thus not suitable for re-use.

If the purpose of the operation is to re-use a broken cable, this can be done provided that it has broken at or near one end. Check before starting that you can afford to lose a little of the overall length of the cable. Clean the broken inner cable carefully. This is best done using aerosol contact cleaner to remove all traces of grease. Next, clean the metal of the inner with fine abrasive paper, and then degrease once more. Heat the cable with the iron, and tin it to bind together the strands of the inner back to the point where they form a tight circular bunch; the object at this stage is to cut away the damaged end of the inner, and tinning the area of the proposed cut will prevent it unravelling when cut.

Work out the distance between the far end of the nipple and the proposed cut, and make a note of this. You will have to remove the same amount of the outer cover to ensure that the effective working length of the complete cable is maintained. Cut off the inner cable, and then withdraw it a few inches down the outer cable. You may like to remove it completely so that both the inner and outer can be cleaned and lubricated prior to the new nipple being fitted.

With the inner removed, the outer cable can now be shortened to suit. Using the side cutters, remove the appropriate amount of the outer cable. The offcut piece will take with it the small metal ferrule which covers the cut end of the outer. You will need to fit a new one to the newly cut end, but with care it is often possible to work off the old one and re-use it. Slide the inner cable back into position, and degrease and clean the tinned end ready for soldering. **Note:** *Remember to refit in the correct sequence any adjusters or stops which are integral with the cable* **before** *the nipple is fitted!* If you have no spare nipples of the appropriate size, you may be able to re-use the old one, provided that it was soldered on originally. Holding the nipple with pliers to avoid burns, heat it up until the solder melts, then shake out the old inner wire. You may need to use a piece of wire or a similar tool to dislodge the strands. While the solder is still molten, blow through the hole to remove excess solder.

If using a new nipple, heat it up with the iron and tin the hole at the centre. Fit the nipple over the end of the inner cable so that about 2 mm protrudes. You will now have to flare out the individual strands into a fan shape. This is a vital stage in the operation, since it is this flared end consolidated with solder which gives the joint strength. If the cable end is left straight it will pull off the end of the inner cable. There are various ways of doing this, but I have found it easiest to clamp a self-locking wrench around the inner cable just below the nipple. Alternatively, clamp the inner wire in a vice. If you are using new, untinned cable, it is quite easy to tease the ends apart, bending them into a fan shape. If the end has been tinned before cutting, heat the cable end first, keeping the solder molten while spreading the ends. The cable end and nipple can now be heated, and solder flowed into the end of the nipple to form a small domed end. Odd strands of inner wire should be smoothed off with a file to complete the job.

Soldering electrical connections

Soldering is without doubt the best way to make electrical connections. Although crimped connections are commonly used these days, the resulting joint is less secure, and a potential trouble spot if corrosion takes place between the wire end and terminal.

Start by stripping back the insulation from the wire to be joined to the terminal. This is best done using a wire stripper tool, though if care is taken you can score through the plastic insulation with a sharp knife blade, taking care not to cut into the copper strands beneath. The insulation can then be slid off, and the strands twisted lightly together to stop them spreading.

Tin the bare end of the wire, keeping it on the bit just long enough for the solder to flow between each strand, but not long enough to melt the insulation. Remember to fit any insulating sleeves or other fittings over the end of the wire at this stage. Tin the corresponding area of the terminal, then bring the two together and apply heat to flow the solder between the two. Take care not to move the two parts as the solder cools. This can result in a 'dry' joint, and poor electrical conductivity.

7 Welding and brazing

Welding is the main method of forming permanent joints between metal parts, and you will find it used extensively on the frame of most

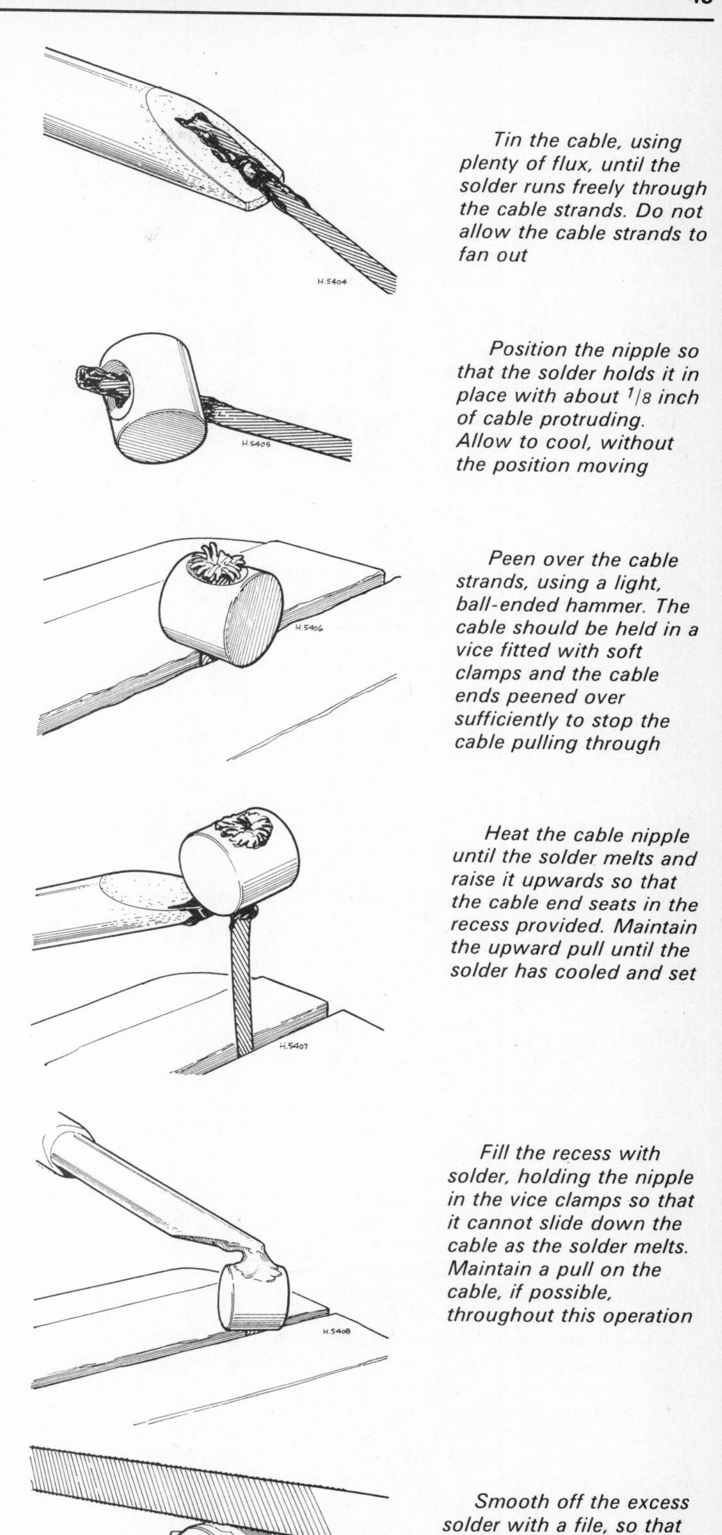

Fig. 3.9 Soldering a nipple to the end of a control cable inner

motorcycles. The distinction between welding and other forms of joining metal, such as glue, rivets, solder and so on, is that the process forms a true molecular bond between the parts, effectively turning them into one piece. This imparts great strength, and it allows a continuous joint to be made, unlike bolts or rivets which place the stress in a few small areas.

All of the many welding methods employ the same basic principle of melting the two parts together in a controlled fashion to produce the required molecular bond as they fuse into one. The methods we will consider are the most common ones, and thus those likely to be accessible for home workshop use. It follows that most of the high technology processes such as laser and electron beam welding will be ruled out of the discussion.

In addition to welding, there is a related process known as bronze welding, or brazing. This is not quite as strong as welding, a bronze alloy rod being used instead of steel, but it is useful for joining thinner materials. This is because the bronze rod melts at a lower temperature than the steel parts to be joined, and the parts are stuck together, rather than fully fused as is the case with welding proper. This means that the joint is not quite as strong as a welded one, but it is a very useful technique for non load-bearing jobs.

Fig. 3.10 A typical lap joint made by gas, arc or Mig welding

Note how the fused area must penetrate both pieces

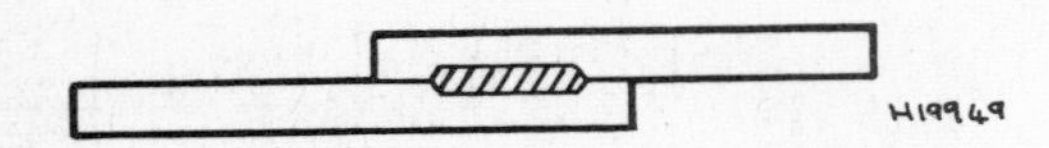

Fig. 3.11 In mass production, spot welding is often used

The two parts are gripped between electrodes and a current passed between them. The metal fuses together at the centre

Arc welding

This is the cheapest route into home welding, arc welding sets being easily obtained at low cost. The welding equipment comprises a metal cased transformer unit, normally with a dial control to regulate the welding current. To this are attached a mains lead and plug, and two heavy-duty output leads. One of these, the earth lead, terminates in a crocodile clip which is attached to the workpiece. The other is connected to an electrode holder, and it is this which does the work.

The electrode is normally a steel rod, coated with a special flux which prevents oxidation of the weld during work. The principle of the process is to select a rod size and current setting appropriate to the metal being welded. The machine is then switched on, and the electrode brought very close to the workpiece. As the tip of the electrode nears the workpiece, an arc is struck between them. This is almost exactly the same as a spark plug in operation, except that the spark is much hotter, and continuous.

The heat generated by the arc melts the rod, and also the pieces of metal to be joined. The molten metal flows together to form the weld, while the flux melts and flows over the surface to exclude air from the weld. Once cooled, the slag formed by the flux is chipped away to reveal (hopefully) a perfectly formed welded joint.

The main advantage of arc welding is its low cost and general versatility. It is a good arrangement for home use, since there is no big outlay on equipment hire, as with gas welding. The range of thicknesses of metal covered by a particular welding set is invariably reflected in its price, but even the cheapest machines will suffice for tacking odd brackets into place or fabricating battery trays and similar items.

The main disadvantage with arc welding is that you will have to spend some time developing your technique before you can start serious work, and it is very easy to get things wrong during the first few attempts. *Note that most home arc welders will not treat thin metal kindly, and will tend to blow holes straight through it.*

Carbon-arc attachments

These devices replace the normal electrode holder with a pair of tapered carbon electrodes, one of which can be slid in relation to the other to regulate the gap between them. In use, the arc forms between the two electrodes and produces a sort of flame, and this can be used to braze thin metal which would be burnt through when using welding rods. This extends the versatility of the arc welder considerably, and some manufacturers include a carbon-arc torch with their sets. The only drawbacks are that it is a difficult process to master; all the problems of welding, plus an electrode gap to regulate. The carbon rods tend to be expensive if you use this method a lot, and the torch itself is bulky, and difficult to use in a confined space.

MIG welding

MIG (Metal, Inert Gas) welders avoid many of the problems of the arc welder, from which they were derived. The electrodes used in arc welding, which gradually burn away in use, are dispensed with. Instead, a reel of wire housed in the casing of the machine is fed by a motor and a capstan along a hose to the welding handset. The speed at which the wire emerges from the welding tip can be varied, as can the welding current. The flux is also eliminated, and instead an inert shielding gas is fed through the same hose to emerge around the welding wire. This prevents oxidation, and does away with the problem of slag formation.

MIG welders allow much thinner materials to be welded, and the shielding gas and lower current settings help to minimise distortion. In addition, the use of different shielding gases and wires permit stainless steel and aluminium welding. The welding current and the wire and gas feed are normally controlled from a switch on the welding handset. This means that much closer control is possible than with arc welding, and that work in confined spaces is much easier.

The only drawbacks with MIG welding are the comparatively high initial cost of the machine, and the fact that brazing is not possible. The relatively low-cost units designed for home or semi-professional use usually use miniature non-refillable gas bottles, and these can become expensive if much welding work is undertaken. It is possible to fit adaptors to take industrial size gas bottles, however, and if necessary, larger wire reel sizes can be used.

Oxy-acetylene welding

This is perhaps the most flexible welding method of all, allowing close control of the flame temperature. The possible range goes from low settings useful for heating metal and freeing seized fasteners, through brazing to welding and even metal cutting. In essence, the system comprises two gas bottles, one containing oxygen and the other a fuel gas, normally acetylene. The gas is passed through regulators and anti-flashback valves to a handset, where the two gases can be mixed in differing proportions to permit various temperatures to be attained.

The main drawback for home use is the high initial cost of setting up a gas welding system. In the UK, the regulators, hoses, handset and nozzles must be bought, and then added to this is the cost of a rental agreement for the gas bottles, plus refilling charges. There are cheaper alternative systems which use disposable gas bottles, but the cheaper versions of these seem to be largely ineffective for welding, insufficient heat being produced to cope with dissipation through the metal. Brazing and soldering are feasible, provided the area of metal is not too great. A more significant drawback is the sheer running costs; it is not possible to get enough oxygen into a small disposable bottle to allow more than a few minutes work, and this soon gets prohibitively expensive. If you feel that gas welding is for you, having tried all the options during a welding instruction course, resign yourself to the cost of a professional welding set – it will be cheaper in the long run.

Less significant problems inherent with all gas welding systems include the risk of distortion, which is always greatest when using gas, because of the heat build-up in the materials. This type of problem is unlikely to be too apparent with motorcycle work, where thin sheet materials are seldom encountered.

Brazing

The brazing process allows two pieces of metal to be joined by introducing a third, in the form of a copper/tin alloy rod. The process requires heat, but far less than is the case with any of the fusion welding processes discussed so far. The two pieces to be joined do not reach melting point, and so there is little risk of burning through or distortion.

The rod melts at a lower temperature and flows between them, forming a reasonably strong molecular bond. This is stronger than soldering, but less strong than full fusion welding, and thus it is not suitable where maximum strength is required. It is, however, very useful where thin steel sections, like splits in mudguards, need to be repaired.

Brazing is normally carried out using oxy-acetylene equipment, together with brazing rods and flux, though as has been mentioned above, good results can be obtained using a simple arc welder fitted with a carbon-arc attachment.

The Lumiweld process

This is a proprietary process for joining aluminium and various aluminium and zinc-based alloys. As such it is of considerable use in the home workshop. Conventional alloy welding methods such as TIG and MIG welding are specialised skilled processes, effectively limited to professional engineering workshops equipped for this type of work. Lumiweld, on the other hand, does not require the parts to be joined to be raised to melting point to allow a full fusion weld to take place. In this respect it is rather similar to soldering or brazing, though the manufacturer claims that a molecular bond is formed.

In use, the workpiece is raised to a temperature of about 730°F (390°C), at which point the Lumiweld rod will melt and flow between the parts to be joined. It should be noted that this is far lower than the melting point of aluminium and related alloys (generally about 1100°F or 595°C) and so there is little risk of sudden melting of the parts being worked on. More significantly, it is normally possible to obtain these temperatures using no more than a butane torch. The Lumiweld technique, which is described in detail in the manufacturer's literature, is fairly straightforward, and is claimed to result in a joint which is four times stronger than aluminium, and harder than mild steel.

Applications for the Lumiweld process include casting repairs, building up stripped threads prior to re-tapping and similar jobs which would otherwise demand professional attention or renewal of the damaged part. It will also cope with zinc alloys, which are normally difficult or impossible to repair. Disadvantages are that the zinc content of the rods means that corrosion takes place where salt is present, and that anodising and heat treatment are not possible, though these are more than outweighed by its general usefulness.

Lumiweld can be obtained from the UK suppliers, Grand Union Products (International) Ltd, who advertise in the motorcycle press.

7.1 Brazing is ideal for light fabrication jobs, such as attaching brackets. Note that the bracket here has been positioned with a piece of brazing rod while it is tacked into place

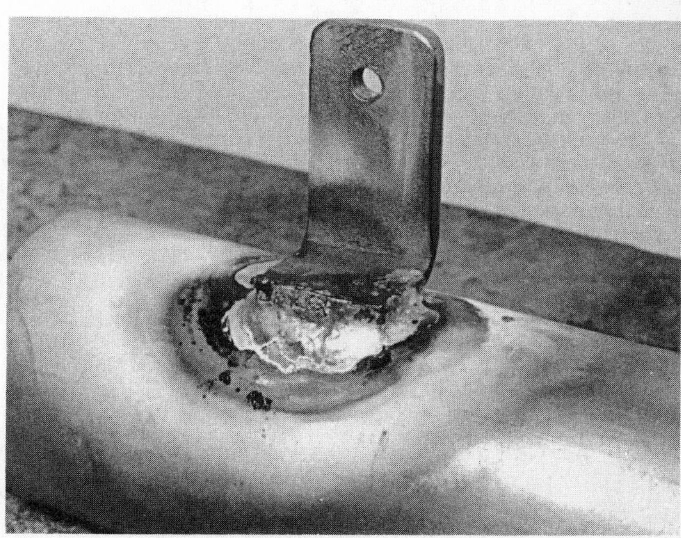

7.2 A brazed joint is reasonably strong (though less so than welding) and less prone than some welding to causing unwanted heat distortion. Note however, that heat may cause discoloration or damage to plated parts

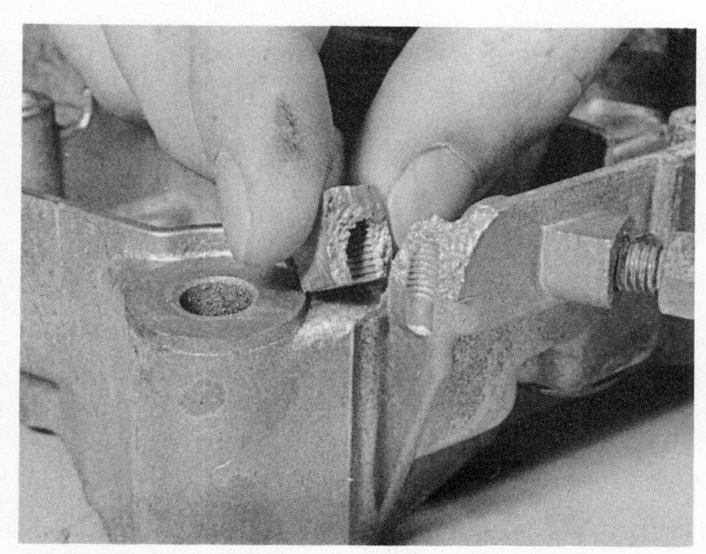

7.3 We used 'Lumiweld' to repair a damaged crankcase boss which had broken away.

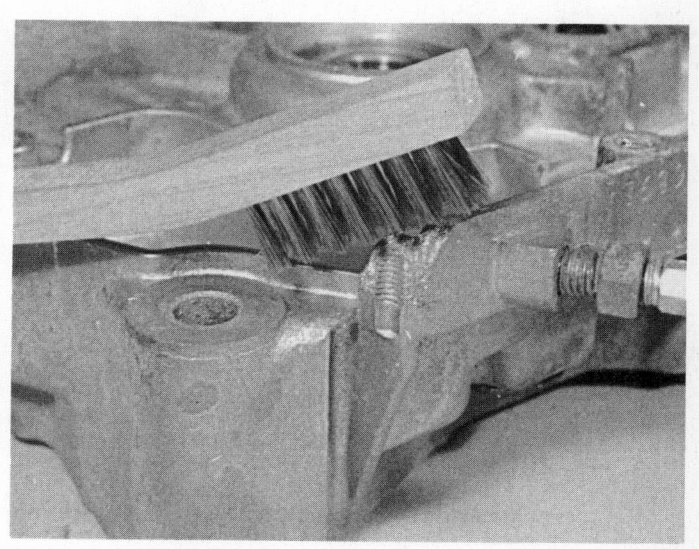

7.4 The faces are first cleaned with a stainless steel wire brush

7.5 The surfaces are heated carefully, and then coated or 'tinned' with some of the Lumiweld rod

7.6 The Lumiweld is agitated using a stainless steel rod to remove oxides and impurities from the metal surface

7.7 The boss is now held in position and 'sweated' into place

7.8 Once secure, the joint can be further improved by filling the cracks around the break

7.9 After cleaning and filing, the threads were cleaned up using a tap. The finished job was both strong and neat

7.10 This section of the crankcase was damaged after the failure of a bearing

7.11 The break was filed into a V-shape to ensure a good bond with the Lumiweld material

7.12 After the usual preparation, the notched areas were filled using the Lumiweld rod

7.13 The finished repair was almost invisible, other than where the light caught the gasket face

To weld or not weld

The decision on whether or not to purchase a welding system is a difficult one. They are undeniably useful, but demand a certain level of skill to be reached before the results are likely to be acceptable. Welding is a skill which can only be learnt by experience, therefore no attempt will be made to turn this section into a DIY exercise. If you feel that you could make use of some sort of welder, try enrolling in a welding evening class at a local college. This will give you a feel for the technique, and should allow you to compare gas, arc and perhaps MIG welding. Your instructor may also be able to suggest the best local supplier for welding equipment and accessories. The type of equipment you buy will depend on the nature of the work you are likely to be doing, but for home use and on the grounds of cost alone, a cheap arc welder is probably the best bet.

Safety

If you do decide to try welding and attend a training course you will be taught about the safety aspects very early on in the course. Again, this is not the time to cover this area in detail, but the following rules should always be kept in mind, even if you are assisting someone else who is welding.

Fire risks are a real problem when welding. Remember to remove all fuel sources, including the machine's fuel tank, before starting work. Work in a well-ventilated area to avoid the risk of a build-up of fumes, and make sure the area is clear of paper or rags which might be ignited by the sparks. Make sure that you have adequate fire-fighting equipment on hand in case of fire.

Eye protection is especially vital when welding. Arc and MIG welders produce an extremely bright arc which can cause serious eye damage unless filtered through special dark glass. Use only an approved full welding mask, and make sure that the glass fitted is of the correct type for your welder. The risk comes mainly from the high ultraviolet levels in the light emitted. You need only look at the arc with the naked eyes for moments to develop a condition called arc eye. This potentially serious condition is extremely painful, leaving your eyes feeling as if they are full of sand or grit. In bad cases, permanent damage may be caused. Do not forget that spectators and pets can suffer from this too, so keep them well away when welding. Oxy-acetylene welding requires a less dark glass, but gas welding goggles should be worn at all times. Never under any circumstances assume that sunglasses will protect your eyes during any welding operation; they won't.

Skin protection will be needed, both to fend off hot sparks, and to protect the skin from the intense ultraviolet light from the arc; you can easily get a nasty version of sunburn from arc and MIG welders. You will need gloves to avoid burns to the hands, and a cotton boiler suit with no cuffs (these could trap sparks). Wear workboots, preferably arranged so that sparks cannot get down inside them. A piece of red-hot slag inside your boot is not at all amusing!

Machine protection is required whatever form of welding is employed, but especially so in the case of arc or MIG welding. If you are welding a bracket or lug to the frame, remember that the welding current will be passing through it, and this can easily damage electrical components. Always disconnect (earth lead first) and remove the battery, and also disconnect the alternator, regulator/rectifier and electronic ignition leads to prevent damage to these components.

A secondary risk is due to the passage of the welding current along the machine's chassis. For example, if the earth clamp is attached to a fork leg while welding is carried out on the main frame, the current must pass through the steering head bearings to complete the circuit. Since this presents only a small contact point, the bearing balls may become welded to the races. Always attach the earth clamp close to the weld area to avoid this risk.

8 Working with plastics

As has been mentioned earlier in this book, plastics are being used for many applications on motorcycles. This presents something of a problem for the home workshop, these materials being difficult to deal with if repairs are needed. The easiest course of action is to renew any damaged parts which are small and inexpensive or perform more than a cosmetic function.

The renewal option becomes less than desirable where the item happens to be a fairing; the cost of buying a new one will be high, and it is worth attempting a repair before resorting to renewal. Many fairings fitted as original equipment will be ABS mouldings, and this material is often used for side panels, seat tail humps and mudguards. Depending on the nature of the damage, it may be possible to repair it by glueing. A simple crack, for example, can often be dealt with using one of the 'superglue' instant adhesives. The type of adhesive required will be dictated by the type of material to be repaired, and in practice this will mean testing the adhesive on a small area to see whether or not it will work on that type of plastic. Unfortunately, it is almost impossible to give firm recommendations here; there are numerous types of plastics in common use, and not all of these can be joined with any one type of glue, so a little experimentation will be required.

Heat welding

The best way to repair many plastics is by heat welding the parts, using a special tool for this purpose. This type of equipment is not readily available, and if you wish to try this technique at home you will have to improvise. The author has had some success using a modified low wattage soldering iron. The main requirement is to reduce the temperature at the bit, and the easiest way to do this is to attach an extended bit in the form of a length of thick wire. This will act as a heat sink, and by progressively reducing the length of the wire tip it is possible to arrive at the desired temperature. This should be just enough to melt the plastic, but no hotter.

Using the improvised tool, melt across the split or join in several places to tack them together, ensuring that the parts are held level. Once you are sure that the parts are correctly positioned, work along the join with the bit, moving it from side to side slightly to help 'stitch' the parts together. Be very careful not to allow the heat to penetrate too far; you need to fuse the plastic together as far as possible, but beware of burning through on the side that will show. With luck, you should find it possible to repair light damage using this method. The join will not be as strong as the surrounding material, but this can be improved if you can find a piece of similar plastic to use as a filler. This should be used in the same manner as welding rod, running the extra material into the weld to reinforce it.

An alternative approach is to attempt a repair using glass reinforced plastic (GRP) methods as described in the following section. The snag here is that GRP may not adhere to some types of plastics. It is worth testing adhesion on a small area first, before carrying out a full repair. If it seems to stick well, then you may as well go ahead.

Plastic windscreens

The clear or tinted windscreen blade fitted to fairings is often a source of problems. Apart from its general vulnerability to knocks, it will eventually get scratched by the road dirt which will build up on it, necessitating an often expensive replacement. The best thing to do here is to avoid scratching it in the first place. It is helpful to use a silicone-based spray polish which will help prevent road dirt sticking to the screen. When cleaning **any** plastic surface, use copious amounts of water to soften the dirt first, then wash it off gently using a sponge or chamois leather.

If the screen is badly scratched, you will probably have to fit a new one, but before doing so it its worth trying to remove the scratches by polishing them out. This is a tedious process, and will work better on some materials than others. Suitable polishes include specialised clear plastic scratch remover and polishes (try a boatyard or caravan accessory stockist) or most liquid metal polishes, like *Brasso*. You can also try paint cutting compounds and haze removers, chrome or aluminium polish, or even toothpaste! Experiment with various grades to see which seems best, working down to progressively finer grades to leave a clear finish. Note that some light scratching will inevitably remain, and if you look through the screen rather than over it this will be dangerous; buy a new screen instead.

Most modern screens are moulded, rather than being produced from flat sheet materials. Any such screen will invariably have a double curvature profile, and this cannot be reproduced successfully at home. Older types used screens with a single curve profile and were usually moulded from *Perspex (Plexiglas)* acrylic sheet. These can be made up at home, provided that great care is taken; the material is brittle, especially when it is cold. A good alternative material is one of the more modern flexible plastics like *Lexan*, though this is usually a lot thinner than the original and may flex too much in use. It does avoid the need to form a curve, however, and will withstand the odd knock and scrape without damage.

The new screen shape can be traced onto the sheet using the old one as a pattern, and is best cut out using a power jigsaw fitted with a fine blade. Alternatively use a padsaw, and take great care not to let the blade snag in the cut or you may split the sheet. It is reasonably easy to mould a single curve in flat acrylic sheet using boiling water. A handy method is to place the sheet in the bottom of the bathtub, and then to pour the boiling water over it until it softens and bends to the curvature of the tub. Make any further adjustments to the curve by hand while it is still pliable, taking care to avoid scalding your hands.

Mark out the fixing holes carefully; they must be drilled accurately or great stresses will be placed on the screen. Using a sharp drill bit at low speed, drill through the sheet **from both sides** to prevent the material shattering when the bit emerges. Check that the holes align correctly, and if necessary file them to make sure that there is no stress on the surrounding material.

Fix the screen using rubber washers or sealing strips so that the screen is allowed to flex slightly. Even better would be to drill oversize holes and then to use rubber 'top-hat' bushes on one side and plain rubber washers on the other; that way the bolts can never twist directly against the screen.

9 Using GRP (glass reinforced plastics) for repair work

As has been mentioned, GRP techniques can be most useful for repair and fabrication work in the home workshop, and GRP bodywork has been used to good effect by the builders of many specials.

Although you will have to devote some time to learning the techniques, you can produce professional results if care is taken.

The most obvious application of these methods is in the repair of fairings or luggage equipment originally made from GRP materials. Many small producers use GRP for this type of work because it allows relatively small quantities of a given item to be made without undue tooling overheads. It goes without saying that in the event of damage, GRP is also the correct material for repairing it.

Other materials can be repaired with GRP too, provided that they are not load-bearing. A good example would be a badly rusted steel rear mudguard. The worst rusting will probably be in the area which shows least, below the seat. If the old paint and rust are removed carefully, the remains of the old mudguard can be built up and reinforced with a GRP skin applied to the underside. Plastics are more difficult to repair; with some types the technique will work well, but others may be damaged by the resin, or the resin may not adhere well enough. In these cases, all you can really do is to test a small area. If the resin will bond well and does not damage the original plastic, then you should be able to use this technique for the repair.

Safety precautions

When working with GRP, note that the resin fumes should be avoided, so work in a well-ventilated area or outdoors. Do not handle the matting without gloves on, or the tiny fibres will become embedded in your skin. Avoid skin contact with the resin, again using rubber or disposible plastic gloves during use. If you need to sand down the hardened material, use goggles and a dust mask to avoid inhaling dust, which will contain fine glass particles and fibres.

Take note of the instructions supplied with the materials you buy. These normally explain safe working practices and any special precautions which must be taken. For example, where accelerator and hardener are supplied separately from the resin, you should keep neat accelerator well away from the hardener until the former has been mixed thoroughly with the resin. If the hardener and accelerator come into direct contact they can explode!

Repairing existing GRP bodywork

Before attempting to repair a split fairing or damaged luggage, have a look at how the original was made up. This tends to be the exact opposite of how you would expect it to be done, rather like starting with the paint and adding the underlying panel! Normally, the part will have been made in one or more sections, each of which will have been laid up in a female mould. The mould is coated with a release agent to allow the moulding to be removed easily, once finished. The first coat applied to this forms the finished outer surface of the moulding and is called the gel coat. Sometimes this surface will have been painted, but more often the resin itself is coloured by adding a pigment to it.

Once the gel coat is completed and cured (the exposed surface never actually gets fully dry while it is in contact with the air, so that it adheres to the subsequent layers) the rest of the laminate is built up. This consists of the glass fibres, in the form of chopped strand matting. Into this is worked the resin, then the air bubbles are stippled and rolled out, more glass mat, and so on until the required thickness is built up. The inner surface is sometimes smoothed off a little by using a fine glass tissue as the final layer. Once the resin has hardened fully, the moulding is freed from its mould and is ready for use once any attachments or fittings have been added. For more complex shapes like full fairings, more than one original moulding is used and these are then bonded together with resin to form the final composite structure.

Minor damage to GRP mouldings is relatively easy to deal with at home, and you can buy all the materials you will need in the form of bodywork repair kits from car accessory shops. If you find that you want to develop the technique for larger jobs, or to make up your own bodywork mouldings, then the materials can be bought in bulk quantities from specialist suppliers or boatyards. The materials needed are as follows:

a) Pre-accelerated lay-up resin (the correct grade of resin, with an accelerator added in the appropriate quantity)
b) Hardener to suit the quantity of resin
c) Body filler, or inert filler powder to be added to the resin to make up your own filler
d) Chopped strand matting (some fine tissue is useful too)

You should also collect together some rubber gloves, a polythene container or a tin or similar for mixing the resin, a 1 in paintbrush, and a mixing stick. Some solvent, such as acetone or a proprietary equivalent, is useful for removing the resin from brushes, but for small jobs it is easier to buy a cheap brush and throw it away after use. If you do use such solvents for cleaning, note that they can be flammable and dangerous if misused. Get advice on safety precautions from the supplier of the particular product.

Cracks, stone chips or scratches in the surface of the moulding can usually be dealt with by filling. Given the amount of road and wind vibration experienced on fairings and luggage it is best to choose a type of filler designed to remain slightly flexible; if it gets too hard it may crack away after a while. Before applying the filler, check that the gel coat in the vicinity of the damage is adhering to the underlying layers. If it is coming away, grind it back until it is sound. Mix the filler and hardener in the proportions recommended by the manufacturer. Apply a thin film of filler using a spatula or plastic spreader, smoothing it to follow the original shape as closely as possible.

Once the filler is cured, sand the surface to blend it into the surrounding material. If you find odd low spots or holes these should be filled, and the sanding operation repeated, finishing with a fine abrasive to leave the surface ready for painting. Note that where the original item was colour impregnated during manufacture, it will be difficult to match this with paint. In such cases it is probably best to sand the entire surface to provide a key for the primer and then spray the whole item.

Small holes and splits in the original moulding can be patched over from the back after bridging the front surface with masking tape to provide a temporary support for the matting. Fix the tape in position, then apply laying up resin with a brush to the back surface, overlapping the sound material to form a strong repair. Cut a piece of matting with scissors, then use the resin-soaked brush to stipple it into place. Make sure that the resin soaks well into the fibres.

Coat the first layer with resin and repeat the laying up process until a reasonable thickness has been achieved. On the final layer, fine mat or tissue can be used to give a smoother finish, though this is not necessary unless it is readily visible. Once the laminate has cured, strip off the masking tape and check the surface. If it stands up in relation to the surrounding moulding, file and sand it flush, then complete the repair using filler to give a smooth surface ready for painting.

Larger holes are repaired in a similar manner as described above, though you will need to contrive some sort of temporary mould to support the matting and resin until it cures. Care must be taken at this stage, since any large discrepancy in levels will be difficult to correct once it is dry. If you have the damaged section, this can be repositioned in the hole and held in place with tape or small metal strips pop-riveted to the surrounding moulding. Lay up fresh layers of matting and resin over the join, then remove the tape or strips from the front when dry. Fill and sand back the crack as described above.

Where a large section is missing, you will have to contrive a mould using hardboard or stiff card, riveting it in place until the repair has been made. This may not be appropriate where the hole covers an awkward double curve or complex shape. In such cases it may be necessary to form a support for the matting *inside* the hole, using this to hold the material in place and stippling the mat into the resin as best you can. Lay up just one or two layers and allow it to harden, then cut away and remove the support, and continue to lay up successive layers from the inside as normal.

Repairing steel pressings

Assuming that they are non-structural items like mudguards, rusted or split pressings can be reinforced and repaired with fibreglass. The success of such repairs is entirely dependent on the standard of preparation. All dirt, paint and loose rust should be removed and the surface wire-brushed until it is bright. A good way to make sure that all loose material has been removed is to prepare the metal as described, then lay up a single layer of matting. Allow the resin to harden until fairly stiff and rubbery, then tear it away from the surface. The resin should bring with it any residual dirt and rust, leaving the surface ready for repair.

Using the same technique as described above, and having taped over any small holes or splits, lay up several layers of matting, working them well into the resin to push out any air bubbles. Once dry, trim off any stray fibres, then prepare and finish the outer surface of the part using filler to cover any small holes or splits.

Repairing rusted fuel tanks

Another use for resin, this time without the glass matting, is in repairing a rusted fuel tank. This technique allows an otherwise expensive item to be salvaged for further use. Start by removing and emptying the tank, taking the normal precautions against the risk of fire. Remove the fuel tap(s), then use an airline to blow the tank dry inside. If you do not have access to an airline, a good alternative is to use a vacuum cleaner hose packed out with rag and stuffed into the filler hole. If possible connect the hose to the cleaner outlet so that it blows rather than sucks. Leave the machine running for a while until all residual fuel has been removed.

Once empty and dry, check the tank for damage, and remove any loose rust or dirt by shaking it. If your ears (and those of your neighbours) will stand it, put a handful of gravel in the tank and shake it around to dislodge as much of the loose material as possible. Investigate any obvious signs of leakage or pinholes, clearing out any damaged metal with a scriber or small screwdriver. Once sound metal has been found, tape over the holes. Before proceeding further, the tap holes must be temporarily blocked off. The emphasis here is on temporary – you need to be able to remove whatever you block them with after the resin has set. Threaded holes can be blocked with a well-greased bolt of the appropriate size, whilst plain holes are best plugged using a piece of wood carved to shape and covered with polythene. Make sure that any plug extends at least an inch into the tank.

Mix up about 1/4 pint of resin with the specified amount of hardener and pour it into the tank. You will now have to try to coat the entire inner surface of the tank with this mixture, turning the tank constantly to ensure that the resin spreads evenly. Once the tank walls are coated, any excess can be allowed to drain to the bottom since it is here that any rusting will have been worse. Once cured, remove the plugs and make sure that they open into the tank; you may have to break away or drill out excess resin from around the plug area. The fuel tap(s) can now be refitted and the tank put back into service with its inner surface repaired and reinforced. Any remaining surface damage on the outside can be repaired by filling and painting in the normal way.

10 Painting and plating: general

The problem of deteriorating paintwork and plating is one with which most owners will be familiar. Although regular cleaning and waxing will reduce the rate at which this deterioration occurs, each new stone chip or scratch provides a starting point for corrosion, and even sound paint will eventually fade and lose its shine.

In the case of bright decorative plating, usually chromium plate, this will last well if looked after, but once it does start to peel or rust little can be done at home to restore it.

In the sections which follow, painting and plating techniques are discussed, together with ways of maintaining them and repairing minor damage.

11 Maintaining and repairing paintwork

Preventative maintenance

Regular cleaning should be considered an essential part of routine maintenance for any machine. Apart from keeping the paintwork in good condition, cleaning will ensure that accumulations of road dirt are removed before they can encourage corrosion to start. In the UK, as in many other countries, salt is spread on the roads during winter to prevent ice and snow building up on the surface. It is inevitable that this will find its way onto the machine, and some salt will remain on the road for some time into the spring, to be carried onto the bike in the form of spray from passing vehicles. If you combine salty water with a fresh chip in the paint film, rapid corrosion of the area will begin, so it makes sense to wash off the salt before this happens.

The process of cleaning a motorcycle should be fairly obvious, though it is worth mentioning one or two details worth bearing in mind.Check the bike over before you start cleaning. If there are grease or oil deposits to be removed, apply a water-soluble degreaser before you start using water for general cleaning. Soak the dirt with water to which a little detergent has been added, and allow this time to soak in and soften the dirt film. You can use hot water to speed this process up. A good method of applying the water, and for rinsing after washing, is to use a pump-type garden spray. These allow the water to be delivered as a jet or as a fine mist.

Once softened, use a soft brush or a cloth to dislodge the dirt, moving it gently over the surface to minimise scratching. Wash the cloth or brush regularly to prevent embedded dirt getting dragged over the surface. Do not forget to clean the areas you can't see, like the underside of the machine. Depending on the type and model of motorcycle, you may find it worth removing the seat, fuel tank and side panels to allow easier access. If you do this, do not omit to protect vulnerable electrical components and the air filter intake from water contamination. When all the dirt has been loosened, wash it off using clean, cold water, preferably using a spray or a hose. To prevent water marks forming as the machine dries off, use a chamois leather to remove water droplets, rinsing and wringing it out frequently.

Washing will take off the film of road dirt, diesel fumes and other deposits which would otherwise attack the paint, but it also leaves the surface open to attack by the same deposits next time you use the machine. You can protect it to some degree by waxing the painted and plated surfaces. Waxes and polishes are available in innumerable varieties and in liquid or paste form. The composition of these waxes varies considerably, some using mainly natural waxes, or mixtures of waxes and silicone, while others use polymer sealants to leave a tough detergent-resistant skin over the paint.

Some polishes include a fine abrasive designed to remove the outer layer of paint, and thus any oxidation of the surface. These are fine if the machine has been neglected for a while and the paint has gone dull, and in some cases can transform the appearance of a bike. Beware of using an abrasive polish on a regular basis, though, because each time you use it, your paint is getting a little bit thinner ....

If the paint has got really dull, or is covered by fine scratches, it can be restored by using a haze remover or even a paint cutting compound. Again, this is an abrasive process which will wear the paint down, and must not be done too often. In particular, be wary of over-use on lacquered finishes, and be very careful near the edges of side panels and the fuel tank; it is quite easy to cut through to the primer in these areas.

Minor damage to frames and fittings

Minor damage, by which is meant stone chips and small scratches in the paint film, can be dealt with at home in most cases. In the case of the frame and its associated fittings, which are normally painted black on most machines, you can get by with touching in the scratches using a small brush. You will need a small paintbrush (about 1/2 inch), some wet-and-dry paper in various grades, a good quality brushing enamel, thinners for cleaning the paintwork and brush, and some metal primer compatible with the enamel chosen.

The make of paint used will in all probability depend upon what you can find locally; not many people use brushing enamel these days, so you may have to ask around to locate a supplier. Most car accessory shops will stock brushing cellulose, together with the appropriate thinners and primer. This will probably be the best choice for touching in existing paint on frames. Another possibility worth considering is the oddly-named *Smooth Hammerite,* manufactured by Finnegans Speciality Paints, and again stocked by car accessory shops and hardware stores. This is a recent addition to their range of tough hammer-finish paints, and produces a normal gloss finish. It can be applied to bare metal, and is touch dry within about twenty minutes, and is thus ideal for repairing stone chips.

Start preparing for painting by rubbing down the damaged area with medium wet-and-dry paper, used wet. Remove any corrosion and cut back to the sound paint, leaving a feathered edge. Apply a coat of primer, (unless you are using the *Smooth Hammerite* mentioned above) overlapping the feathered edge slightly, and allow it to dry for the recommended length of time.

Once the primer is dry, use fine wet-and-dry paper to flat down the surface, removing any brush marks or runs and leaving the surface keyed ready for the top coat. Remove any residual surface dust using a clean rag moistened with thinners. Using a clean, small brush, apply the top coat, carefully brushing out the brush marks and taking care not to apply too thick a film, which could cause sagging or runs on vertical surfaces. If a second coat is needed, flat the first coat again using fine wet-and-dry paper.

Damage to tanks, side panels and mudguards

On most machines, the main areas of decorative paintwork will be found on these areas, though it can be assumed to cover other items finished in the main colour. The best way of dealing with scratches and chips will depend upon the type of finish used by the manufacturer. With a simple paint finish, you can adapt the procedure described above for the frame, but you will have to find a suitable colour matched paint. Many dealers stock aerosol touch-up paints in a range of colour-matched shades for the models they stock and supply.

If the damage is simply a scratch or small chip, the easiest method is to use a child's paintbrush to touch in the damage. Spray a small amount of the paint into the can lid, and then carefully apply this along the scratch using the brush.

More extensive damage is rather more difficult to deal with, and the paint will have to be sprayed on. If you intend to do any form of paint spraying note that you should pick good conditions in which to do it. The air must be dry, and reasonably warm, or a poor finish will result. Do not work in a dusty or badly ventilated area, and on no account use any form of heater; this would pose a serious fire risk, and will tend to stir up any dust. You can try working outside if the weather is warm, dry and still. This does bring its own problems, in the form of flies which seem to find wet paint irresistible. Wherever you decide to work, remember to take precautions to avoid spray drift landing anywhere it would not be welcome, wear a clean overall, and use a dust mask to keep the atomised paint out of your lungs. Materials needed include various grades of wet-and-dry paper, aerosol spray cans of colour-matched top coat and primer, masking tape and newspaper.

Start your preparatory work by removing the part to be worked on, and remove all traces of wax or polish using a clean rag dipped in thinners. Using progressively finer grades of wet-and-dry paper, again used wet, rub down the affected area until the damaged paint edge is blended into the metal below. It should be noted at this stage that any decals in the vicinity may have to be renewed, in which case order replacements well in advance. The old decal may peel off, but more likely you will have to sand it away.

Use masking tape and paper to protect the surrounding area from overspray, rubbing the edges of the tape down smooth with the thumbnail. Following the paint manufacturer's directions, mix the primer by shaking its can for the specified length of time (and don't cheat!) then spray on the primer coat. Start spraying on the masked area, positioning the spray head 6 – 8 inches from the surface, and moving the can smoothly and horizontally across the spray area and onto the masking paper before releasing the spray head. Gradually move down, applying the paint in overlapping bands until the area is covered. Do not be tempted to apply too much paint; several thin coats are preferable to one thick one, which will probably result in runs.

The primer coat(s) should now be flatted using fine wet-and-dry paper, having allowed sufficient time for drying. Wipe over the surface with a thinners-moistened rag to remove dust. The top coats can now be applied. With these, the need for careful application is vital if runs are to be avoided. If a run does occur, let the paint dry fully, then remove the run with abrasive paper before continuing. Once the final coat has been applied, let it dry and then remove the masking tape and paper. There will almost certainly be an overspray line where the masking tape was, and this can be removed once the paint has hardened fully. Use a fine cutting compound to smooth away the excess paint, blending the new and old coats together. Any decals can now be applied, and the paint surface waxed for protection.

Lacquered and metallic finishes

Many current machines are produced with a plain or metallic lacquered finish, often in more than one colour. Whilst the effect of this is attractive, it does pose problems when the inevitable scrapes and scratches occur. As a general rule, it is not possible to touch in small areas of metallic paint without odd effects being caused by the direction in which the metallic particles lie. The only real solution is to respray the entire part. This is even more the case where a clear topcoat or lacquer has been used; refinishing the entire panel or part is the only satisfactory solution.

Whilst it is possible to obtain acceptable results from spray cans over small areas, the consistency of the finish is likely to be variable over large areas. This means that you may be able to get away with spraying a tank or side panel, but an entire touring fairing may be rather too much of an undertaking. If the fairing is a plastic moulding, rather than a GRP one, you will need to use special paints which are compatible with the plastic used in the moulding process. Ordinary paints may react with the plastic, or tend to flake off. On some of the larger models with complex integrated bodywork, paint refinishing becomes a complex operation, and is probably outside the scope of the humble aerosol can. In such cases as those outlined above, you are strongly advised to leave the work to a professional. The only other option would be to set up your own spraying facility, although the cost of a compressor, spray gun and related equipment is unlikely to be a viable prospect for occasional use.

If you should decide to undertake this sort of work yourself, be prepared for a considerable expenditure on equipment, and an investment of time in learning spraying skills. You may be able to find a training course for this purpose at a local college. You will also need to establish the type of paint used originally so that a compatible product can be obtained. A local paint stockist will be able to advise, if you take in a sample part and explain what you are attempting to do.

The basic procedure for dealing with larger areas is broadly similar to that described above, only bigger and thus more demanding if a good finish is to be achieved. A simple summary of the process would look something like this:

a) Remove and clean the part to be sprayed
b) Assess the type of paint originally used, and choose a colour-matched alternative. Seek advice from your paint supplier on this point. To ensure an accurate colour match, take a sample of the existing paint when ordering the new paint
c) Use a spirit wipe to remove all traces of old polish (especially where silicones may have been used)
d) Wash down with warm water and detergent, then rinse with clean water
e) Sand down and fill any damaged areas
f) Fill any paint defects with a suitable stopper
g) Flat the paintwork using medium (P240 to P400) grade wet-and-dry paper
h) Check that all paint edges have been feathered
i) Apply any necessary masking
j) Apply the primer coat (a special type may be needed on some plastics and on aluminium parts)
k) Flat the primer when dry using fine (P400 to P1200) grade wet-and-dry paper
l) Apply the top coat(s) noting that the masking pattern needs to be worked out beforehand if the part has more than one colour.

From the above you will probably appreciate that paint refinishing at this level becomes rather complex. Alternatives to attempting the work yourself or sending the whole job to a professional include doing your own rubbing down and flatting, saving a good deal of time and expense at the spray shop. You could even apply the primer yourself, but agree this beforehand with whoever is going to do the final spraying work. In the case of side panels or other smaller parts, you could even consider buying a new (or even secondhand) replacement.

12 Frame refinishing: the alternatives

Sooner or later the frame on most machines will have got to the stage where touching up is insufficient to keep corrosion at bay, and in some cases this could get to the point where the corrosion made the machine unsafe. This is especially true of machines which use a pressed steel spine frame, or fabricated sections welded to a conventional tubular frame. To avoid reaching the stage where the frame is in need of renewal, it will be necessary to refinish the frame, and thus halt corrosion.

Preparation

The decision to have a frame refinished is not one which should be taken lightly. You will have to strip off **every** part from the frame. This means not just the engine and wheels, but the rear suspension, front forks and steering head, and the entire electrical system. It goes without saying that you should only contemplate this sort of restoration work if the machine warrants it. If the machine is rare or valuable, or if you intend to keep it for some time, then fair enough, but it would not be worthwhile on an average machine nearing the end of its useful life.

Once all fittings have been removed, the frame should be examined closely for signs of damage, corrosion or similar problems. If any repair work is required, this should be carried out now, rather than after paint removal has been carried out.

The bare frame should next be cleaned with degreaser, and the remaining old paint removed. You can do this at home if you so wish, using a chemical paint remover or a blowlamp and scrapers. Both approaches require the use of suitable safety equipment, in the form of gloves and eye protection. The risk of burns from the blowlamp may seem obvious, but do not forget that chemical paint strippers are highly caustic, and can cause serious damage to skin and eyes. Whichever method you choose it is a long tedious job, and best avoided unless you have nothing better to do for a few days. There may not appear to be much to a frame, but it is surprising just how long it takes to remove all traces of paint and corrosion by hand.

Blasting

A better solution by far is to have the frame stripped by blasting. In this process, compressed air is used to blast the frame with either abrasive grit or glass beads, removing all traces of the old paint and also any rust. The frame, swinging arm, stands and any other parts to be refinished should be completely free of grease and road dirt prior to blasting. You should also note that the blasting process will damage machined surfaces, so drive out any bearings or bushes which would otherwise be destroyed.

When you take the parts for blasting, specify either bead blasting, or **fine** abrasive grit blasting; coarse grit will leave the metal too rough for painting. Make a list of all the parts you take so that you can check that they are all there when you collect them. The blasting process should result in a perfectly clean frame with a fine matt finish to the now-exposed metal. Check that the blaster has got into all the corners and recesses – it is easy to miss a few. It is important to apply the new paint finish on the blasted parts as soon as possible, preferably on the same day. If you leave them longer, surface rusting will occur, and even if this is not apparent the potential for further corrosion will be lurking under the new paint finish. Before rushing home and applying primer, however, read through the alternative finishes discussed below.

Frame finishes

In addition to the various conventional paint finishes available, you can also have the frame finished in several other coatings, many of which offer a number of advantages over normal air-drying paint. The most common choices are listed below. In practice you will probably have to settle for whatever can be obtained locally, epoxy powder being a fairly common process, and well suited to our needs. It is worth enquiring about companies offering this type of finish when dealing with the blaster; many have reciprocal arrangements with a local powder coating specialist.

Stove enamels were once the traditional frame finish, and these gave a tough and reasonably hard gloss finish. These paints are normally baked on at fairly high temperatures during manufacture. Traditional stove enamels have largely been superseded by modern synthetics, some of which require baking in much the same way during original manufacture. The biggest demand for *refinishing* paints is from the car trade, and it is no surprise that these high-bake products are not normally available, for the simple reason that vehicle refinishers have no use for any process which would require all the trim to be removed to avoid it burning or melting in the paint oven.

Cellulose finishes are commonly used for car and commercial refinishing work, and are easy to use, relatively safe and air-drying. They are less than ideal on frames and stands of motorcycles, however, since they chip easily and are likely to lose their initial shine quite quickly.

Two-pack paint finishes work in much the same way as epoxy resins, that is they harden chemically rather than by air-drying. They give a good finish which is quite durable, but have the great disadvantage of being isocyanate-based products. This makes them toxic, and special breathing apparatus must be used when spraying them.

Epoxy powder coatings are used in general manufacturing, where an even and tough finish is required. Fortunately for us, epoxy powder is also an ideal finish for motorcycle frames. The frame is hung up in a booth and an electrical charge passed through it. The spray gun applies a coating of powder which clings to the entire surface, attracted to the electrostatic charge in the frame. This tends to even out the coating as it goes on, and also helps to avoid any thin or missed areas. Once coated, the frame is baked in a high temperature oven, where the powder melts, flowing out over the surface of the frame in a continuous smooth film. The resulting finish is quite acceptable cosmetically, and very tough. The only disadvantage may be that a wide range of colours is not normally available, though this will depend on the range stocked by the local expert in this type of work. This will usually be dictated by the nature of the main part of his business.

Nylon coating is a similar process to epoxy powder coating, except that nylon is used as the coating medium. The resulting film is less smooth and inclined to be duller than with other finishes. The main advantage is that it is relatively resistant to stone chips in use. The drawback is that it is difficult to find a paint to touch in any damage that does occur, and like epoxy powder, is available in a limited colour range.

13 Specialist paints and finishes

Various other paint finishes and coatings will be found on most motorcycles, either as purely decorative elements or as protection for the underlying metal. In some cases these coatings are used simply because they cut down production time and costs.

An example of this can be seen on most fork legs and engine covers. Where alloy castings are easily visible, the traditional approach was to grind the rough cast surface smooth, and then to buff the metal to a smooth, bright finish. The resulting surface was attractive to look at, and had a secondary advantage in that it was fairly easy to maintain in this condition, regular polishing restoring the shine if it became oxidised.

As the manufacture of motorcycles became more mechanised and labour costs rose, this type of hand finishing work becames less realistic, and alternative ways of finishing these areas were introduced. In some cases, the casting is sanded to remove the rougher parts of the surface, and then a coating of aluminium-coloured paint applied. This gives a bright and reasonably smooth finish and is often used on fork lower legs and cast alloy wheel rims. A similar process is often applied to the engine covers, where a lacquer coating is often applied. Alternatively, a heat-resistant satin black finish can be used, with detail areas like the edges of fins being polished up as highlights.

From our point of view, this treatment of castings can present problems once it becomes damaged and chipped. This sort of damage is inevitable during normal use, and spoils the appearance of the machine. Most owners will be familiar with the problem; the paint around the stone chips allows oxidation of the metal underneath, and this then spreads, flaking off the surrounding paint. Once this occurs to any great extent, the only solution is to remove the old coating and start again, a process which will require the removal of the part(s) affected. Even where the original finish was not coated, general wear and tear or neglect will leave it looking discoloured and pitted, especially where road salt has been allowed to speed the corrosion.

The intricate surface shape of most castings will make the removal of the old coating difficult using normal chemical paint strippers, but it can be done this way. You will need to allow a good deal of time, and have to hand a selection of scrapers, wire brushes and steel wool to help dislodge the softened paint. If you use this method, be sure to wear overalls, rubber gloves and goggles to avoid getting the corrosive stripper on the skin or in the eyes. Once removed, you will have to deal with the underlying corrosion, which will appear as unsightly pitting on the metal surface.

The time needed for preparation and the corrosion problem can be dealt with by having the part bead blasted. This process, unlike abrasive blasting, is safe for use on soft alloy and will dispose of both the paint and the corrosion, leaving the surface with a bright matt finish. An additional advantage is that the blasting process works equally well on fairly rough cast surfaces, and thus can be applied to items like the crankcases on European models, maintaining the original finish. If possible, choose aqua-bead blasting in preference to a dry bead blasting process; the resulting finish is finer and smoother. The main point to remember is that any surface that you want to protect from the blasting process will have to be masked off thoroughly, and simply applying masking tape is often not enough. The tops of fork legs can be plugged with rag or large corks, while in the case of engine castings, plug any small openings, and screw the cover down onto a wooden base using woodscrews through the normal fixing holes.

Once the corrosion and the old finish have been removed, you will have to decide on how to refinish the part. In some applications, it may be most effective to choose a simple polished surface. This will entail smoothing the surface with progressively finer grades of abrasive paper, and then using a buffing mop to polish it to a bright finish. The success of this process is somewhat dependent on the composition of the alloy, but with the exception of zinc alloys it will generally work quite well.

If you wish to duplicate the original finish, the choice of coatings is important. The matt black paints and clear lacquers available in aerosol cans, unless specifically designed for these applications, are unlikely to be tough enough, and may flake off in time. High temperature exhaust paints can work on the engine covers, but are not very resistant to wear and tear.

The best option is to consult a local metal coating specialist. You may find that the blasting company will be able to recommend a local company. Explain in detail what you want to do, and see if they can offer a suitable process. These companies specialise in applying various coatings to manufactured parts, and will probably have a coating which will suit your application. Many of these may require high-temperature stoving, so check that this will not damage the component, and make sure you remove any fittings which might be affected.

14 Electro-plating

Many parts on motorcycles are finished by electro-plating, either as a decorative element, or simply to protect the metal from corrosion in use. The most common decorative finish is chromium plating, and this will be found on almost all machines. Common applications include exhaust systems, headlamp rims, wire-spoked wheel rims, and some control levers and pedals. Chrome plating is also found on items like the front fork stanchions, but in this type of application its main use is to provide a hard, smooth working surface. This type of chroming is normally described as 'hard chrome' or 'industrial chrome' to distinguish it from the decorative processes.

Less cosmetically-important items are often given a protective electro-plated finish. Fasteners, for example, are given a bright zinc or cadmium coating, and black chrome has become a popular choice for coating exhaust parts, an application in which it outlasts most paint-based finishes.

With one exception, which we will look at a little later, electro-plating is a specialised process, totally outside the scope of the home workshop. There are two simple reasons for this; plating equipment is very expensive, bulky, and needs skilled handling, and the chemicals used in most plating processes range from very unpleasant to potentially lethal. For these reasons, we will not delve into the intricacy of the subject in this book – if you want to know more it would be better to find a specialist work on the subject.

At its simplest, the principle behind all electro-plating processes is to set up a bath containing suitable salts into which the part to be plated is suspended from a conductive wire. Also in the bath is a lump of the metal to be plated; chromium for example, again connected electrically to the plating transformer. When the process is started, a current flow from the piece of chromium (the anode) passes through the bath to the item to be plated, carrying with it a tiny amount of the anode metal, which is then deposited onto the part in a thin layer.

In practice, things are not that simple. For a start, the biggest part of chrome plate is actually nickel. This is built up onto the underlying base metal, which will have first been through numerous stripping, cleaning and polishing operations. It is these polishing stages which determine the quality of the finished work, and any small pit or scratch will show through the plating.

Once in the nickel bath, the plating process is started, the temperature, plating current, pH balance and numerous other factors having first been checked carefully. In due course, after a few thousandths of an inch of nickel have been deposited, the part is removed and rinsed before inspection. After any necessary polishing of the nickel surface, the chromium layer is added. Hard though it is to believe, the chromium layer will be only 0.000005 in thick. Despite this, the chrome is what gives the part its blue-white sparkle, and prevents the normal dulling of the underlying nickel layer.

Cleaning and maintaining plated parts

From the above description of the plating process it will be appreciated that plated surfaces, though fairly tough, will not withstand too much wear before corrosion can occur. The thin layer of chromium is hard and quite durable, but once worn through the underlying nickel layer is exposed, and it is the oxidation of this layer which produces the greenish-white deposits which form in corners of plated parts exposed to attack from road dirt and water. Chrome plating which is in good condition can be cleaned and waxed in exactly the same way as painted parts, and if treated with care will last well.

Re-chroming

Once surface rusting is seen it is fair to assume that the plating is well on the way to failure. The best that can be done at this stage is to remove surface corrosion using a fine metal polish. Application of a wax polish, or even a clear lacquer, will slow subsequent corrosion, but will not prevent it.

If it has become badly speckled with rust, the part will either have to be renewed or replated. Often with small items it is quicker and easier just to fit a new one, or to visit a local breaker in the hope of obtaining a used part in better condition. In the case of wire spoked wheels, the normal course of action is to have a new rim built onto the hub, usually with new spokes. Unless the machine is very old or has an unusual rim size, it is not normally worth having rusted rims polished and replated. Bear in mind that plated steel rims can often be replaced with aluminium alloy ones – check with your local wheel builder to establish the best choice for the replacement parts.

You should check around locally before sending parts for replating. Often there will be one local firm which has a good reputation for plating motorcycle parts, whilst others may not be so reliable. With plating work, preparation is the key to a good job, and you may not find out if this stage has been skimped on until your newly-plated parts start to peel some time later. As a rule, you should contribute nothing to the preparation yourself, apart from carefully cleaning and degreasing the parts before taking them to the platers. **Note:** *on no account should parts for plating be blasted; the resulting surface will need a great deal of polishing to get it smooth again, and on some parts the thickness of the metal may make this impossible.*

When you deliver the parts for plating, take with you a list of the items taken, together with your name and address and daytime telephone number. Keep a duplicate of the list yourself so that you can check that everything is there when you collect them.

It is not generally worth having exhaust system parts replated, and many platers will not accept them anyway, due to the risk of their baths becoming contaminated. Remember that exhausts often rot through from the inside; if the plating has gone, it is safe to assume that the underlying metal is heading that way. The best cure for rusting exhausts is to purchase a replacement system made from stainless steel.

It is feasible to plate many other items on the machine, though the cost of preparing items not originally plated will be high due to the large amount of polishing work needed. Chromium plating does tend to weaken steel parts by making them brittle, however, and consideration should be given to the safety aspects before sending parts like wheel spindles to the platers.

It is also possible to chromium plate aluminium alloy, as you may have seen on custom bikes. Be warned that you cannot guarantee plating to stay on aluminium for long – it will tend to flake off after a year or so. The plated layer will also tend to act as a heat barrier, and for this reason is best avoided where good heat radiation is desirable.

Cadmium plating

This is a good way of refinishing fasteners and odd brackets and fittings, being reasonably cheap to have large quantities plated. The finish obtained will be a fairly bright satin silver colour wtih good resistance to rusting. If you decide to have the frame refinished, it is a good idea to send the various fittings away for cadmium plating at the same time. Prepare the parts by carefully cleaning them and remove any surface rusting. Given that the finish will be matt anyway, you might consider having the parts bead blasted before plating.

Before you decide to have all the casing screws replated, consider replacing them with stainless steel Allen screws. These will never rust, and are far more practical than the standard cross-head types. If you intend to keep the machine for any length of time they are well worth spending money on.

Home plating kits

As has been mentioned, it is possible to carry out nickel plating at home using a kit of materials especially designed for this purpose. This type of kit is best suited to those who restore older machines, where the original finish on plated parts may have been nickel originally, and you should bear in mind that, unlike chrome plating, nickel will require regular cleaning and re-polishing. It is possible, however, to use nickel plating to build up worn parts for re-machining, and this may have uses in some applications.

The kits contain the nickel anodes, chemicals and full details of how to set up your own plating bath. You will also need a power supply in the form of a battery charger or 12 volt battery, a suitable small tank, a bar from which to suspend the parts for plating, some bulb holders, bulbs and electrical cable. In addition, a small hydrometer and a thermometer are required. Complete nickel plating kits and individual items can be supplied by the manufacturer, *Dynic Sales,* whose address can be found in the motorcycle press.

The plating process is essentially a scaled-down commercial plating system and can give excellent results if the instructions are followed carefully. The bulbs and bulb holders are used as a method of regulating the plating current, and you must calculate this having worked out the surface area of the item to be plated. Note that as supplied, the standard kits allow only small parts to be dealt with, but you may be able to expand your system to take larger parts if you find this necessary.

Anodising aluminium parts

Anodising is a technique for stabilising and sealing aluminium surfaces, often with the addition of a dye in a range of basic colours, which is used to impart a translucent colour to the metal. It is frequently used on new parts such as aluminium wheel rims and brake line fittings, and can be an effective cosmetic touch as well as good protection for the metal.

Most platers will carry out anodising work, though they may not be able to deal with items as bulky as a wheel rim. If you are carrying out a full rebuild, you might consider this technique for use on small parts and fittings.

Chapter 4 Engine and transmission related tasks

Contents

1 Introduction

Maintaining, overhauling and repairing faults on most reasonably new machines is a relatively straightforward matter, requiring a reasonable toolkit, the necessary parts, a Haynes Owners Workshop Manual and some common sense. Few detail components will require very much testing; it is more common now to make a few simple checks to see whether the part or assembly is working, and if not, to fit a new unit.

With some items it is possible to obtain replacement parts, an example being carburettor components. In other cases, though, the assembly is treated by the manufacturer as a single renewable item. Good examples of this approach are the small starter motor fitted to many mopeds and commuter models, instrument panel assemblies, alternator and flywheel generator stators and so on.

As this philosophy is pursued by the manufacturers, the opportunity to keep a motorcycle running cheaply is diminished as the availability of small replacement parts decreases. This might sound exploitative, but in fact it is an economic reality of modern manufacturing techniques; the cost of stocking minor parts, and the expense involved in terms of dealer time in fitting them rule out this approach when it ceases to be cost effective.

The concept of renewable assemblies, rather than renewable parts, makes sense while the machine is reasonably new and in good general condition, and is probably the most economic way of approaching the problem of repairs, both from the point of view of the manufacturer and dealer, and that of the machine's owner. After several years of normal use, however, the logic of this approach is less clear, as the machine begins to suffer the inevitable general wear which comes with age.

The manufacturer would obviously prefer that you disposed of the machine at this point and purchased a new one, and eventually every machine will get to the stage where it is no longer economic to continue using it. The stage at which this happens is difficult to assess, being dependent upon a number of factors, not least of which is how much you want to spend in terms of cash and time. The only general rule which can be applied is that a machine which is dealer-serviced and maintained will become uneconomic earlier than the owner-maintained equivalent, for the simple reason that you do not have to pay for your own time when working on the bike.

In this Chapter we will be looking at some of the more usual problems which crop up when working on the average engine/transmission assembly, which cannot be included in the specific Owners Workshop Manual for a particular model. For detail information on dismantling and reassembly procedures, however, you will need to refer to the appropriate workshop manual, where you will find these areas covered in depth, together with the necessary specifications and clearances.

2 Repair assessment

An important aspect of any overhaul is assessing the overall condition of the engine unit and then armed with this information, deciding on the best course of action. If the machine is fairly new and just out of warranty, for example, the failure of a single component with no attendant damage to the rest of the engine is easily assessed; just purchase the new part and fit it. In the case of older machines the situation is less straightforward. You must assess the nature of the immediate problem, just as you would with a newer machine, but in addition, you should also check the rest of the unit in detail to see whether it is worth carrying out the repair.

It is usual to find that when the engine is dismantled in response to an obvious problem, say worn valve guides, the rest of the engine is nearly worn out too. There would be little point in fitting new guides if the valves, bores, pistons and crankshaft bearings are worn out, and you will have to work out whether a full overhaul would be warranted if the machine in general has seen better days.

This might seem a rather pessimistic attitude to take, but it is a necessary one if you are to avoid wasting money. However fond you may be of that particular machine, if it is likely to cost more to repair than the bike is worth, it is time to consider cutting your losses and looking for a newer model. Alternatively, if something disastrous has happened to the engine of an otherwise sound machine, think about obtaining a secondhand engine unit from a motorcycle breaker. Although you may be taking something of a chance regarding the condition of the new engine, this approach can be a great deal cheaper than attempting to repair the old one if wear or damage is extensive.

3 Sudden component failure

If something serious goes wrong with the engine or transmission while the machine is in use, the chances are that it will be readily apparent to the owner when this occurs. Any major component failing in use will invariably produce some sort of knocking, rattling or vibration, even if the engine continues to run. Should this happen, pull in the clutch immediately to prevent further damage. If the engine is still running, use the kill switch or ignition switch to stop it, and coast to the side of the road. Try to find a way of getting the machine out of the way of passing traffic – a gateway or layby if possible, then set about trying to diagnose the fault.

Where the engine stops abruptly with accompanying ominous rumblings or other noises, it is a fair assumption that something unpleasant has occurred. It is also probably safe to assume that little more can be done there and then; you will have to resign yourself to further investigation once the machine has been recovered. There are, of course, exceptions, and you should make a quick preliminary check on the roadside in case the fault can be dealt with easily there and then.

If the problem looks like being a serious failure, the first thing to do is to get the machine home so that you can check it in the relative comfort of the workshop. The next thing to consider when trying to assess a sudden major failure is the most likely source of the trouble. Think about where you were at the time, how hard the engine was working, whether it was being revved hard, and whether there were other signs such as unusual noise or vibration just before the problem happened. If you have an oil pressure warning lamp (four-stroke models) was this lit when you first noticed the problem? If the engine stopped suddenly, did you notice any slowing down of the machine just before it did so? Try to get any such symptoms clear in your mind at the earliest possible opportunity. Whether you intend to repair the machine yourself, or ask a dealer to do so, such information is invaluable, pointing to the general area of the failure before dismantling commences.

4 Overhaul assessment

Whilst the engine and transmission assemblies of modern motorcycles will run happily for many thousands of miles, given reasonable treatment and regular servicing, the time will come eventually when an overhaul will be necessary. It is impossible to define the point at which this will occur, and you will either have to assess this for yourself, or ask your local dealer for his opinion. More often than not overhauls are prompted by a number of symptoms. You may, for example have noticed high oil consumption or leakage and have been prepared to accept this condition for some time. If the engine then begins to get unusually noisy, however, you might feel that the time has come to check the whole unit over and put right these and any other problems.

It is difficult to establish the cause of a problem on a machine which is in need of general servicing; excessive play due to infrequent adjustment and failure to take up normal free play can cause a good deal of noise. It follows that regular services will make any fault diagnosis easier, and will also highlight developing faults at an earlier stage.

In the sections which follow we will consider some of the problems that you are most likely to come across so that you are better able to make the decision to proceed with an overhaul. There are, inevitably, many other things which may go wrong with your machine, but the following should give you an indication of the checks you should make on faults in general when assessing their seriousness.

5 Lubrication system

Excessive oil consumption – two-stroke engines

This problem is mostly of interest to owners of machines with four-stroke engines, where oil is circulated around the engine. Two-stroke engines are lubricated by a total loss system in which oil is fed to the engine internals either by a pump, or as a fuel/oil mixture via the carburettor. In each case, the excess oil is burnt off after reaching the combustion chamber, and excessive oil consumption will produce dramatic smoking from the exhaust of the machine.

If the level in the oil tank of a pump-fed system drops faster than normal, the probable cause is maladjustment of the oil pump cable or a damaged or worn pump. The first thing to check is that the pump control cable is in good condition and correctly synchronised to the carburettor. The procedure for doing this varies widely, and you should consult the appropriate Owners Workshop Manual for details.

Some pumps allow the volume of oil delivered on each stroke to be regulated, and whilst this setting should not normally require alteration, it should be checked in cases of high consumption. Once again, instructions for checking and setting the pump stroke will be found in the relevant Owners Workshop Manual. In other cases, the pump may not be adjustable and will have to be renewed.

There is very little to go wrong with a premix system, apart from getting the fuel/oil ratio wrong. Check the mixture ratio in the owners handbook or Owners Workshop Manual and revise this if you are using the wrong quantities.

What if you find the typical symptoms of excessive oil consumption, such as heavily smoking exhausts and oil-fouled plugs, but the level in the tank drops at the usual rate? This is quite informative, since the only other way the engine can burn oil is to draw it from the reservoir of lubricant held in the transmission area of the crankcase. It does this by drawing oil through worn crankshaft seals, and you should monitor the transmission oil level carefully to see whether it drops over a week or so. If this is the case, the crankshaft oil seals are leaking, and it is likely that the machine will be running poorly and that it is reluctant to start, due to the loss of crankcase compression through the seals. You now know that new seals are required. You may well find that the main bearings will also need renewing, because play in the bearings is often the cause of damaged seals, though they may just be worn out through sheer age and the effects of heat on the sealing lips.

Two-strokes have another way of disposing of oil, through leakage. This is usually loss of the transmission oil from a seal or gasket joint, and will result in the outside of the crankcase becoming coated in a mixture of oil and road dirt. This problem is common to four-strokes as well, so refer to the rest of this section if you have oil leak problems with a two-stroke engine.

Excessive oil consumption – four-stroke engines

Most current four-stroke designs are wet-sump, ie the reservoir of oil is held in a space at the bottom of the crankcases from where it is pumped under pressure around the engine and also to the transmission assembly which shares the same cases. (Older designs use a dry-sump system where the oil is held in an external tank. It is circulated by pump around the engine and then returned to the tank. In this system, the transmission is lubricated by a separate oil bath arrangement.)

Since the oil is under pressure, the passages and joints through which it passes must be sealed to contain it. Oil can also be lost through being burnt in the engine, and in the case of four-strokes this points to wear in the cylinder bores, pistons and rings, or valve and valve guide wear.

To deal with leakage, you must first identify the source of the leak. Start by degreasing the engine thoroughly. Once cleaned, the exact point of any leakage can be established by running the engine. If the leak is serious, it will show up after a few minutes running on the stand. Note that care should be taken not to overheat the engine by allowing it to run with the machine stationary for too long a period.

With a less serious leakage you will have to ride the machine for a few miles to provoke the problem; some leaks may only occur when the engine is at full operating temperature. The leakage will show up quite readily in most cases, though bear in mind that oil will be swept back from the source of the leak by the airflow past the engine.

Check over the engine carefully and note where the oil is coming from, having removed any minor covers (eg sprocket covers) as necessary. Leakage around seals will be fairly obvious, as will most gasket leaks. There are a few exceptions however. Most Japanese four-stroke engines pass oil through passages in the cylinder block and head to lubricate the cam and rocker gear. The joints at each gasket face are usually sealed by O-rings. Sometimes, oil can leak at the crankcase to barrel joint or around the base of a holding stud. From here it works up the stud to emerge at the head to barrel joint surface, giving a misleading impression of the location of the problem. Bear this in mind, and remember to check for such migrating leaks during the overhaul.

Finally, not all oil leaks are due to simple seal or gasket failure. Poor assembly of the engine or incorrect tightening of components or covers can be responsible. You should also check all engine breather arrangements; these must allow air in the crankcase to be displaced as the engine runs. If the breather system gets blocked, pressure in the crankcase will cause oil to be forced out of otherwise sound joints and connections. The same applies to the more sophisticated recirculating systems where oil vapour is redirected from the breather into the air filter chamber to be burnt before expulsion into the atmosphere.

If the loss of oil is not due to leakage, the engine must be burning it. This is almost always a good indication of general engine wear. Worn or damaged cylinder bores or piston rings will cause high oil consumption, the oil film on the cylinder wall being drawn up past the rings on the induction stroke and then burnt. There are many causes of this type of problem, but all indicate the need for a top-end overhaul at the very least. Most problems of this type relate directly to worn or damaged pistons, piston rings or bores, and may be as simple as a stuck oil ring or as complicated as a cracked or deeply scored piston. Less obvious areas include worn plain big-end or main bearing surfaces; these can allow excessive amounts of oil to be thrown onto the cylinder walls. The piston rings are unable to cope with the high volume of the oil, which finds its way past them and into the combustion chamber. Problems of this nature can go unnoticed for some time before exhaust smoke becomes significant.

Worn valve guides and valve guide seals allow oil to be drawn down from the cylinder head area into the combustion chamber, where it is burnt. One indication of this is a brief cloud of exhaust smoke at the moment the throttle is opened after the engine has been on overrun down a hill; the oil builds up in the combustion chamber and then burns off as the engine begins to pull once more. Get someone to follow you on a short ride and check whether this is happening rather than attempting to monitor the situation with the rear view mirrors.

There are several other causes of oil burning, though these are less common and less obvious than those mentioned above. If too much oil is added to the crankcase, burning of the excess may occur. This problem is easily resolved by draining off the excess oil to restore it to the correct level. Very occasionally, an oil pressure relief valve may stick, causing the oil system pressure to become excessive. If noticed early enough, the valve can be cleaned to restore normal operation, but if left for any length of time, seals, O-rings or gaskets may be damaged by the abnormally high pressure. Finally, the accumulated debris in the oil system of a neglected engine may block passages, causing either pressure leaks or oil starvation in localised areas. Even where oil starvation does not occur, the abrasive contaminants in the oil will cause rapid engine wear. This highlights the essential nature of regular oil and oil filter changes.

6 Engine/transmission noise identification

Engine noise

Engine noise usually develops as a result of excessive clearance due either to wear, or the need for routine adjustment of certain wear-prone areas of the engine/transmission assembly. An example of the latter would be the valve clearances. These need to be checked and adjusted periodically to ensure that the small clearance required during normal running is maintained; if the gap is too small the valve will be unable to close fully and will burn out. If the gap is too large, the extra clearance will result in the noisy operation that most owners will be familiar with.

All engines produce some mechanical noise during normal operation, this being an inevitable by-product of all mechanical devices. What we are concerned with are abnormal noises which indicate an internal problem. Becoming attuned to the various normal engine noises is an essential first step in being able to spot the abnormal ones, and this requires experience and a degree of familiarity with a particular engine. If you ride one machine on a regular basis this will happen subconsciously; you will find that you can estimate your road speed quite accurately before you check the speedometer, for example, and this is largely due to an association of a particular engine note with that road speed. In the same way you will automatically notice a new noise quite readily. The skill comes in associating that noise with its probable cause, and this will require a certain amount of practice.

Describing even the most common engine noises is not easy – we simply do not have the vocabulary to do so concisely. If you ask several skilled mechanics to describe what a worn big-end sounds like, each will give a slightly different description. This is simple due to the fact that any one person can only describe what they think the noise most resembles, and you may not agree with their subjective impression. However you care to describe a noise, if you hear a worn big-end on several occasions you will recognise it again in the future. Bearing the above in mind, let us consider some of the more common engine noises and attempt to describe them.

Big-end bearing wear is caused by excessive clearance in the big-end bearing due to wear or bearing failure. It is normally described as a knocking or rattling sound, and increases with engine speed. It is likely to be loudest at low to medium engine loadings, especially during the transition from idle speed to acceleration. If the cause is general wear, the noise is likely to be most evident when the engine is first started, diminishing somewhat as the oil pressure in the lubrication system increases.

Main bearing wear or damage produces a similar noise to big-end wear, but is usually a duller rumbling sound. The noise will be less affected by engine speed or load than would a worn big-end bearing, but may be unnoticeable at idle speed. The intensity of the noise will probably increase progressively as engine speed rises, and it is often possible to feel the resultant vibration through the footrests.

Piston and bore wear is characterised by piston slap, a light metallic rattle caused by the piston rocking in its bore at the top and bottom of each stroke. It is likely to be most noticeable when the engine is first started and will diminish as the piston warms up and expands. If you suspect piston slap, try pouring a little engine oil into each cylinder through the spark plug hole. Refit the plug and start the engine. The oil will tend to damp out the worst of the noise for a few moments, after which it will be scraped off the cylinder walls and the noise will return. Another good indication can be gained by performing a compression test. Low compression is a good indicator of bore wear, and will often confirm your diagnosis of the noise. On four-strokes, add a little oil to each bore and repeat the compression check. If the reading improves, bore wear is confirmed, but if the same reading is found the compression loss may be due to leakage at the valves.

Gudgeon pin and small-end bearing wear normally results in a light rattling sound. Two-stroke engines are often susceptible to this type of wear or damage, particularly where the lubrication to the small-end has failed.

Valve train wear can be difficult to pinpoint due to the numerous possible areas in which wear can develop. You should check first of all that the valve clearances are set correctly to eliminate this as the cause of the noise. If the noise is unchanged, check for wear between the rocker arm and its spindle (pushrod ohv and some ohc engines), the camshaft bearings or bushes, cam lobes, cam followers or tappets and the camshaft drive. In the camshaft drive area, check the driving chain, paying particular attention to the tensioner mechanism. It is not unknown for these to fail or jam in use, and the problem can be recurrent and difficult to resolve with some models.

Pinking and pre-ignition are related conditions which produce abnormal engine noise due to incorrect combustion. These problems normally arise when the wrong fuel grade or fuel/air mixture is present, if the ignition timing is set incorrectly, or where the compression ratio of the engine becomes abnormally high due to carbonisation of the combustion chambers. Other contributing factors are using the wrong grade of spark plug and general overheating of the engine. Although there is a subtle difference between the two faults, they are closely interrelated and give rise to similar symptoms; namely a light metallic-sounding rattle often referred to as pinking (or pinging in the US).

The problem is usually most evident when the engine is hot and being worked hard, and will often vanish when the machine is not accelerating. Without direct comparison it is not easy to distinguish between these conditions and mechanical faults such as wear in the small-end bearing which produces a similar sound. Whatever the actual cause, this type of noise indicates the need for careful examination of the combustion chambers and pistons. If left unresolved, detonation can damage the pistons and overload the hard-pressed big-end and small-end bearings. In extreme cases the piston crown may become holed or collapse under the excessive pressure.

Pinking, or knocking, can occur as a result of over-advanced ignition timing, or an excessively high compression ratio. Part of the mixture charge in the combustion chamber burns normally, but the remainder is then provoked into exploding due to compression pressures. Many engines will suffer slight, barely audible, pinking, but if the problem becomes severe, damage can result.

Pre-ignition is caused by ignition of the fuel/air mixture at the wrong point in the engine cycle. This can be due to accumulated carbon deposits on the piston, combustion chamber and spark plug or by overheated components.

Transmission noise

Transmission noises are due to wear or damage of the transmission components, which for the sake of convenience we will use to describe anything within the crankcases not covered above. There are innumerable possible causes, and the most likely ones will depend upon the machine in question. Some of the more common general problems are as follows:

Clutch and primary drive noise can result from general wear and tear, or from mechanical failure. Noises can range from whining primary gear teeth to rattles or knocking from a worn or damaged clutch. On some four cylinder models, even carburettor imbalance can be enough to cause snatch and a resulting rattle from an otherwise sound clutch assembly!

Gearbox noise is generally confined to one ratio, though it can apply in all gears in cases where the input shaft or output shaft or bearings are involved. As a general guide, input shaft related problems will vary with engine speed, whilst those relating to the output shaft will be controlled by the road speed of the machine. Do not forget that the whole of the final drive will be involved here, so check that the drive chain or shaft is not causing the problem.

Generalised rumblings or roaring noises are often attributable to worn bearings, whilst regular clicking or rattling noises are indicative of chipped or broken gear teeth. Whining noises can often occur where general wear of the gear teeth has occurred and will build up slowly, rather than appearing suddenly.

Other noises can be produced from almost anywhere, and are largely dependent on the model producing them. For example, where balancer shafts are used on single-cylinder or twin-cylinder models, there are extra bearings to wear out. Similarly, some machines may carry jackshafts to drive the alternator, or minor gears to control oil pumps.

Locating the source of noises

Provided that you can provoke the noise while the machine is stationary, there are a number of ways of attempting to work out the general area of its source. This can be difficult if you just listen unaided, because sound can be transmitted through the cases to emerge some way from the source. The simplest method is to run the engine then place a large screwdriver against the engine at various points. If you place your ear against the handle, you will be able to hear the mechanical rumblings produced by the engine. By moving the tip around you should get a good idea of the location of the suspect sound.

To refine this process, use a length of fuel pipe as a makeshift stethoscope, or better still, buy a real stethoscope. These are often sold in engineering shops or tool shops for this very purpose.

7 Engine/transmission unit: dismantling and checking

If the preliminary checks described above have failed to pinpoint the cause of a problem, or if you have confirmed your worst fears by identifying the source of an ominous noise, you will have to prepare to dismantle the engine/transmission unit to allow detailed examination of the affected parts. To do this you will need some source of reference giving the dismantling and reassembly procedures and also the relevant specifications and torque wrench settings. Either the manufacturer's service manual or our own Owners Workshop Manual for

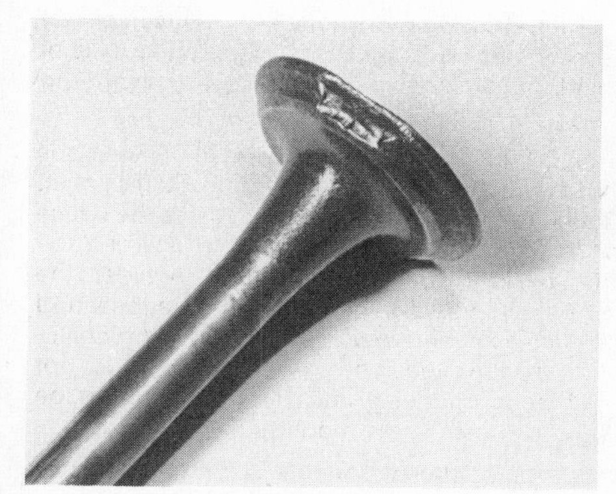

6.1 Burnt out valve faces and valve seats often result from inadequate valve clearances

6.2 Detonation damage can collapse or hole the piston crown, often near the spark plug electrodes

6.3 In more severe cases the piston may be completely destroyed as in this example

6.4 Nicks in the surfaces of the teeth, caused by metal particles getting caught between the gears

6.5 Watch out for etched markings like this when renewing gear pinions

your machine will prove a necessity, but do not expect to find sufficient information in the owners handbook supplied with the machine; this does not cover overhaul or repair work.

The manual will show you what degree of dismantling is required for a particular job, and whether you need to remove the complete unit from the frame to allow access. The extent of the dismantling work will also depend upon the nature of the fault. If, for example, an internal part has broken up in service, it would be risky not to strip the entire unit so that any residual debris can be cleaned out. If you fail to do this, metal particles may be dislodged at a later date, wrecking the engine again.

Be sure to clean the exterior of the engine/transmission unit thoroughly before dismantling commences. This will avoid road dirt getting inside the unit, and will make it a lot easier to work out exactly what went wrong. Look carefully at each part as it is removed, making an initial check for damage. Look out for pieces of metal or swarf which might give you a clue as to what has failed and why. The location of any such debris will be a key factor, so approach this methodically. If you spot something which looks wrong, stop and work out whether this has any significance before proceeding; once you have removed it and forgotten its exact position it may be too late to make this sort of deduction.

The dismantled parts should be laid out in sequence on a clean bench top, small parts being placed in marked containers so that they can be refitted in their correct locations during reassembly. This is particularly important with items like valves, which will have bedded in to a corresponding guide and valve seat. If you have room, leave everything laid out in this way for now. Failing this, place groups of components in boxes and stack them to one side. Used cardboard cartons will provide useful temporary storage of this type, and can be thrown away when you have finished with them.

Before you start cleaning the individual parts, look at them closely for further signs of wear or damage, eg the build-up of carbon on pistons, plus any scorch marks or discoloration below the rings will give a good indication of the condition of the bores and rings without the need for direct measurement. Check the sludge deposits which will have formed at the bottom of the crankcase. Is this just dirty oil, or does it contain metal particles? If so try to find out where they came from.

At the end of the dismantling process you should have formed a fairly clear picture of the condition of the engine, having noted any obvious failures and also the visible indications of normal wear. The next step is to clean each part thoroughly to remove oil and dirt, placing each part on a clean surface after doing so. If parts are to be left for any length of time, put them in closed containers or cover them with clean cloth to prevent them from getting dirty again.

Using the specifications in the manual in conjunction with the measuring equipment described in Chapter 2, you can now set about making a detailed assessment of the engine's condition in readiness for any overhaul or reconditioning work. Obvious wear or damage will make this unnecessary; eg a badly scored bore surface will mean that the engine will have to be rebored and new piston(s) fitted, so there would be little point checking the old pistons or rings. It is a good idea to make a written note of any wear or damage found during this process. Later, you can sit down and consider the best course of action, and decide which aspects you can deal with at home and which require specialist help. The rest of this Chapter deals with the more common problem areas and suggests the best way to approach them.

Sometimes it is necessary to apply heat to a part that needs to be dismanted or removed, more often than not when dissimilar metals are in direct contact with each other. Since each metal will have its own rate of expansion, when heat is applied one will expand more than the other, so that parts that were once a tight fit can be separated quite freely.

The manner in which the heat is applied requires careful consideration because if it is applied too locally as, for example, by a blowlamp or a welding torch, it is only too easy to cause the component to distort permanently, rendering it useless. Whenever possible, an oven should be used to bring the whole component to an even heat before any attempt is made to separate it into its component parts. If this is not practicable, the entire component should be heated by a flame, moving the heat source backwards and forwards, up and down, so that there are no hot spots. Remember to keep naked flames away from any fuels or similar combustible products and to take great care in handling the heated parts. To avoid the risk of serious burns, wear protective gloves or use some other form of protection to avoid contact with the hot surfaces. When the dismantling work has been completed, allow the individual parts to cool slowly in the open atmosphere. Do not attempt to force cool them as this too may lead to distortion.

Special care is needed when heating alloys as too much heat will allow them to melt, with disastrous results. This also applies to large castings, such as cylinder heads and crankcases, which will distort with ease unless adequate care is taken. Never use undue force to separate any components. If heat alone will not permit separation with gentle pressure, there is another reason why this will not occur, which must be investigated.

Never use a naked flame to separate components cast in Elektron, a mixture of magnesium and aluminium alloys sometimes used in an attempt to reduce weight on competition models. Magnesium burns very fiercely when it catches fire, which explains why it was used extensively in wartime incendiary bombs. Once started, a magnesium fire is very difficult to put out by some of the more common types of fire extinguisher.

When reassembling components always pay particular attention to the recommended torque settings and especially tightening sequences. This is particularly important when refitting the cylinder head as incorrect tightening could cause distortion of the gasket face.

7.1 Scoring of the piston skirt is evidence of a partial seizure. Light damage can be carefully filed smooth, though a new piston will be a more satisfactory repair. Establish the cause of the seizure before reassembly

8 Bearings and bushes

Plain bearings and bushes

Most current engine/transmission units use a wide variety of plain bearing surfaces, either in the form of simple bushes to support lightly-loaded shafts, or in the form of plain bearing inserts on crankshafts. They rely on a lubrication film between the bearing surface and the rotating part to provide support and prevent wear, the shaft being carried on a thin film of oil at fairly high pressure.

In the case of most four-strokes, the surface can be damaged by contaminants in the oil finding their way past the filter and scoring the bearing surface. In extreme cases the hardened surface of the shaft will also be damaged. The problem is often caused by neglected oil filter changes; the filter becomes choked with deposits, and to maintain circulation a pressure-sensitive bypass valve opens, allowing unfiltered oil to circulate. This is better than nothing, but the oil will get progressively dirtier, and damage to the bearing surfaces is almost inevitable if the situation continues for long.

With two-stroke transmissions, and those of many four-strokes, lubrication is by oil bath, the rotating parts flinging oil around the casing and onto the bushes and bearings. Again, neglecting regular oil changes allows the oil to become contaminated with particles (there is no provision for filtering on two-strokes and some four-strokes) and degraded by moisture.

General wear of plain bearing surfaces is difficult to assess in the average home workshop, even armed with the necessary clearance specifications. The tolerances required are very fine, and high-precision equipment would be needed to check them. It is, however, usually possible to gain some impression of the stage of wear by visual assessment. In the case of plain bearing inserts used in main and big-end bearings on most four-strokes, you will be able to see quite easily where the bearing layer has worn through to the copper-coloured backing metal, and any discernible movement is usually a fair indication that the bearing has completed its useful life. Minor bushes can be checked by feel, and once again any movement felt will generally indicate the need for renewal.

A major problem area can be found on many Japanese four-strokes, where the surface of a casting is often machined to form the bearing surface. No renewable insert is fitted, so when the surface wears, the whole component is effectively scrapped. Typical examples are to be found where camshafts run direct in the head material. This is a common problem area usually found on high mileage engines, and the original cause of failure is very often nothing to do with the camshaft. What happens is that bearings elsewhere in the engine (notably the crankshaft main and big-end bearings) become worn, and eventually the capacity of the oil pump is outstripped by the rate of leakage around these bearings.

This results in a loss of oil pressure beyond this point in the lubrication system. The supply to the cylinder head is diminished, and the otherwise sound camshaft bearing surfaces are rapidly destroyed. If you examine an engine to which this has happened, you will find that the bearing surface furthest in line from the oil pump fails first, the condition progressing back along the oil circuit. The owner may have noticed that the engine seems to be making a little more valve train noise than normal, but when the engine is finally dismantled it is not uncommon to find the cylinder head and camshaft(s) effectively destroyed.

This example illustrates two things. First, the average Japanese engine works reliably and performs well **only** as long as all engine tolerances are within the specified range. Once one or more of these limits are exceeded, even by a small amount, rapid wear can follow. There is **no** safety margin designed into the engine. Secondly, you will appreciate the need for fastidious maintenance, regular oil changes, and above all, the need to resolve any unusual noise immediately.

Worn or damaged bushes are fairly easy to deal with. You will need to remove the old bush, using one of the extraction methods described in Chapter 2. Assuming that the corresponding shaft end is sound, a new bush can then be fitted. In a few cases you may need to ream the new bush to size, though many are ready to use once pressed into place. In a very few cases, the bush must be line-reamed, which means locating the axis of the reamer from another reference point, usually another bush on the opposite side of the casing. This is a difficult job to carry out at home, and in such cases you will need to get help from a good engineering shop.

If the shaft end has become worn or damaged, you have a choice of either renewing it together with the bush, or getting the old one reconditioned. Again, this is a job for a professional engineering workshop. It may be possible to re-grind the bearing surface of the shaft if it is not too worn. A new undersized bush will then have to be made up to suit the new diameter. Alternatively, the shaft end could be turned down to remove the damage, the surface then being built up by plating, or by a process called metal spraying. The reconditioned surface is then ground back to its original size and can be used with the standard bush. If you are faced with this sort of choice, first get a price for a new shaft and bush, then armed with this information, seek the advice of an engineering company, who will be able to advise you whether reconditioning would be cost effective by comparison. In cases where the machine is obsolete, or parts are unobtainable for some other reason, reconditioning may be the only option open to you.

In the case of split plain bearing inserts, which are normally used for crankshaft main and big-end bearings and on the camshaft bearings of some models, renewal is relatively straightforward. New bearing inserts are selected according to the manufacturer's specifications. Often the journals will be graded according to tolerance, and marked with a code letter or number. This is used to select the insert from a range of sizes to obtain the required oil clearance. The inserts are usually identified by a dab of coloured paint.

The selection procedure described above will work provided the journal itself is unworn. If you detect free play after fitting new bearing inserts of the correct grade, and thus need to check journal wear, a product called *Plastigage* will prove invaluable. This consists of a graded deformable plastic strip. A piece of the material is laid across the journal, and the bearing cap tightened down, squeezing the strip flat between the surfaces of the journal and the bearing insert. The bearing is then dismantled again and the width of the strip read off against a scale on the *Plastigage* envelope. This will indicate the actual oil clearance of the bearing, and any irregularity of the strip will give a good visual indication of taper or unevenness in the journal. *Plastigage* should be obtainable through larger motorcycle dealers. If one or more journals are worn or damaged, specialist attention will be required. This is covered in the following section.

Journal ball and roller bearings

These will be found in most engines, and are almost invariably used in two-strokes where a high-pressure lubrication system is not available. It is almost impossible to measure wear in this type of bearing, and an assessment of its condition will have to be made by feel. Whilst there will be a little play evident when the inner race is rocked in relation to the outer race, no discernible radial play should be evident. The bearing should turn smoothly and evenly, with no signs of roughness or tight spots. This can be checked after cleaning the bearing in solvent, but note that a dry bearing will invariably produce a little noise when spun. Apply one or two drops of clean motor oil to the cleaned bearing. If grittiness or noise is evident, the bearing should be renewed. On seriously damaged bearings it may be possible to see pitting or imperfections of the rollers or balls, and in the bearing tracks of the races. Any such blemish means the bearing is no good.

Needle roller bearings are often used in two-stroke small ends, where they can survive the meagre lubrication available. When these break down you will usually find cracked or damaged rollers, a broken cage, and in bad cases, indentation or pitting and scoring of the connecting rod small-end eye and the piston gudgeon pin. If only the bearing is damaged, it is permissible to renew this alone, though if the surfaces on which it runs are damaged these too will need attention. In

8.1 A strip of *Plastigage* is placed on the journal and the bearing is assembled, taking care not to turn the shaft

8.2 When dismantled the width of *Plastigage* strip will indicate the amount of bearing clearance

8.3 By lining up the markings on the *Plastigage* package, the exact clearance can be read off on the printed scale

the case of small-end bearings, this may mean fitting a new crankshaft assembly.

Similar bearings will be found in other areas of some engines, often pressed into blind bores in the crankcase. If inspection shows these to be worn, remove the old items using a slide-hammer and bearing extractor, and use a tubular drift of the correct size to tap the new one into place. Otherwise, do not disturb the bearing – removing it from the casing will almost certainly destroy it.

Roller big-end bearings may be encountered, and again almost every two-stroke engine uses this arrangement. The bearing assembly includes the crankpin, and this is usually in the form of a pressed-up assembly forming part of the crankshaft. With this type of crankshaft assembly there should be a specific amount of axial (side-to-side) movement of the connecting rod between the flywheels, but no discernible radial (up-and-down) movement. Axial play is normally measured with feeler gauges, whilst it is normally sufficient to check the radial play by feel. If you feel the need to measure this, set the crankshaft up on V-blocks and take a measurement using a dial test indicator (dial gauge). Be sure not to confuse any side play felt as being radial play.

If wear is discovered, some manufacturers offer an exchange crankshaft service for their models, though it is now increasingly common to simply purchase a new crankshaft outright. It is theoretically feasible to rebuild a roller big-end, but only if the replacement big-end assembly can be obtained. Note that this may well include the connecting rod. The process of pressing apart a crankshaft, assembling it with a new big-end and then trueing the flywheels should be considered a specialist job, and thus best entrusted to an engine rebuilding company. Some motorcycle dealers will be able to undertake this work, or arrange to have it carried out on your behalf. As a general rule though, a damaged big-end tends to indicate the need for a new crankshaft assembly.

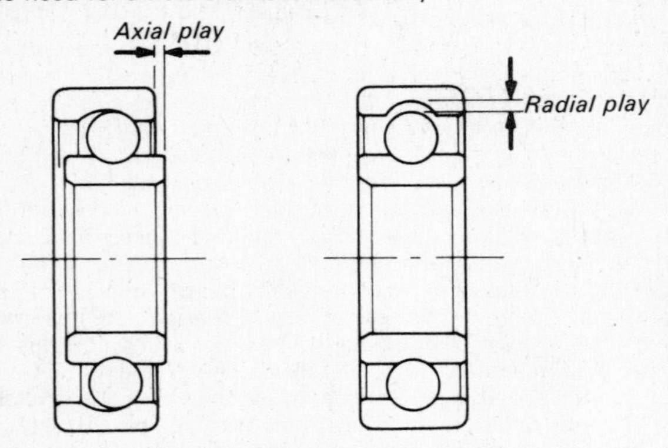

Fig. 4.1 Axial play and radial play in ball bearings

This is best checked by feel. If you can detect excessive movement, renewal is necessary

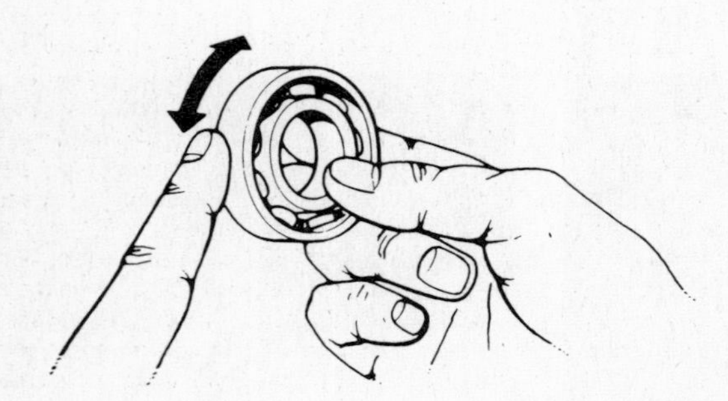

Fig. 4.2 Spin a ball or roller bearing to check for tight spots or roughness

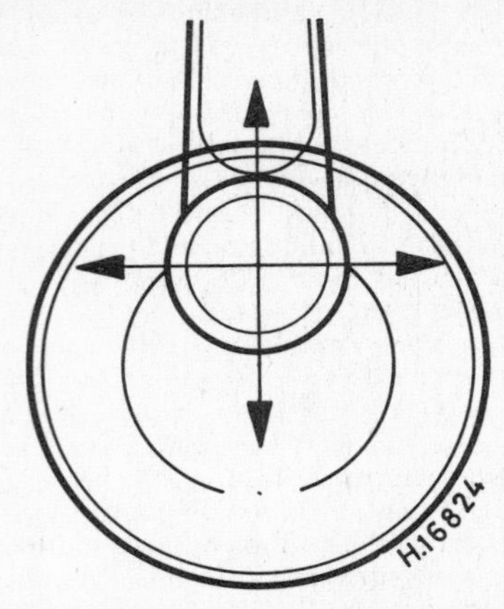

Fig. 4.3 Checking big-end bearing radial play

Checks for free play in the bearing as shown. To do this accurately, use a dial gauge

8.4 Most two-strokes use caged needle roller small-end bearings like this (though beware of uncaged rollers during dismantling!)

8.5 Small-end rattles are usually caused by wear of the bearing, though damage of this type will also cause a certain amount of noise! The owner of this machine was lucky to catch things at this stage, before the connecting rod broke completely

8.6 The scored and pitted centre section of this gudgeon pin shows the damage caused by a seized small-end

8.7 Result of a wrecked small-end bearing. Broken rollers were swept up though the transfer ports to the combustion chamber ...

8.8 ... where they were hammered into the piston crown and cylinder head

8.9 Axial (side-to-side) play in the big-end bearing is measured using feeler gauges. There should be no radial (up-and-down) play

8.10 With some pressed-up flywheel assemblies it is necessary to check various dimensions after reconditioning

9 Crankshaft and camshaft: reconditioning

As has been mentioned, it is possible in some cases to salvage an otherwise scrap crankshaft or camshaft assembly in the event of severe wear or damage. This is very much a specialist task requiring a great deal of skill and some very expensive machinery. If the damage extends to a non-renewable bearing surface, eg camshafts which run direct in the cylinder head, it may be possible to carry out a repair rather than have to renew the cylinder head, camshaft and camshaft caps.

Repair work of this nature entails building up the damaged casting by aluminium welding. A new bearing surface must then be machined into the cylinder head, using a surviving sound bearing face to locate from. It is not possible to generalise about the viability of such repairs; in some cases it can be done, in others it would be impossible or would cost more than a new head. The only constructive advice in such cases is to price the cost of new parts, and then get a comparative assessment from a reputable engineering company. It should be stressed that this is very much specialist work, and some engineering firms may be unwilling or unable to undertake this sort of job. Ask around locally for the name of a reputable company, and seek their advice.

Renovation of plain bearing crankshafts is, superficially, a simple matter of taking the crankshaft to an engine reconditioning specialist for grinding, and it is customary for the engine manufacturer to offer a range of special undersized bearing inserts, just as oversized pistons are available for reboring a worn engine. Unfortunately, the majority of motorcycle manufacturers no longer hold with this practice, and do not list undersized inserts. Today it is assumed that the crankshaft should last for the useful life of the machine, and on this basis there is no requirement for re-grinding and the fitting of undersized inserts.

This means in practice that you will have to purchase a new crankshaft, unless you can track down a specialist offering their own, unofficial, reconditioning service. Many such companies exist, and advertise in the motorcycle press. They usually specialise in one or two areas of reconditioning work, or on a particular model, and some will even undertake more major conversions, such as converting damaged cylinder heads to accept needle-roller camshaft bearings. Whilst most of these companies are quite honourable in their intentions, and provide an invaluable service, bear in mind that any manufacturer's warranty will be invalidated by such an unauthorised modification. That said, it cannot be a bad thing to make an expensive and unavoidable replacement part a viable reconditioning prospect, and such conversions go a long way to compensate for the general lack of interest by manufacturers in the subject of durability in their products.

The best way to reclaim a crankshaft with damaged journal surfaces would be to have the journal built up by plating or metal spraying and then ground back to the original size, as described above. Done properly, the crankshaft will be as good as new, and you will be able to use the original bearing inserts.

Metal spraying may sometimes allow a damaged small-end bearing surface to be reconditioned, though if there is enough metal around the eye, you could consider having the eye bored oversize and a bush fitted. Once again, the best course of action is to take the offending part to a specialist for evaluation.

Camshafts can be treated in a similar fashion to that described for crankshafts, and a good engineering company will be able to build up and re-machine both the bearing journals, and where necessary, the cam lobes. Camshaft lobe re-profiling is a fairly common procedure, and can also be used to alter the cam lift characteristics for a particular engine. This is useful where the engine is being tuned for competition purposes, but make sure you deal with someone who knows the engine concerned, particularly its limitations in racing applications.

In cases where catastrophic engine damage has occurred, such as broken or bent connecting rods or valves etc, it is essential that you get the components associated with the broken parts checked carefully. This means getting them crack detected, and where necessary straightened, stress relieved and reground. For most practical purposes on motorcycle engines, it will be quicker, safer and probably cheaper to fit new parts. If necessary, consult your engineering specialist for advice on this point. They will probably be able to assess quite easily whether it is worth attempting to repair the damage. Remember that the crankcase or cylinder head may well be damaged too. Take them with you together with the damaged shaft so that the parts can be checked together. Finally, remember that damage of this nature can be far more extensive than it looks; if the engine has blown in a big way it is probably time to start scouring the local breakers for a new unit.

9.1 Crankshaft re-grinding (car-type shown). Often the unavailability of suitable undersize bearing inserts will require the journals to be built up before they are ground back to size

10 Cylinder head: reconditioning

Initial preparation

The amount of work involved in cylinder head reconditioning depends on the engine type; eg a single cylinder two-stroke being far easier to deal with than a dohc four-stroke four. In the case of four-strokes, the valve and camshaft components must be removed before cleaning and inspection can begin. Remember that each valve assembly must be placed in a marked container so that it is refitted in its original port and guide. Camshaft caps and any cam follower or rocker arm assembly should be stored in a similar manner, and for the same reason.

Once the head is stripped to the bare casting, cleaning work can begin. The object is to remove all traces of carbon deposits without causing damage to the soft aluminium alloy. If possible, it is best to have the deposits removed by dipping the head in a specially designed solvent cleaner. This is a commercial tank process using fairly strong chemicals and is not normally available for home use for safety reasons. If you can find a local engineering company with this facility, however, it is worth paying to have the head cleaned in this way. It will save you a good deal of time and effort, and will remove deposits inside ports and other awkward areas. If you decide on this method and can arrange it locally, try to get the piston(s) and cylinder barrel cleaned at the same time. Note that the chemicals will probably remove any paint coating, and that all seals and any plastic parts must be removed or they too will be dissolved.

If the head casting is generally well-used, and especially where the outside surfaces are ingrained or corroded, consider having the casting bead blasted. Again, this method of cleaning is extremely efficient and well worth paying for. Before taking parts for blast cleaning, make sure that they are thoroughly degreased; dry carbon deposits are quite acceptable, but oil and grease will be received with little enthusiasm by the prospective blaster. On ohc four-stroke models, or any machine with bearing surfaces on the head casting be sure to mask off these areas. Where camshafts are held by caps, cut a length of wooden dowel or plastic tubing to the same length as the camshafts and clamp this in place with the bearing caps. Cover valve guide bores by plugging them or by fitting plastic caps.

Be absolutely certain to specify glass bead or aqua-bead blasting only; on no account allow alloy castings to be abrasive blasted. After blast cleaning, **all** traces of residual dust and beads must be removed. Use a high pressure airline to blow through ports and oilways, and clean thoroughly in a degreasing solvent afterwards. Any small residual deposits of beads left unnoticed will invariably get dislodged after the engine is rebuilt, and the damage which can result can be quite dramatic.

If you are unable to have the head cleaned chemically or by blast cleaning, you will have to resort to the time honoured (and time consuming) procedure of scraping off the carbon deposits. The success of this procedure is directly proportional to the amount of effort you are prepared to put into it. You will need various scraping tools, and these will have to be made up to suit. Choose materials which are slightly softer than the alloy to avoid accidental scoring of the surfaces (particularly the gasket faces), avoiding the temptation to use an old screwdriver or chisel. Clean out the ports on four stroke engines using steel wool or Scotchbrite pads. Other useful equipment for this process includes small rotary brass wire brushes which can be used in the chuck of a power drill. Better still, purchase a flexible drive which can be fitted between the drill chuck and the brush. This will allow you to get into ports and other recessed areas otherwise inaccessible to a power tool.

Inspection of the bare cylinder head

When cleaning has been completed, the head casting can be checked for wear or damage. Normally, this consists of a careful visual check for cracks, and in the case of four-strokes, checking the condition and security of valve seats and guides. If all this is in order, attention can be turned to checking for warpage of the gasket faces, and checks and measurements of all wear-prone areas such as valve guides, seats and bearing surfaces. To carry out these checks you will require the manufacturer's service specifications and wear limits, and once again you will need either the factory service manual or the relevant Haynes Owners Workshop Manual for the model concerned.

Crack and leak detection

If you have reason to suspect other damage, it is a good idea to arrange to have the head crack detected. This again is a specialist engineering process, and one which may prove difficult to arrange locally. Despite this, it is well worth having done if you are dubious about the condition of the head; there is little point spending money reconditioning a head which later turns out to have hairline cracks or other fundamental defects present.

The normal crack detection processes use a fluorescent dye which penetrates the cracks and will then show up during inspection under ultraviolet lighting. This method can be used to good effect on most engine components. Whilst it will identify even hairline cracks which might normally be invisible to the eye, it will not draw attention to other sources of leakage, such as pressure leaks around studs or valve seats. A related process is magnetic crack detection which uses iron powder instead of a fluorescent dye. This system requires the component under inspection to be magnetised, and is thus of no use on non-ferrous parts such as aluminium alloys, brass or bronze.

A traditional method of crack detection in the normal workshop (and the only one easily adaptable for home use) is to apply a film of kerosene (paraffin) and ether in equal proportions to the surface to be tested. This is then left to soak in for a few seconds, before blowing it dry with compressed air. The ether is a volatile substance, and any which has sunk into a crack will exude from it as it evaporates, making the crack easily visible. The most convenient source of ether is in the form of one of the ether-based 'easy-start' aerosol sprays.

Pressure leak testing is another useful process, and this will show up defects other than simple cracks, such as the pressure leaks mentioned above. The chosen process will in all probability depend upon the system and equipment available at the engineering company entrusted with this work.

Checking for warpage of gasket faces

Warpage of the gasket faces can occur due to incorrect tightening down of the head fasteners, or less commonly, through extreme conditions such as an overheated engine getting suddenly soaked in cold water. Most manufacturers prescribe warp limits for major engine components, and these should be referred to before making the check.

The easiest home method is to place the casting, gasket face down on a clean surface plate. Ideally, this should be a specially-made ground steel plate, specifically designed for this type of work. A reasonable alternative is to use a sheet of plate glass (not modern float glass) and a good source of this would be an old mirror or similar. Check by sighting across the glass surface that it really is flat, with no visible rippling of the surface.

The gasket face is placed against the surface plate, making sure that both surfaces are completely clean – even a small piece of grit will give a false reading. Look for signs of daylight between the surfaces, and measure any discernible gap using feeler gauges. If the gap exceeds the manufacturer's warp limit (usually about 0.05 mm/0.002 in) it will be necessary to have the surface machined to correct the distortion.

Slight distortion can be corrected at home, using a sheet of abrasive paper taped to the surface plate or glass sheet. The gasket face is rubbed in a circular motion on the abrasive paper until the distortion is eliminated. Lift the head regularly and check the appearance. High spots will show up as matt grey areas, and these will gradually get larger as the material is levelled. Eventually the surface should show an even matt appearance over its entire surface. Note that where the surface is generally flat, with high spots around stud or bolt holes, this indicates over-tightening of the cylinder head fasteners in the past.

Where the head is too big to fit on the abrasive sheet, or where the warpage is severe, it is better to have it surface ground by an engineering company. A popular type of machine for this type of work is shown in the accompanying photographs, though there are many other ways of machining the surface flat again. Whichever method is used, it is essential that the minimum of metal is removed. Excessive grinding or milling of the head surface will increase the engine compression ratio, and this in turn can cause pinking problems. In extreme cases and on certain four-stroke engines, the reduced clearance between the top of the piston and the valve heads when fully open can cause the two to come into contact, often with devastating results. It is sometimes preferable for this reason to entrust the work to a specialist. This will allow the machine operator to take clearance measurements, and where necessary, carry out corrective machining to other parts.

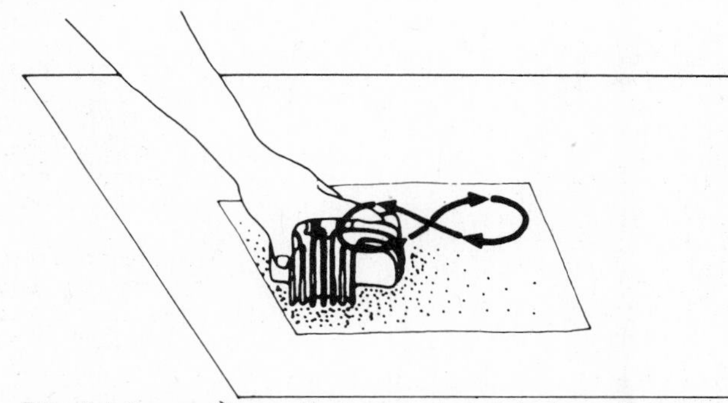

Fig. 4.4 Correcting cylinder head gasket face distortion

Minor distortion can be corrected by moving the gasket face in the pattern shown on a sheet of abrasive paper. Note that the surface used must be flat, and that it is helpful to tape the abrasive sheet in position

11 Valves, valve seats, guides and springs: overhaul

After the head has been cleaned, checked and pronounced sound, attention can be turned to the valve components. For obvious reasons, two-stroke owners may skip this section. Many of the measuring and checking procedures described can be undertaken at home. Alternatively, you could take the entire head assembly, after initial cleaning, to an engine reconditioner for measurement and any necessary repair work to be carried out. Without wishing to discourage anyone wanting to attempt as much of the reconditioning work as their workshop facilities allow, it is often more cost-effective to have this type of work done professionally. With cylinder head work, minor defects can have a significant effect on the efficiency of the engine, so if you intend to do any work at all it is worth checking **all** aspects of the condition of the head and related parts.

If you decide to have any professional help in dealing with this (or any other) area of the engine rebuild, try to supply as much specification information as possible to avoid wasting time and

10.1 A simple method of checking for distortion is to use a straightedge in conjunction with feeler gauges

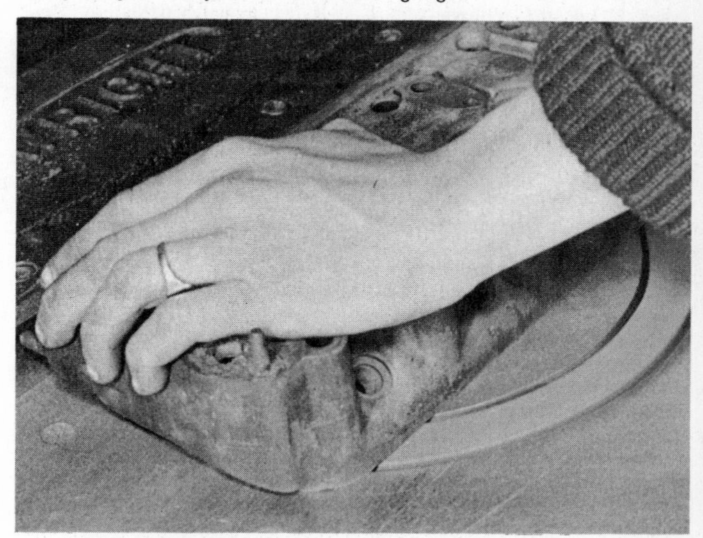

10.2 Most engine reconditioning specialists will undertake surface grinding to restore warped gasket faces

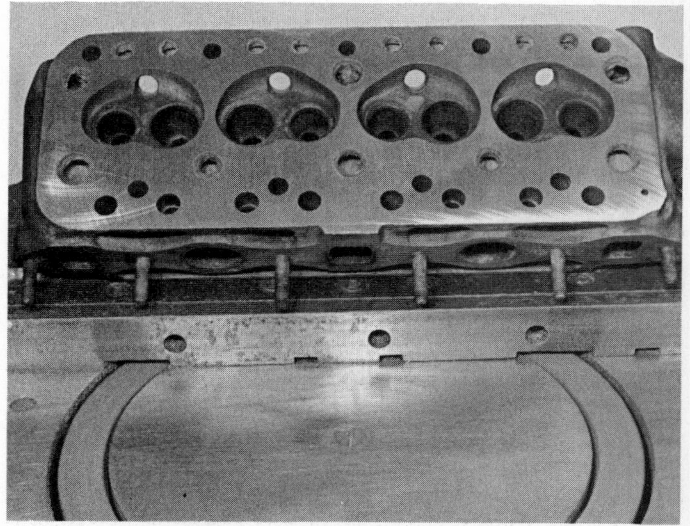

10.3 This shot shows a (car) cylinder head after grinding. Any distortion is now removed, leaving the gasket face smooth and true

perhaps necessitating guesswork. Remember that the majority of engine reconditioning work is carried out on car engines, which normally have far greater tolerances. The more accurately the parts can be checked, the better the result is likely to be, and few reconditioners engineering shops will have all of the required information to hand.

The main areas to be checked are the valve face and the corresponding surface of the valve seat, the valve stem and guide, and the valve springs. The first job is to clean off carbon deposits from the various components to allow close inspection and measurement. Clean off the accumulated carbon from the valves by scraping and with abrasive paper, taking care to avoid scratching or scoring the valve contact face or the stem. One way to speed this process considerably is to clamp the valve stem in the chuck of a bench-mounted power drill or a lathe. A blunt scraper can then be applied very carefully to the valve to skim off the carbon. Take great care if you employ this method, both to avoid catching clothing or fingers in the rotating chuck or valve, and to avoid injury from flying debris – use grinding goggles to protect the eyes. A polished finish can be achieved using progressively fine grades of abrasive paper.

Carbon in the valve guides and around the valve seats should have been removed during the head cleaning process. Any residual carbon can be removed from the guide bores using specially designed tools in a drill chuck.

Examine the valve and valve seat contact faces for wear or damage. Any indication of pitting or burning of these faces will require remedial work, or possibly renewal of the valve. Major valve defects are best dealt with by renewal, but in the case of the valve seats this is usually not possible. The valve seats can be cleaned up by recutting them, a process described below, but if this is necessary (as evidenced by pitting of the contact area) you should first check the condition of the guide and valve stem, as their condition will affect the sequence of operations. The valve contact face too can be reconditioned if necessary.

Check each valve stem for wear, which will show up as a dull area on an otherwise highly polished surface. If you have a micrometer, check the valve stem diameter in various positions, comparing it at the most worn point with an unworn section. Most manufacturers will give a maximum wear limit figure for the stem, and if worn to or beyond this, a new valve will be needed. The manufacturer's service manual or the relevant Haynes Owners Workshop Manual will give details of the various areas requiring measurement, together with the specified dimensions and service limits. A typical example is shown in the accompanying illustration.

If the valve is within limits, check the fit of each valve in its guide. In many cases, the manufacturer will quote a maximum clearance figure, often assessed as the amount of 'wobble' found at the valve stem end. This can be checked using a dial gauge. If the only clearance specified is a maximum bore diameter, this will have to be checked using a bore gauge, and this is best entrusted to someone suitably equipped – it is not worth buying this type of equipment for occasional use.

Valve guides

If the guide proves to be worn it will have to be renewed. This will entail recutting the seat to the new guide bore, so it is best to let a professional take care of the entire process. It is perfectly feasible to carry out the removal and refitting work at home, provided that you have or can make up special piloted drifts to support the guide during fitting, and have access to the necessary reamers to take the bore of the new guide to its finished size. In some cases the cylinder head may have to be heated in an oven to allow the guide to be removed and fitted without the risk of damage to the bore in which it sits. If this approach is specified in the workshop manual, it should not be ignored. A normal domestic oven, or even immersion in near boiling water will often suffice, but follow any directions regarding temperatures and heating methods carefully. One method to avoid is any form of heating with a flame; the localised nature of the heat may distort the head, and the relatively low and abrupt melting point of aluminium alloys can prove embarassing at the very least!

The old guide is driven or pressed out of the head, and the new one fitted in its place. Where the guide locating bore in the head has become loose, it may be necessary to have a guide with an oversize outside diameter made up and fitted. On no account should you even consider the barbaric process of knurling the existing guide to obtain a good fit; this will ruin the head and is rarely successful. Equally, using bearing fitting or locking compounds is no real solution to this problem. A new guide should be made up using the same material as the original, its outside diameter being determined by the extent of damage to the head bore, and the strength of the surrounding area. Where there is insufficient material around the guide bore in the head, it may be necessary to build up the damaged hole by aluminium welding and then machine it back to accept a standard size guide. The machining required will have to be of a high standard if the original valve angle is to be duplicated exactly. If required, the valve guide bore must now be reamed to the finished size to give the specified clearance.

Valve seats

Once the guide is in place, the valve seat face will have to be recut to the new guide bore, the axis of which will inevitably have changed slightly. This is best done using a professional seat cutting machine. This will normally use a rotating tungsten carbide cutting tool or a grinding wheel. Note that the cheap hand-operated milling cutters sold in car accessory and tool shops for this purpose will be found to be totally useless for motorcycle work. This is due to the use of Stellite or other very hard materials for the seats on air cooled engines, and to the work-hardening effects of heating and cooling in use. Often the seat material will turn out to be harder than that of a cheap cutter.

If the valves are to be reused, it is preferable to have them refaced to suit the reconditioned valve guides and seats. This will ensure that the contact surface is regular and of the correct width and angle. Another type of specialised grinder is used for this work, and in skilled hands will ensure an accurate and gas-tight seal at the valve, dispensing with the need to lap in the new surfaces by hand. Some reconditioning companies will carry out a pressure check of the finished work as a matter of course, as well as ensuring that the valve heights are correct for each assembly. This is vital, since if the valve sits too high or too low in its seat it may prove impossible to obtain the correct clearance adjustment on the assembled engine.

This last consideration applies equally where valves have been lapped in by hand on frequent occasions, using grinding paste. It should be kept in mind that each time this is done, some of the valve seat surface is ground away, and if done to excess, the valve will become recessed into the seat, or 'pocketed'. Not only does this affect the height of the valve stem and thus the valve clearance adjustment range, but the pocketing effect makes the path that the incoming mixture and the outgoing exhaust gases must take tortuous and inefficient. More seriously, it is not possible in some cases to renew a damaged valve seat, and so excessive valve lapping can effectively ruin the entire cylinder head. For this reason, lapping valves in by hand must be done only occasionally and very gently, using a fine compound only. If pitting or burning of the faces has occurred, this should be corrected by machining as described above.

The problem of valve seat renewal is difficult to deal with where the manufacturer does not recommend or supply parts to facilitate this procedure. The only choice other than buying a new head is to have a new seat made up specially and fitted into the existing head. Once

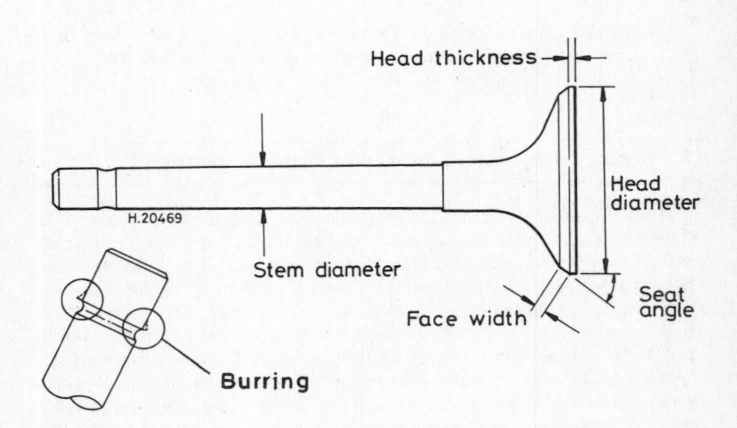

Fig. 4.5 Valve measurement points

Note that any burring around the collet groove should be removed before the valve is withdrawn from its guide, or the guide bore may be damaged

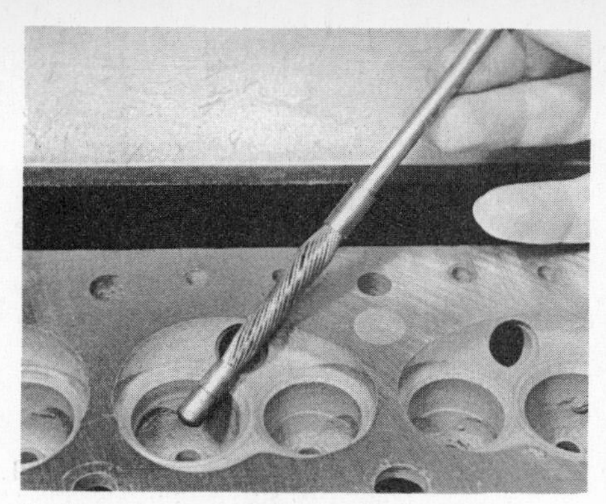

11.1 Remove carbon deposits from the valve guide bores using brushes, or true clearances will not be found

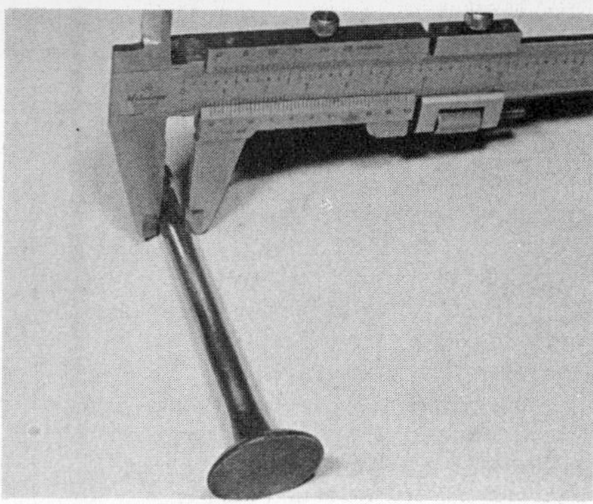

11.2 Compare worn and unworn areas of valve stem. Vernier calipers will give a reasonable indication ...

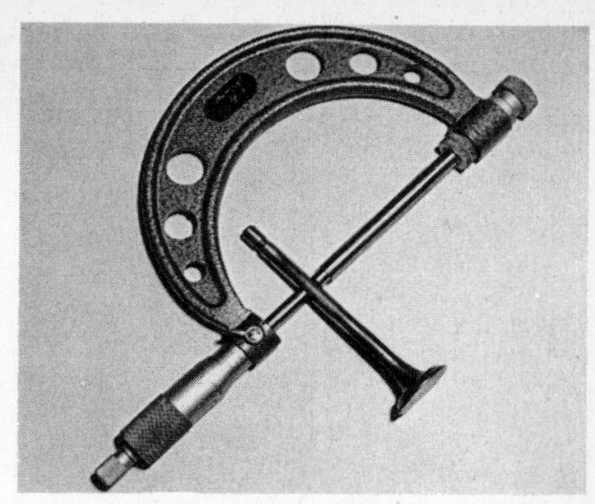

11.3 ... but a micrometer will give more accurate results

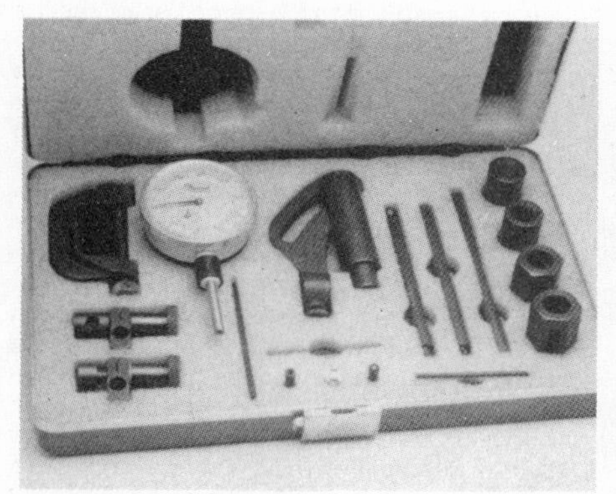

11.4 For direct measurement of valve guides a set of bore gauges will be required

11.5 The gauge is used as shown to take a direct measurement of the bore diameter ...

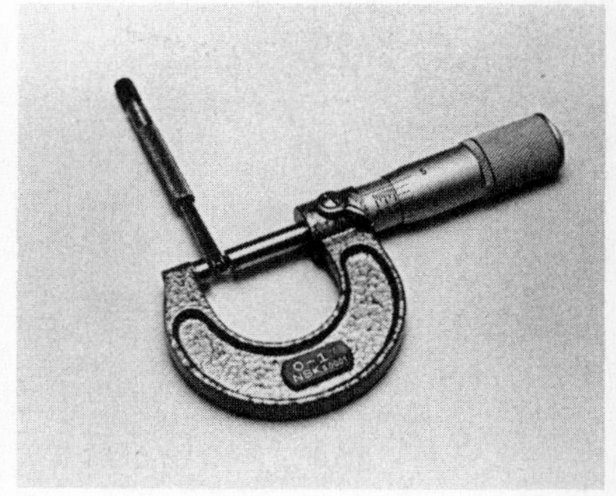

11.6 ... and the resulting reading read off using a micrometer

11.7 After new guides have been fitted, the guide bore should be reamed to size as shown

11.8 The cylinder head is clamped in place awaiting valve seat re-cutting.

11.9 The tungsten-carbide cutting tool simultaneously cuts the three different angles which form the seat face

11.10 Compare the re-cut seat (centre) with its neighbours

11.11 This machine is used to re-grind valve faces to remove pitting and restore the correct profile

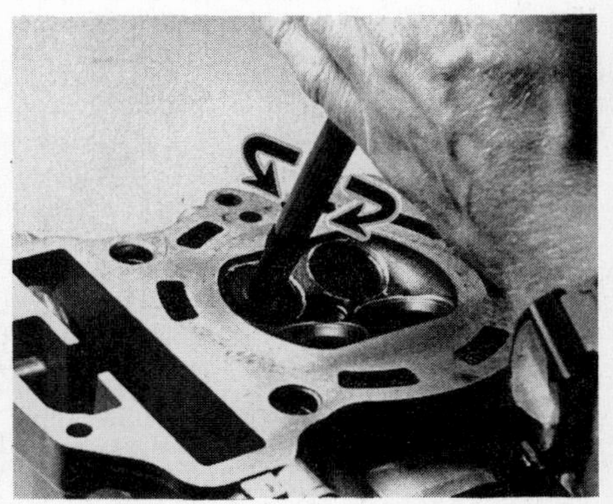

11.12 Most owners will be familiar with valve lapping. The tool is twisted to and fro between the palms of the hands

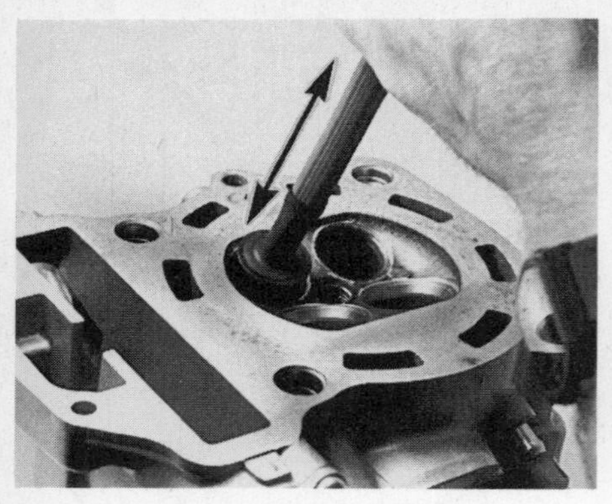

11.13 Periodically, the tool and valve should be lifted, turned and then lowered to redistribute the paste

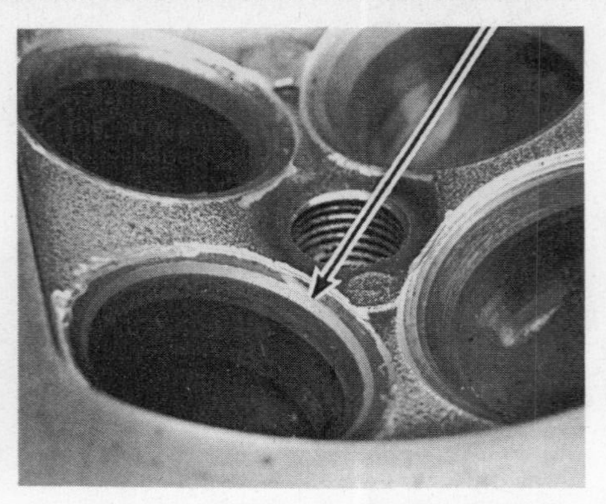

11.14 Re-cut the seat if after a few minutes lapping the surface is not restored

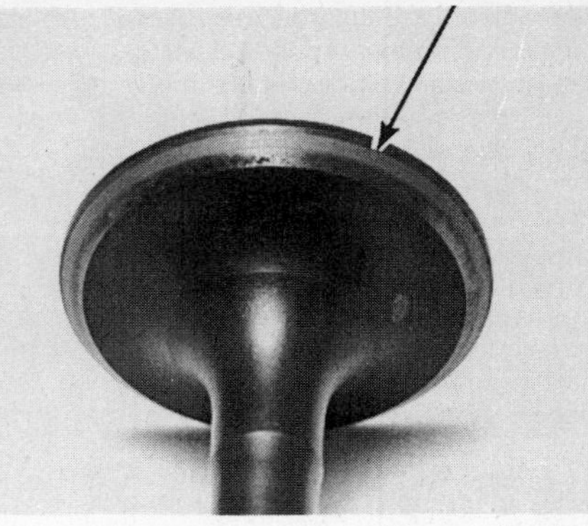

11.15 The contact area should be confined to a narrow band at the centre of each seat, with no residual pitting or grooving which might allow leakage

again this is specialised and precision work, and is best left to a skilled professional. Seat renewal will be needed if the valve has become pocketed so much that normal clearances cannot be maintained, or where the old seat has loosened in the head.

Valve springs

The valve springs are used to close each valve after it has been opened by the cam lobe. On most models, two concentric springs are fitted to each valve, the coils normally being wound in opposing directions. It is also common for the springs to be progressively wound, with tighter coils at one end than at the other. This should be checked during dismantling and the correct direction of fitting observed during assembly. This information usually appears in the workshop manual, but do not rely on it being available; always check this yourself during dismantling.

Initial checks should be made, looking for any indication of cracks at any point of the spring coils. Such defects will mean that the spring should be renewed as a matter of course, and common sense dictates that the remaining springs should be renewed as a precaution against their subsequent failure.

After a period of use, the effects of heat will begin to weaken the springs. This may not become apparent in normal use, but if the machine were to over-rev as the result of a missed gearchange, it is possible that the weakened springs might allow the valves to strike each other. When valves get tangled in this fashion, both are wedged open, and the next time the piston nears top dead centre, bent valves and a holed piston usually result. Less dramatic symptoms include valve bounce, where maximum engine speed, and thus power, is restricted by the lazy response of the valves due to the weakened springs. This can usually be heard when it occurs.

To avoid these occurrences, the springs should be checked during the overhaul. Many manufacturers specify the minimum pressure of each spring when it is compressed to a specified length. Whilst this is a good way to assess condition, it requires a special test rig to do so. Other ways of specifying service limits on springs is to define a minimum free length, the spring being due for renewal if it has become permanently compressed to this limit. This is more easily checked by simple measurement, using a steel rule or a vernier caliper.

You should also check that the springs are still true, renewing them if distorted. This is easily checked with a set square, or by rolling the spring on a flat surface. In the absence of the necessary service limits for the valve springs, it is worth renewing them as a precautionary measure on a high mileage engine. The relatively low cost is more than offset by the comparative cost of damage which could be caused by just one defective spring.

11.16 The contact width is important, and some manufacturers specify a maximum figure beyond which re-cutting is prescribed

12 Camshafts, cam followers and rockers: overhaul

The majority of current four-stroke engine designs use a single overhead camshaft (sohc) or double overhead camshaft (dohc) layout. Depending on the design, movement from the cam lobe is transmitted directly to the valve stem through a bucket-shaped cam follower, or indirectly through rockers. Normally, dohc designs opt for the more efficient bucket-type cam follower, though where more than

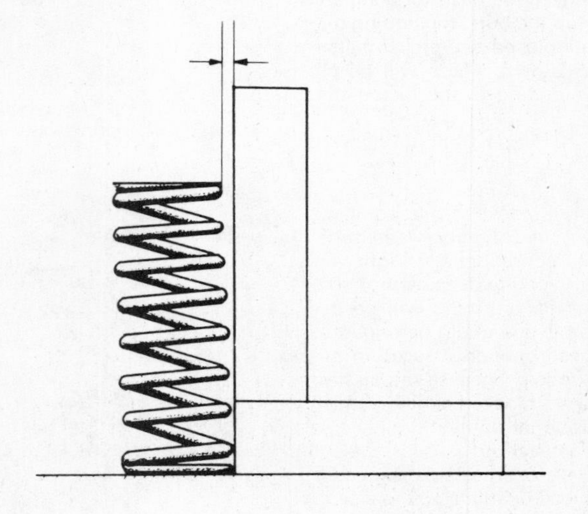

Fig. 4.6 Checking the valve springs for squareness

Renew if clearance exceeds manufacturer's service limit

two valves per cylinder are used it is sometimes necessary to use rockers to allow independent adjustment of each valve clearance setting within the confines imposed by the valve layout.

In the case of pushrod-operated overhead valve (ohv) designs which are still in use on some models, the camshaft(s) reside in the crankcase. Reciprocating motion is transmitted to rockers housed in the cylinder head by means of pushrods.

The general check of the camshaft lobes and bearing surfaces have been discussed earlier in this Chapter, and reference should be made to this when assessing wear and damage. Most manufacturers specify service limits for bearing surfaces and lobe heights, and these can be measured using a micrometer. This will give an accurate picture of the degree of wear of these areas, though in practice a visual check for scoring, scuffing and other damage to the surfaces is likely to be more revealing. You will also need to check all other associated parts for wear or damage, as detailed in the workshop manual. Wear in cam followers, rocker mechanisms and the camshaft drive and tensioner can create a lot of mechanical noise, and it is as well to eliminate this at an early stage in the overhaul.

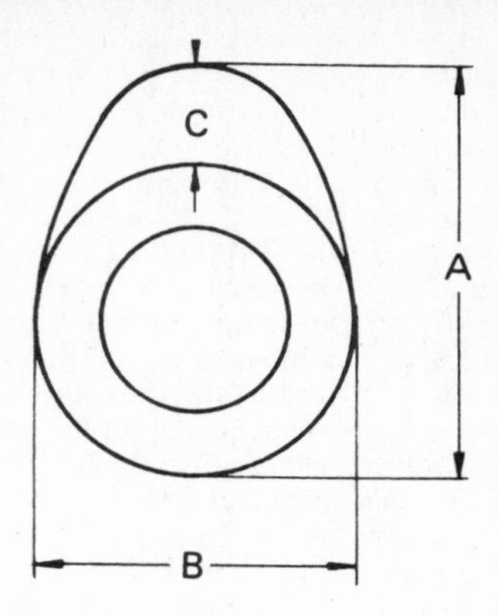

Fig. 4.7 Camshaft lobe dimensions

A Lobe height
B Base circle diameter
C Cam lift

13 Cylinder bore and piston: overhaul

Cylinder bore examination

When examining the cylinder bore(s) and piston(s), attention should be focussed on the condition of the bore surface before consideration is given to that of the piston. This is for the simple reason that if the bore is found to be worn or badly scored, you will have to have it rebored to suit an oversized piston, and it follows that further attention to the existing piston would be pointless.

Before we go any further, it is worth mentioning plated alloy cylinders, such as *Nikasil* and similar processes. With this type of bore, the aluminium alloy surface is coated with a harder deposit to provide the cylinder surface. This is highly resistant to wear or abrasion, but once it does become damaged, no reconditioning is possible. The only action you can take is to buy a new cylinder barrel or block, complete with new matched pistons. You will find this type of bore only on a few machines, notably Moto Guzzi and BMW.

After thorough cleaning, examine the bore surface closely, looking for signs of score marks or other damage. This will be made easier if a light source is shone through from the far end of the bore. Towards the lower end of the bore you will find an unworn area denoting the extent of piston travel in this direction. You will need to use this part of the bore as a reference point for further inspection and any measurement.

Careful examination of the unworn part of the bore will reveal a smooth machined surface, and you should be able to see a diagonal or diamond pattern formed by the fine honing marks left after the bore was originally finished; these fine scratch marks are intentional, and provide a good basis for comparing the rest of the bore surface. As you move up the bore, the honing marks will become more faint, eventually being replaced by vertical marks where the piston has worn the bore surface down. These will be most evident on the parts of the bore corresponding to the piston thrust faces, at 90° to the gudgeon pin bore.

At the upper limit of piston travel, a distinct wear line or ridge will be found, a few millimetres below the top of the bore, denoting the upper limit of travel of the top ring. With a few exceptions, bore wear will be greatest at this point, again at 90° to the gudgeon pin axis. Before checking for wear, satisfy yourself that the bore is in good general condition. Very light vertical lines are normal and acceptable, but any scoring deep enough to allow the loss of compression pressure means that reboring will be needed. Damage of this sort is almost invariably due to dirt getting into the engine either because the air filter has been omitted or used in a damaged condition, or in the case of four-strokes, because the oil has not been changed regularly. Another cause is a blocked oil filter, which will have caused the bypass valve to open and let unfiltered oil circulate.

Whatever the cause of the contamination, the dirt inside the engine will have been trapped between the piston or rings and the cylinder walls, each particle acting as a cutting edge and wearing grooves into the wall. Where scoring is severe and deep on the thrust faces of the piston and on the corresponding area of the bore, partial seizure due to overheating or inadequate lubrication may have occurred. In such cases the piston skirt area will be badly scored, with areas of the skirt material smeared down the piston. Sometimes this will have caused

12.1 Many ohc models use bucket-type followers containing small shims to permit valve clearance adjustment

12.2 A code number electro-etched into each shim indicates its thickness (although this can always be checked using a micrometer)

the piston rings to become trapped in their grooves. Once again, if damage of this nature is evident, reboring will be needed.

To make accurate measurements of the bore wear, you will need either a bore micrometer or dial gauge, or a telescoping bore gauge used in conjunction with an outside micrometer, though a rough check can be made using a piston ring and a set of feeler gauges. Most manufacturers quote a service limit for bore taper, this representing the difference between the unworn bore area below the limit of piston travel, and the area of maximum wear, usually just below the upper wear ridge and at right angles to the gudgeon pin axis. This is usually in the region of 0.1 mm (0.004 in), though you should check the exact figure in the workshop manual for the machine being worked on. Accurate measurement will require checks of the bore diameter to be made at several places along the bore to find the area of maximum wear. From this is subtracted the diameter measured at the unworn area to give the amount of overall wear or taper.

If you do not have the necessary equipment for this check to be made, use one of the piston compression rings to gain an approximate indication of the extent of wear as follows. Place the ring into the bottom of the bore, using the edge of the piston skirt to position it square in the bore. Using feeler gauges, measure the end gap between the two ends of the ring, and make a note of the reading. Now repeat this procedure, this time with the ring positioned just below the wear ridge near the top of the bore. Subtract the smaller end gap figure from the larger, then divide this by three to give a rough wear figure. If this exceeds the manufacturer's service limit for bore wear or taper, it is time for a rebore. Note that it is advisable to have your findings confirmed by an expert before taking further action.

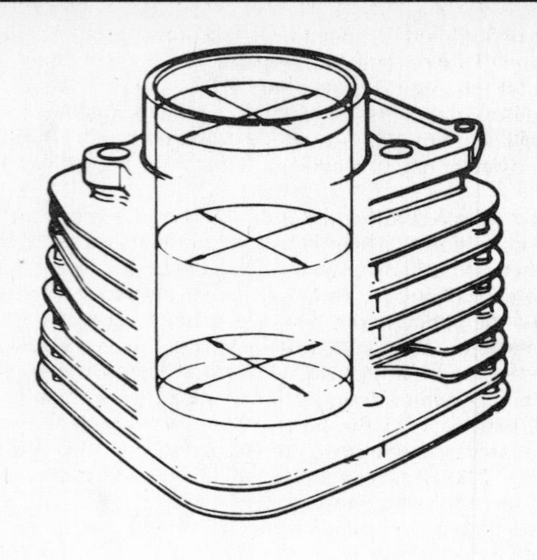

Fig. 4.8 Cylinder bore wear measurement points

13.1 If you find damage looking anything like this, put away the measuring tools! It is clear that a rebore and a new piston will be needed

13.2 This photograph shows clearly the wear ridge near the top of the bore. It denotes the upper limit of travel of the top ring, and is a good indication of the need for reboring

13.3 Bore wear can be checked accurately using a dial-type bore gauge

13.4 Alternatively, use a telescoping bore gauge and read off the measurement with a micrometer

13.5 The traditional 'home workshop' method is to compare ring end-gap readings taken at worn and unworn areas

Some manufacturers specify a figure for maximum ovality, or out-of-round. This requires several measurements to be made along the bore in each position taking one measurement in line with the gudgeon pin axis and one at 90° to it. It is only feasible to make this type of measurement using direct and precise measuring equipment: the ring-and-feeler gauge method will not give an accurate indication.

Piston and ring examination

If bore condition is acceptable, and the original pistons are to be re-used, they should be checked carefully for serviceability. If the pistons are still in place on the connecting rods, remove the circlips which retain the gudgeon pin, displacing the pin to free the piston from the connecting rod. If the pin proves tight, remove any carbon from the ends of the piston bore. If this fails to allow removal, warm the piston with near-boiling water, taking care to avoid scalding your hands. The piston will expand with the heat and the pin should then push out easily. As a last resort, make up a drawbolt arrangement to remove the pin (see Fig. 2.35); never resort to drifting the pin out or you risk damage to the connecting rod and bearings.

Continue preparation by removing the piston rings, taking care not to break them during removal. As each ring is removed, it should be checked for markings. Many rings have a letter or other marking near one end, and this normally indicates the upper face of the ring. It is important to fit rings in the correct groove and the right way up; some rings have a slight taper section. To be on the safe side, mark each ring as it is removed to indicate its position and upper surface. A spirit-based felt marker is ideal for this.

It is possible to purchase ring removal tools which minimise the risk of breakage. A cheap alternative is to use several thin feeler gauge blades, working them under the ring at equidistant points around its circumference. With the ring lifted just clear of the ring groove in this way, it can be slid off the piston with ease. The traditional method is to spread the ring ends with the thumbs and then slide the ring off the piston. This last method requires a good deal of care if the ring is to be removed in one piece; piston rings are usually of cast construction and thus very brittle. The oil ring on four-stroke models comprises two scraper rails held apart in the groove by a flexible steel expander. The rails are removed individually, and then the expander is peeled out of the groove.

Some models, especially two-strokes, use Dykes or other ring sections. Note the way that these are arranged, and if necessary make notes to act as a guide during installation. On occasions, two similar-looking rings will in fact differ minutely in section, or one may have a plated working face. All such factors should be noted, bearing in mind that you may not be reassembling the engine for some time if much work is needed, and it is very easy to forget such details.

Clean off all accumulated carbon from the piston crown using a blunt scraper and/or brass wire brushes. The ring grooves must also be cleaned out, but great care must be taken to avoid enlarging them. This demands patient and careful scraping, using whatever improvised tool you can find or make up. If you have a broken piston ring to hand, the end of this will make a good scraping tool. If you can get the pistons dipped in a carbon removing solvent tank, all the better.

Assuming that there is no obvious damage which would make further examination pointless, check the piston carefully for signs of cracking, especially around the bosses inside the skirt. Any hint of cracks will mean discarding the piston as a precaution against its complete disintegration later. If you have any doubts, have the pistons crack detected (see Section 10).

At the same time, check the condition of the gudgeon pin bores in the piston bosses. If these have become worn or damaged through seizure, or if there is any play to be felt between the pin and the bores, the piston should be considered unserviceable. If you attempt to re-use the piston in this condition, the engine will be noisy, and the repeated impact allowed by the play can cause the bosses to fracture and break up.

The areas to be measured to check for wear will depend on the service information supplied by the manufacturer. Often a piston-to-bore figure is quoted, and this will have to be checked by direct measurement of both components. It should be noted that pistons are usually slightly oval in section, the widest diameter being at the thrust faces, or at 90° to the gudgeon pin axis. Wear will occur on the thrust faces, and is generally most significant near the base of the skirt due to the rocking action of the piston as it passes TDC and BDC.

Wear will also be evident in the rings and in the ring grooves. Wear of the working face of each ring where it bears upon the cylinder wall is

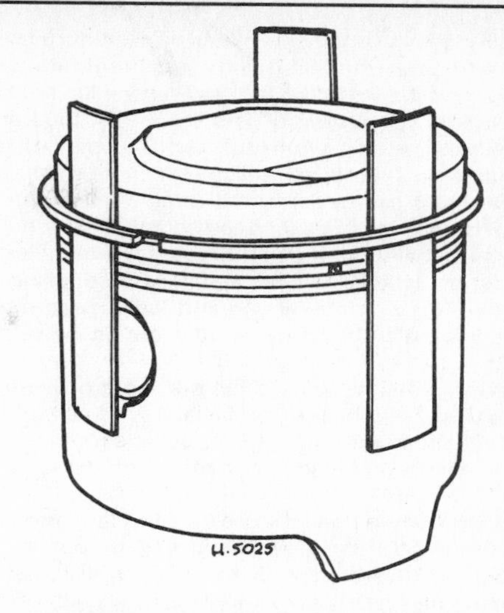

Fig. 4.9 Method of freeing gummed piston rings

13.6 A pointed instrument will be required to remove gudgeon pin circlips. Pack the crankcase mouth with rag to catch dropped circlips

13.7 If money is no object, it is possible to purchase ring-removal tools to reduce the risk of ring breakage. Alternatively, work thin metal strips beneath each ring to allow it to be slid off the piston

fairly obvious, and this is usually checked by inserting the ring in the unworn area at the base of its bore and measuring the ring end gap with feeler gauges. If the installed end gap service limit is reached or exceeded, renewal will be required. Less obviously, the ring top and bottom surfaces will become worn due to the continual reversal of thrust as the engine runs. Most manufacturers quote a service limit for ring thickness, and this can be measured using a micrometer.

Perhaps surprisingly, the rings tend to become thinner more quickly than the relatively soft alloy of the ring grooves. Nevertheless, the ring groove widths should be examined and measured using a vernier caliper, or by checking with a new ring and feeler gauges. If the ring groove has become enlarged, there is little option to renewing the piston.

It is theoretically possible to machine out the grooves to accept a thicker ring, but this presupposes the availability of suitable rings; the motorcycle manufacturer will certainly not be able to supply them and would probably not recommend this method of reclaiming a worn piston.

Fitting new rings would appear to be an easy and inexpensive way of getting more mileage from a part-worn engine, but in reality this practice is fraught with problems. First of all, whilst the additional depth of a new ring may help locate it in an unworn section of the ring groove, the outer edge will lack proper support where ring groove wear has occurred. This can allow the ring to 'flutter' in the groove and this will often cause rapid failure. The unworn area at the back of the ring groove can often be seen as a step in the sides of the groove.

Worst still is the problem of the wear ridge at the top of each bore. This is **always** present on a used engine, and the old rings will have worn to accommodate it. If you fit new rings, the likelihood is that the top ring will strike this ridge and cause the ring to break up. This problem is avoided by removing the wear ridge. This can be done by using a de-ridging tool, or by honing the bore until the ridge is gone. Given the need for the part-worn bore to be honed before new rings are installed, honing the ridge away at the same time is probably best. Unfortunately, once the honing work has been carried out, the bore may well have gone beyond the service limit, so such techniques are often of limited use. If you are considering this sort of work it is worth checking with an engine reconditioning specialist before it is carried out.

14 Arranging for reboring or other machining work

Having examined and measured the cylinder bore(s) and piston(s) as described above, if wear or damage has been found it will be necessary to arrange a rebore. Most dealers will have an arrangement with a local engine reconditioner or small engineering works for dealing with this type of machining, in which case you should make arrangements to deliver the cylinder barrel or block to the dealer, who will then supply the necessary pistons and get the work carried out.

If you intend dealing direct with an engineering company, or if you want any other machining work carried out, you should first work out exactly what you require to be done. You may also have to supply pistons, or any other new parts which are to be fitted. If the chosen company deal specifically with motorcycle repair work they may have their own source of replacement parts, but often it will be quicker and easier if you supply these yourself. In the case of pistons, you will have to check the existing bore size, calculating from this the next oversize for the new pistons.

When doing so, make sure that the proposed rebore will remove all existing damage and order pistons accordingly, checking the available oversizes in the workshop manual. As a general rule, piston oversizes are supplied in 0.25 mm (0.010 in) increments, and there are usually four oversizes available; +0.25 mm, +0.50 mm, +0.75 mm and +1.00 mm. These steps are chosen so that normal wear or damage will be removed by reboring to the next oversize, but you must be certain that this is the case before ordering.

If, for example, the engine has a nominal bore size of 50 mm, and this is worn to 50.30 mm at the worst point, there is little point ordering +0.25 mm pistons; reboring to this size will not remove the damage, and you will have to go up two sizes to +0.50 mm. On multi-cylinder

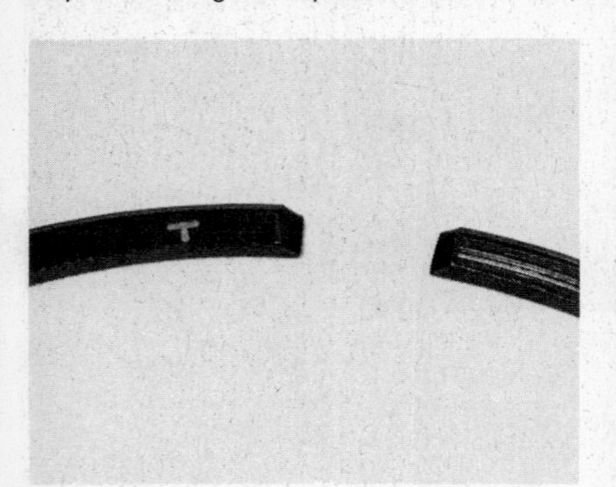

13.8 As each ring is removed, note any markings denoting the top surface of the ring

13.9 In this case there will be no need to bother with piston measurement – the top ring has broken up, destroying the ring groove and thus the piston

13.10 This two-stroke piston has suffered a major seizure, the rings becoming welded to the cylinder wall and tearing out of the piston

13.11 Damage caused by the ring locating pin working loose and being forced out of the ring groove

13.12 Piston wear can be measured using a micrometer, the measurement normally being made at 90° to the gudgeon pin boss

13.13 Check for wear in the ring grooves by inserting **new** rings and checking the clearance with feeler gauges

engines, the deciding factor will be the most worn bore, the remaining bores being taken out to the same size as this. You must also take into account whether the engine has been rebored previously, and if so, whether further oversizes are available. This will be indicated after the bore(s) have been measured, though in some cases the oversize markings on the piston crown will be visible after cleaning. If the engine has been bored to its maximum oversize on an earlier occasion, you may need to purchase a new cylinder barrel or block and start again with new standard pistons. Alternatively, it may be possible to fit sleeves to the old block to allow it to be restored to its nominal bore size. Decisions of this sort will require expert advice.

If you are arranging the work through a dealer, or if the engineering shop agree to take care of the whole process, you can leave them to make the required calculations. Do check whether they require specification information, and if required remember to supply this when you take the work in. In particular, provide details of the piston to bore clearances so that the bores can be finished to suit the new pistons.

Where you propose to fit new rings in an existing bore, it is well worth getting the bore honed, both to provide the cross-hatched surface needed to allow the rings to bed in normally, and to remove any wear ridge from the top of the bore. Honing is carried out using a special tool comprising abrasive stones held in either a flexible or a fixed honing tool.

A flexible hone is spring-loaded and this ensures that the correct pressure is applied automatically to the stones as the tool rotates in the bore. This type of tool will remove the glaze from an existing bore, but is less good at removing the wear ridge. Another popular type of tool for the removal of glaze is the *Flex-Hone* and similar devices. These look rather like large bottle brushes, the end of each bristle terminating in an abrasive tip. Once again, this type of tool will not remove the wear ridge. Ridge removal requires the use of a de-ridging tool prior to flexible honing, or the use of a fixed hone which is capable of removing the ridge as part of the honing process.

The honing operation, which will also be required to complete a new bore surface, is designed to finish the surface correctly so that the bore is able to retain a certain amount of oil on its surface during use. This is achieved by leaving a fine cross-hatch pattern in the bore surface, and it is this which allows the rings to bed in correctly to give a good seal. The cross-hatch pattern, which will be visible if you catch the light on the new bore surface, should have the honing marks intersecting at about 45°. Although it might seem an insignificant process, the quality of the final honing has a lot to do with the success or failure of the whole reboring operation. Poor honing will cause excessive oil consumption, poor ring sealing or even abnormal ring wear.

The general remarks above apply just as much in the case of other machining work. You will have made your own checks and measurements when you decided what needed doing, and so will have a good idea of the repair work required. Even so, be guided by the engineer who will be carrying out the work; it is likely that they will have done this sort of work many times in the past, and thus will have the experience to be able to suggest alternative methods of achieving the desired result. Always take with you as much specification information as possible. Whilst this may not be needed, it is better to be sure that it is available than to leave the engineer to guess at clearances. Try to work out in advance exactly what needs re-machining throughout the entire engine, and take all of the work in at once to avoid delaying the rebuild. While these parts are being dealt with, you can turn your attention to any remaining preparatory or repair work, as detailed in the following sections.

15 Damaged and broken fasteners

However careful you are, it is highly probable that at some stage you will be confronted by a damaged fastener. This can range from a casing screw with a chewed-out head to a broken off bolt or stud. All have something in common; they are difficult to deal with, and require a high degree of care and patience if the problem is not to be made worse. This section describes some of the more common problems and suggests methods of dealing with them.

Damaged casing screw heads

This is probably the most common problem of all. The universal

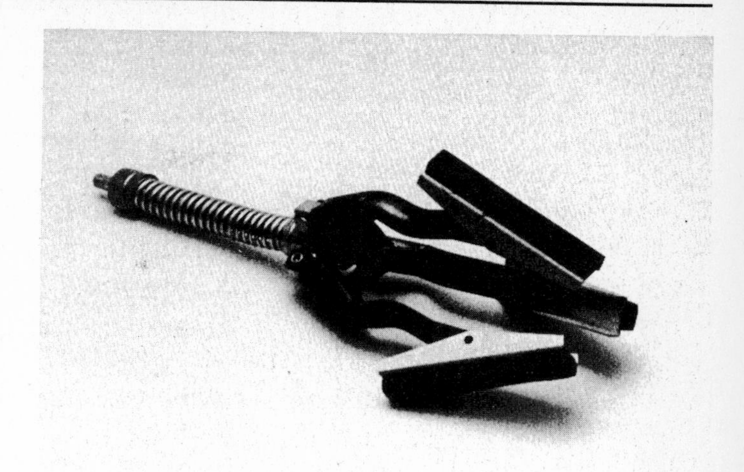

14.1 A simple flexible hone like this will produce the correct surface on a newly-machined bore, or will remove glaze from a used but unworn bore surface

14.2 A fixed hone like this will be needed if it is necessary to remove an appreciable wear ridge from the top of the bore

14.3 In the foreground is a typical small boring machine. The cylinder barrel is clamped to the underside of the surface plate and cannot be seen in this shot

casing screw is invariably tight, and this combined with a ridiculously soft cross-head means that damage is almost inevitable unless great care is taken when removing them. If you are confronted with this problem it is essential that you take remedial action before the head gets too badly damaged, or you will be in real trouble.

The first thing to do is to examine the damaged head closely. If it has the remains of the driving tangs still evident, albeit distorted, you may not have too much trouble. Using a parallel punch, tap the centre of the screw so that the tangs are pushed down into the screw. With luck this will return them to a vaguely cross-head shape, even though a screwdriver will not fit them any more.

You will now need an impact driver and a range of cross-head bits. Using a similar screw head in good condition, find the bit which most closely fits the profile of the head. Hold the bit against the damaged head, and tap it into the head so that it fits snugly. Now fit the bit into the impact driver, having checked that it is set to twist anticlockwise. Push the tool hard against the screw head and tap the end sharply. With luck, the screw will loosen and can then be unscrewed. If the attempt fails, and access permits, you can try hacksawing a slot across the head. If you can cut a slot it should be possible to use a normal slotted bit in the impact driver to remove the screw.

Where the above methods fail to free the screw, you may have to resort to drilling out the head of the screw. The remains of the head should suffice in locating the tip of the twist drill, though if necessary you can centre punch the head at the exact centre. Choose a twist drill slightly larger than the diameter of the screw shank, then drill into the head until it comes free of the shank. The cover can now be removed, leaving a short piece of the shank protruding from the case below. Apply a few drops of penetrating fluid and allow this to work for a few minutes. Grip the end of the shank between the jaws of a self-locking wrench and remove the remains of the screw.

In cases where the head has been damaged due to the screw being seized in the crankcase, or where the screw has broken off, refer to the text below for further information on dealing with this type of problem. Whatever method has been used to remove a damaged screw, on no account re-use it or you will have similar or worse problems on the next occasion that removal is required. It is well worth considering replacing the entire set of casing screws with Allen screws, preferably in stainless steel. These can always be removed easily using a hexagon key, and the stainless versions will not suffer from corrosion problems.

If you choose to fit plated steel screws, either of the original type or Allen-headed, it is worth guarding against problems in the future by applying a thin coat of a copper-based or graphite grease to the threads. This will help prevent corrosion of the threads by excluding moisture, and will also act as a lubricant to ease future removal.

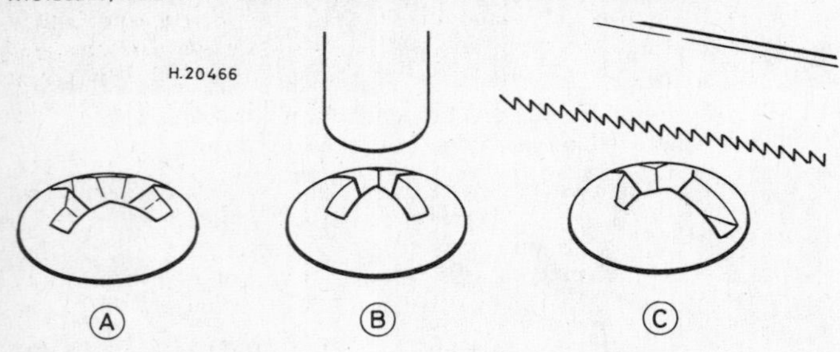

Fig. 4.10 Removing damaged cross-head screws

A The cross-head screws favoured by most Japanese manufacturers are prone to damage unless the correct type of screwdriver is used

B It is sometimes possible to re-form the head by judicious use of a punch

C If access permits, cut a screwdriver slot across the head to allow a flat-bladed screwdriver to be used

Freeing seized fasteners

Casing screws, bolts and other types of fastener which have become seized in their threads will have to be tackled with care; it is all too easy to end up with part of the shank still seized in its thread after twisting off the head. The usual cause of seized fasteners is corrosion of the threads, and if this gets bad enough it is possible for the threads to become firmly stuck together. Another possible cause is that the

14.4 After the bore has been machined to size by the cutting tool (centre of picture) the bore is then honed to give the correct surface finish

screw has bottomed in the hole, causing the threads to distort and jam in position. This can be due to the wrong length of screw being fitted, or to accumulated debris at the bottom of the hole. In such cases, removal by normal methods may not be successful.

Start by applying penetrating fluid around the threads. This is likely to be more successful if the fastener can be persuaded to move, even slight movement indicating that there is enough of a gap around the threads to allow the fluid to penetrate. It is sometimes useful to try tightening the fastener very slightly and then backing it off again; this may work where attempts to unscrew it have failed. Using an impact driver may help by shocking the fastener enough to get slight movement.

Leave the penetrating fluid to work for as long as possible. If you can, let it stand overnight before further attempts at removal are made. You may also care to try applying a little heat to the area; the differing rates of expansion between the steel and alloy may assist in breaking the bond between the threads. Be very careful how you do this; direct heat from a flame can cause the alloy to melt very suddenly, and there is also a risk of distortion. It is safer to use a heavy-duty soldering iron applied to the screw head. Alternatively, a hot air gun of the type used for stripping paint can be used, or the area treated with near boiling water (take care to avoid scalding yourself).

Once you succeed in getting the fastener to move, start unscrewing it carefully. If it becomes tighter, this will be due to the corrosion locking the threads at one point. Do not continue forcing the fastener, but instead apply more penetrating fluid, screw it in by a turn or two to dislodge the corrosion, and then try again. Repeat this process until it can be removed completely.

Removing broken off fasteners

This can be a difficult undertaking, the chances of success being dependent on the original reason for the breakage. If it is a simple matter of an otherwise sound fastener being over-tightened and snapping off, the chances are that the remainder can be removed reasonably easily. If on the other hand a fastener has snapped off because it was corroded in place or bottomed in its hole, you still have to deal with the seized threads, but without an easy method of turning it.

The first task is to mark the **exact** centre of the sheared fastener with a centre punch. This is easier to do if you first file the broken end flat, but in some situations this will be impossible. It is vital that the punch mark is exactly at the centre of the fastener thread, so take time to ensure this. It is worth noting that the shock from the centre punching may help matters; the impact itself might help loosen the fastener, and the compressive action can deform and thus loosen the threads a little.

You will now need to drill a small pilot hole through the fastener, and this is best done using a bench or pillar drilling machine. Where this is impractical, you will have to do the drilling work freehand, taking great care to ensure that the drill bit is kept perpendicular to the fastener.

You must also take extreme care that the twist drill does not snap off as it breaks through at the end of the fastener; you have enough of a problem already without having to extract a broken off drill bit! Try to gauge the likely length of the fastener, comparing it to a neighbouring undamaged one, and ease off drilling pressure as the bit nears the end of the fastener.

If all goes well, the next step is to try to remove the screw with a screw extractor. These devices (also known as 'easy-outs') are available in sets. There are two basic types, one of which employs a tapering left-hand thread. The tool is turned anticlockwise, which screws it into the pilot hole and, hopefully, then unscrews the remains of the fastener. The second type has a thin tapering square or cross section shank which is tapped into the hole to locate it, and is then turned to remove the damaged fastener. Either type will work on most occasions, provided that the head was not originally sheared off **due to seizure**. If the fastener is really stuck, the extractor may only succeed in spreading it outwards, making it even more securely stuck. Worst still is the risk of the hardened extractor breaking off in the pilot hole, making matters far worse. By all means try these tools, but bear in mind the drawbacks outlined above.

Where mechanical extraction methods fail, and assuming that you have not snapped off an extractor in the attempt, you will have to resort to drilling out the remainder of the fastener. Use drill bits of progressively larger diameter, finishing with one at the core diameter of the damaged thread. If you are very lucky, this will leave just the thread remaining in the hole, and it is sometimes possible to peel this out of position to leave the tapped hole relatively unscathed.

As a last resort, it will be necessary to drill out the fastener complete, together with the thread in the casing. Before you do this, check the appropriate drill size for a wire thread insert tap for that thread size. Drill out the hole, and then either fit the thread insert yourself (see Chapter 2) or have this done by a motorcycle dealer equipped for this work. Other methods, requiring specialist attention, include aluminium welding to fill the hole, followed by drilling and tapping to the original size, or making up and fitting a solid thread insert.

How to deal with snapped-off drill bits, taps and screw extractors

As you will appreciate, this sort of situation is extremely difficult to resolve, and best avoided if at all possible. If the worst should happen, there is very little chance of coping with the problem at home, and even small engineering companies may be hard pressed to tackle the removal of hard steel tool fragments from a seized and broken off fastener in a soft alloy thread.

The only guaranteed method of removal is by spark erosion, or electrical discharge machining as it is sometimes known. As the name suggests, the process works by eroding the metal with a high intensity electrical current discharged from an electrode. This is a specialised process, and you may have some difficulty in finding a local company who can offer this service. Once located, find out what will be required by way of preparation, and make arrangements to have the work carried out.

Dealing with corroded studs and through bolts

This problem is normally confined to those models where a fastener is exposed to the elements along part of its length, although in some circumstances corrosion can occur even where the fastener is fully covered. Particularly vulnerable are cylinder head holding studs, where these pass through open finning at the front of the cylinder block, and similarly, engine mounting bolts, where a single long bolt or stud passes through two or more short lugs. Typical examples of the latter arrangement can be seen on Moto Guzzi engine mountings.

In countries where the roads are salted to prevent icing during winter, the deposited salt, water and repeated heating and cooling inevitably lead to corrosion of these fasteners. Whilst this can be reduced or eliminated by coating the shank with copper grease during installation, this advice is of little comfort when you encounter a well-rusted example for the first time.

The best method of removal will have to be worked out to suit the particular situation, but in all cases start by using copious amounts of penetrating fluid, leaving this to work in for as long as possible. The location of the component, and that of surrounding parts will determine whether it is safe to apply heat; if it can be done without causing damage then it is worth trying. In the case of engine mounting bolts, try drifting the bolt out. Start by unscrewing the nut, but leave it

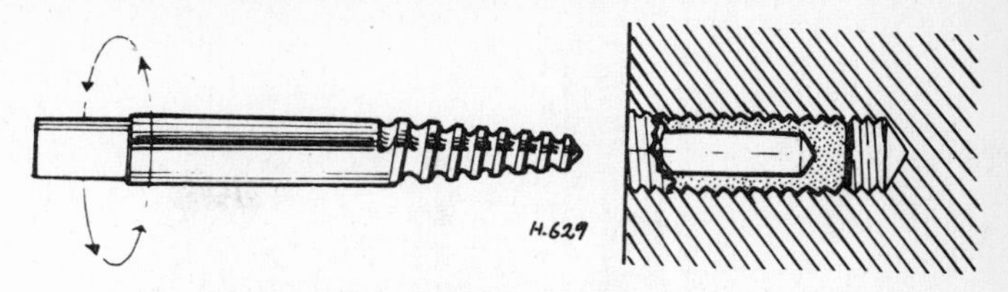

Fig. 4.11 A screw extractor in use

The extractor is screwed anticlockwise into the hole drilled into the centre of the broken stud or bolt. Further turning locks the tapered extractor thread in place and unscrews the broken fastener. One problem is the risk of the wedging action of the extractor spreading the end of the fastener, jamming it in the hole. There is also the risk of the extractor breaking off if excessive pressure is applied

in place on the end of the thread to prevent the bolt spreading when it is struck. If you succeed in moving it even slightly, apply more penetrating fluid, then try turning the bolt until it begins to loosen.

Look out for this sort of problem during normal overhauls. Even if a bolt or stud is not seized in place, watch for signs of corrosion which might get worse if ignored. Apart from applying a film of copper grease to minimise future corrosion, consider having the affected parts blasted and then cadmium plated to avoid the formation of corrosion at a later date. Better still, try to replace the affected bolt or stud with one made from stainless steel.

Methods of removing studs

Studs will be encountered in a number of areas on most machines. Basically a length of steel threaded at each end, the stud is intended to be screwed into a casting, where it normally remains for the life of the machine. If you need to remove it for renewal or for any other reason, the lack of a hexagon head will prove a distinct disadvantage.

Provided that the stud is not too tight, the simplest method of removal is to fit two nuts to the exposed threaded end. The nuts are then tightened against each other to lock them, the stud then being unscrewed by turning the locked nuts. This method should be used only where the stud is reasonably easy to move; if it is unusually tight, or if a thread locking compound has been used to retain it, the upper thread may well strip before it can be freed. If it feels tight, abandon the attempt and use a tool to remove the stud safely.

It will come as no great surprise to learn that the best tool for removing studs is a stud extractor. This is normally a device with a square driving end which can be attached to socket set lever bars and extensions. The stud is passed through a hole in the tool. When it is turned, a hardened and knurled roller grips the stud, providing a means of extraction.

Less desirable methods include the use of self-locking or stilson wrenches. These will usually do the job, but be careful not to inflict too much damage on the stud if you have to use this method.

When fitting studs, check that the threads are sound and completely clean, then apply one or two drops of semi-permanent locking compound. This will prevent unintentional unscrewing of the stud.

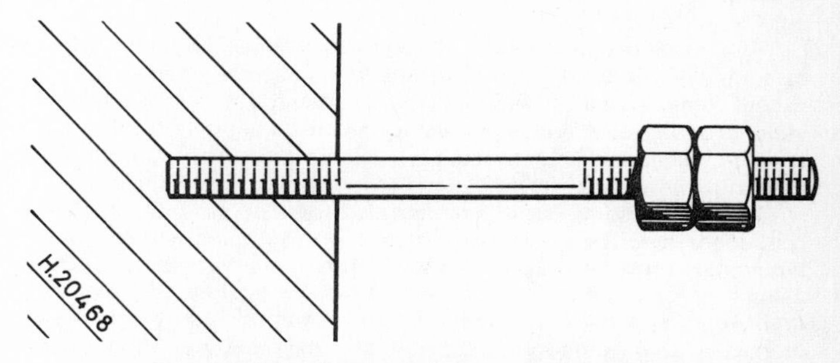

Fig. 4.12 Removing a stud using locked nuts

Provided that the stud is not unduly tight, two nuts can be locked together as shown to permit the stud to be unscrewed using a spanner

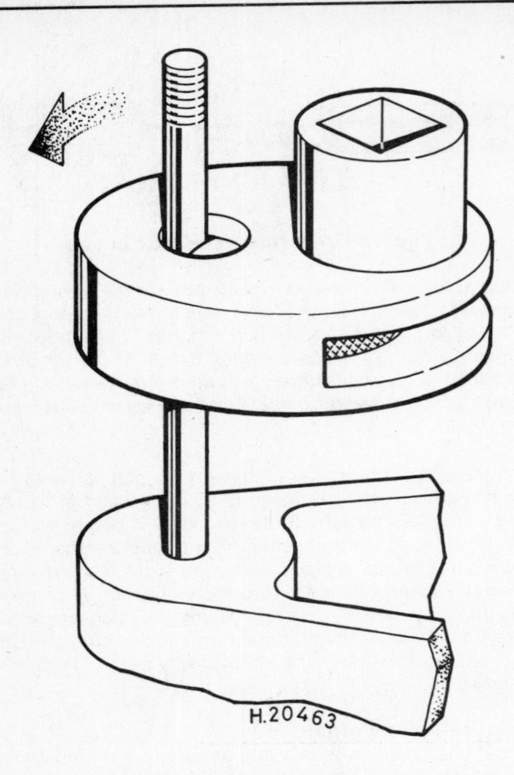

Fig. 4.13 A commercially-produced stud extractor

The square-drive knurled wheel inside the tool grips the stud which is passed through a hole in the tool body. Further movement unscrews the stud

16 Damaged castings

The procedure for making effective repairs to casting damage will depend on the nature of the damage and its location. Impact damage to an unstressed outer cover, perhaps resulting from the machine being dropped, is annoying but relatively easy to deal with. Internal fractures of the crankcase assembly due to stress or catastrophic failure of some internal engine part are an entirely different matter, and you will have to seek expert advice to find out whether repair would be viable. In many instances it will be easier and a lot safer to renew the damaged casting.

A cracked outer casing can sometimes be dealt with at home, using an epoxy resin filler or putty. The cracked area must be absolutely clean and free from all traces of oil or grease. Apply the putty or filler according to the manufacturer's directions, trying to follow the final surface shape as closely as possible (the resin sets very hard and sanding it is no fun). The finished repair can be disguised by refinishing the cover.

The other normal method of repair is by welding, and where castings are concerned this is definitely a job for an expert. Do not expect a miracle cure; castings can be difficult and unpredictable to weld, even in expert hands and using the best equipment. At best, the finished result is likely to be serviceable rather than cosmetically pleasing.

Although casting repair is effectively of limited success, and ruled out for the most part from home workshop use, there is one process which may prove effective. This is a form of aluminium braze welding using a propane torch and special rods known by their trade name of *Lumiweld*. Kits of these rods are available for home and small workshop use. Although this technique may not be adequate for serious structural repairs to the crankcases, it will be quite adequate for use on broken lugs or fins. The special rods used have a lower melting point than aluminium alloys, and so there is little risk of uncontrolled melting as is the case with alloy welding. For more information, refer to Chapter 3.

17 Engine seizure

A seized engine is the usual result of lubrication failure or serious overheating. The overheating problem may have been caused by lubrication failure, though in other cases, a condition such as an abnormally weak mixture may have caused it, and the subsequent failure of the lubrication film. In other words, overheating and loss of lubrication may either have been the cause or the symptom of the failure.

In most cases, it is the cylinder bore and piston areas which are most at risk; if the lubrication film breaks down for any length of time, the piston and cylinder wall come into direct contact, and the resulting friction generates a tremendous amount of localised heat. The heat and friction will cause galling of the piston surface, and quite often significant amounts of the piston material will be smeared along the length of the piston thrust faces, trapping the rings in their grooves. Eventually the condition will become so severe that the piston may end up virtually welded into the bore. At this stage, if not before, the engine will stop abruptly.

Seizure can also occur as the result of component failure. In the case of the bore, ring breakage, the dislodging of one of the ring location pegs (two-strokes only) or a displaced gudgeon pin circlip are typical causes, and are usually attributable to failure to check these areas during a previous rebuild, or to the re-use of old parts in the latter case. Bearing surfaces, too, can suffer seizure. These include small end-bushes or bearings, especially on two-strokes where lubrication is less predictable in this area than is the case with four-stroke models.

The crankshaft big-end and main bearings may fail due to loss of lubrication, either because of oil starvation, or oil pressure loss from bearings that have become too worn. On two-strokes, failure of the pumped supply may be the cause, or even reduced lubrication due to loss through worn main bearing seals. In all of the above examples, the cause of the problem will require the engine to be stripped for examination before the exact reason for the failure becomes clear. In each case, seizure will be the final outcome of neglected services or the failure to notice a general deterioration of the engine condition in time.

In use, the first signs of impending seizure are a slight increase in engine noise, accompanied by a loss of power. This will often seem puzzling at first, almost as if a brake has been applied. If you spot the problem at this stage and declutch and stop the engine immediately, you may have acted in time to prevent the more serious total seizure. Even so, there is likely to be a certain amount of damage to the piston skirt at the very least, and the engine should be dismantled for examination.

The first thing to do is to try to establish the cause of the seizure. If this is something obvious, such as running out of oil on a two-stroke, allow the engine to cool and check whether it can still be turned over. If so, get the lubrication system working again, and then ride home very gently, stopping at once if you notice any further problems. Where the cause of the problem is obscure, the engine is making unusual noises, or the seizure is total, it is preferable to have the machine recovered by trailer.

In cases of total seizure you may have problems in getting the engine apart. Applying penetrating oil around the top of the piston after the cylinder head has been removed may help to free it off, but often this will not work. In such cases, try dismantling the engine around the cylinder barrel/piston assembly if you can. The crankshaft assembly and the cylinder barrel or block can then be taken as a unit to the bench for further attention. Where this is not possible, you will have to try to separate the piston from the bore in position.

With any serious seizure, you should accept that the piston at least is likely to be of little further use. The best that can be done at this stage is to try to avoid further damage to other engine parts. It may be possible to dislodge the piston by passing a length of timber down from the top of the bore and using this to drive the piston out. It will help if an assistant pulls upwards on the barrel or block.

In really extreme cases, you may have to demolish the piston so that the crankshaft can be freed, hopefully without damage to the crankshaft assembly. The exact method used will have to be decided according to individual circumstances, but will inevitably have to be fairly barbaric. Try to separate the crankshaft assembly from the crankcases if this is at all possible, leaving the crankshaft with the seized piston and barrel or block attached.

One possible approach is to chain drill around the piston crown to detach it from the seized skirt, remembering that the hardened steel gudgeon pin will have to be avoided during this operation. Once the crankshaft is freed in this way you will be left with the skirt area to be cut or chiselled out of the bore. This should be done with care, trying not to cause any more damage to the bore surface.

A less dramatic method is to have the seized piston pressed out of the bore, assuming that it is possible to gain access to allow this to be done. It is not easy to carry this out at home, and this job is probably best left to an engineering company.

Unless the seizure was only partial, the piston will have been damaged too much for re-use. Very light scuffing of the skirt area can be rectified with judicious use of abrasive paper or fine files, but if material needs to be removed in this way it is unlikely that the piston will give the correct clearance. In such cases it is always more effective to renew the piston assembly. Seized rings, where the ring land material has become smeared over the rings, trapping them in their grooves, will require careful freeing off, scraping away the smeared alloy until they can be removed and the grooves cleaned up properly. In cases of full seizure, there will be little alternative to reboring and the fitting of a new piston. In all cases of seizure, make certain that you establish and rectify the cause of the failure before the engine is rebuilt and run.

17.1 This crankpin illustrates the effects of big-end seizure. The bearing has broken up, smearing metal around the pin

18 Spline and keyway damage

Splines and keyways are frequently-used methods of locating parts on the ends of the various engine shafts, common uses ranging from alternator rotors, which are usually keyed to the tapered end of the crankshaft, to components like the kickstart ratchet block, which will normally be splined to the end of its shaft. In general, the spline or key acts as a method of location, the component being secured in place with a nut or circlip. In the case of the gearbox shafts, though, the gears are allowed a controlled amount of lateral movement for selection purposes, this being limited by circlips or shoulders machined into the shaft.

Where the component is fixed in position, wear is unlikely to take place unless it is not secured properly. If this happens, the component will chatter in relation to its spline or key, wearing both parts. A common example of this can be found on many kickstart levers; if not tightened securely, the splines of both the shaft and the lever will wear quite rapidly. If even slight wear is discovered it should be dealt with before it can become worse.

In the case of keyways, remove and examine the key. If it is worn, try fitting a new one, but check that this will prevent further movement. If the keyway is worn, it is best to have the area built up with weld and a new keyway machined into the shaft end. An easier, though less satisfactory, cure is to use one of the specialist products like *Loctite Quick Metal*, to make up the free play in the keyway. If you do this, be sure to avoid getting the product on the taper surface, or you may never get the joint apart again!

Splines are less easy to deal with by welding, and in these cases it is preferable to use *Quick Metal* or a similar product to eliminate free play, provided that the spline supports a fixed component. With worn kickstart or gearchange splines, check both parts for wear. If only the pedal or lever splines are worn, you may be able to resolve the problem by renewing the worn part. If the shaft splines are worn too, you can try using *Quick Metal* to avoid the need to remove and renew the shaft to which it is attached. Remember that you will have to allow for the need to remove the pedal or lever occasionally, so avoid any product which will produce a permanent joint. A traditional method of repair, which avoids the need to renew the shaft, is to fit the lever or pedal in its normal position, and then to drill a hole along the length of one spline, leaving a semi-circular slot in both parts. A hard steel taper pin, available from engineering suppliers, can then be tapped in place to locate the pedal or lever, and the pinch bolt tightened to secure it.

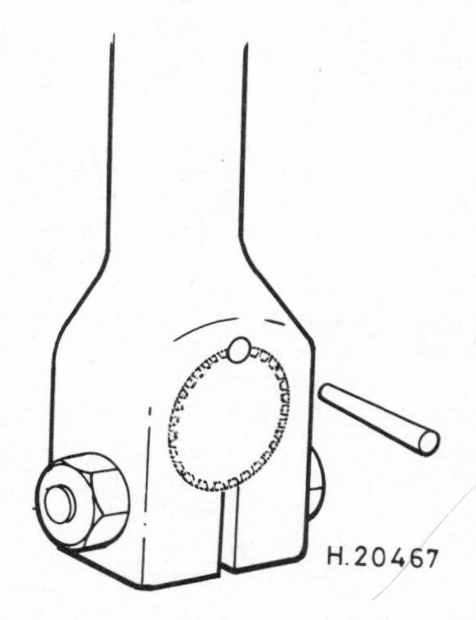

Fig. 4.14 Small taper pins can be a useful way of securing splined kickstart levers to a worn shaft, after drilling a locating hole along the spline axis

18.1 No amount of 'Quick Metal' would make this keyway serviceable again. The only alternative to a new crankshaft would be to fill the damage by welding, and then to re-machine the keyway

19 Clutch and gearbox: overhaul

If the splined clutch centre becomes worn, the clutch plates will tend to chatter against it, causing indenting of either the splines or the slots in the outer drum. Damage of this type can often be responsible for erratic clutch operation, symptoms such as unreliable engagement or dragging occurring as the plates catch in the indentations.

In the case of the clutch outer drum, it is usually worth filing the drum slots flat and smooth, provided that the damage is not too severe. This will usually restore normal operation. In cases of heavy damage, however, the drum should be renewed.

The clutch centre is less likely to suffer, but should problems arise it is almost impossible to correct by filing. In such cases the clutch centre should be renewed, together with the plain and friction plates where these have become sloppy or badly worn. Similar problems can occur in the case of wear between the clutch centre and the shaft to which it is attached, although this is normally prevented by the centre being secured in place by a retaining nut.

In the case of gearbox shafts, or other instances where a sliding component is involved, wear will necessitate the renewal of the shaft and/or gear concerned; it is not possible to repair this sort of damage economically, and unless the shaft or gear is unobtainable, it is not worth having the damaged splines welded and re-machined.

Wear or damage in the transmission components is often indicated by the symptoms which brought about the overhaul in the first place; either gear noise due to general wear and tear, or a specific problem like poor gear engagement or the machine jumping out of gear. Very occasionally, a particular model may have a known design fault in some area of the transmission, though such defects are normally resolved early on in the production run, and affected machines repaired under warranty.

Imprecise gear changes or the machine jumping out of gear are often indicative of worn dogs on the gear pinions. Check these for wear or damage, looking for rounded off or deformed leading edges. The best course of action in such cases is to renew the damaged parts, though it may be possible to have the dogs built up with weld and then filed and ground back to shape.

Erratic gear selection can also be attributed to worn or bent selector forks, worn grooves or worn selector drum tracks. These should be checked visually for obvious damage, and the fork and groove widths measured and compared with the specified dimensions. If wear of this type is discovered it is normally best to renew the gear(s) or the selector fork(s) as required.

Worn or chipped gear teeth will usually be obvious on inspection. This type of damage can occur as a result of general wear, abuse or neglected oil changes. The cost of building up and recutting the gear teeth make repairs impractical; fit new gear pinions, preferably as pairs.

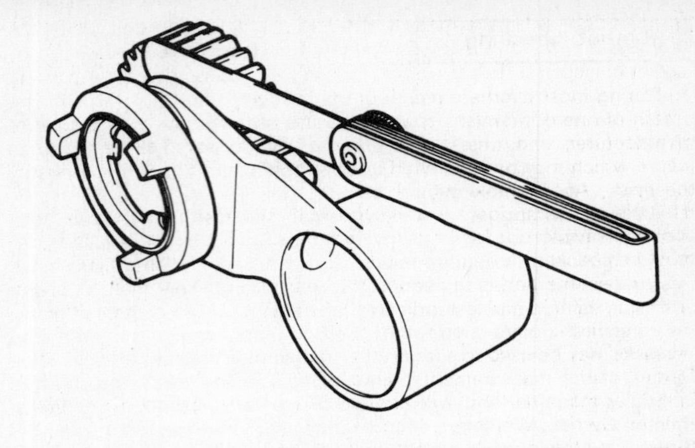

Fig. 4.16 Gear selection problems can be caused by excessive play between selector forks and grooves

Check by measuring clearance using feeler gauges

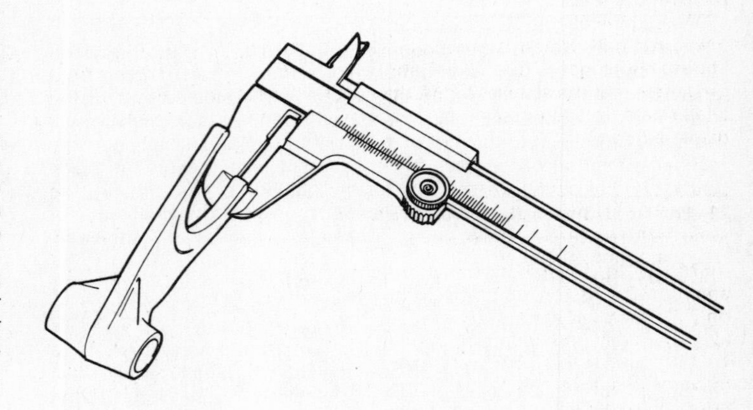

Fig. 4.17 Selector fork end thickness check

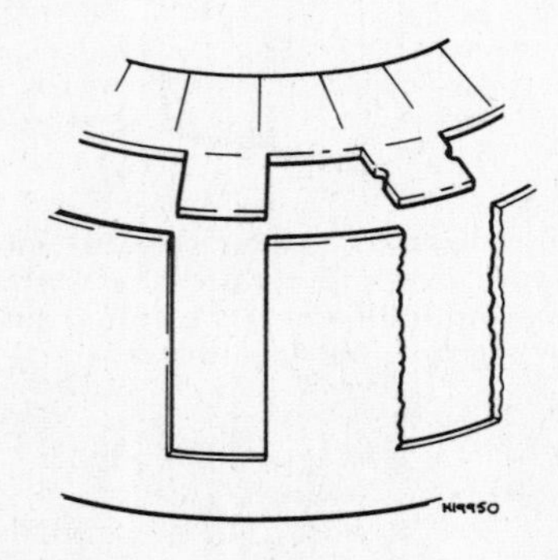

Fig. 4.15 Comparison of worn and unworn clutch outer drum slots and friction plate tangs

19.1 Damage caused by passing a bar through the tangs of the outer drum to prevent clutch rotation

19.2 This sort of damage is unusual, though not unknown on certain models with a design fault in the transmission

19.3 Worn dogs on transmission gears are the usual cause of the machine jumping out of gear

20 Runout: checking

During most overhaul procedures it will be necessary to check the runout of one or more shafts against the service limits specified by the manufacturer, and this is even more important after a component failure which may have resulted in abnormal loadings being applied to the shaft. An example would be after the failure of an engine or transmission component leading to the locking of the drive train; the sudden heavy torque loadings may have caused shaft damage, and this must be checked during the repair.

The best method is to support the shaft between centres in a lathe or similar, using a dial test indicator (dial gauge) to check for runout at the specified points along the shaft. Where this is not practical, V-blocks may be used to support the shaft at its ends or by its bearings. Details of the measurement points along the shaft are specific to a particular machine, and will usually be described in the appropriate Haynes Owners Workshop Manual, or in the factory service manual, together with allowable runout figures. The accompanying illustration shows a typical example.

Simpler items such as clutch pushrods or valve stems can be checked for straightness using a surface plate or a sheet of plate glass. Place the rod on the surface plate and roll it along, noting any runout present. This can be checked more acccurately by sliding feeler gauges beneath the rod.

As a general rule it is not practical to attempt to straighten any machined part which has become bent in service. It will be almost impossible to get it true again, and even if you succeed, there is no guarantee that the stresses left in the metal will not lead to a fracture at a later date. In such cases it is always best to play safe and renew the damaged part.

21 Endfloat and shimming: checking

Many of the shafts and other components on motorcycle engine/transmission units are assembled to fine tolerances to allow only a controlled amount of endfloat. This ensures that meshing parts align correctly and keep to a minimum the noise and vibration which would occur if this play were excessive. In many instances this free movement is governed by machining tolerances, and if the service limit figure is exceeded, renewal of the worn part(s) is indicated.

Where shims are used to control endfloat, it is essential that care is taken during dismantling to ensure that each shim is placed over the shaft to which it belongs, in the correct relative position. This is especially important in the case of some European models. In the case of the bevel-drive Ducati engines, for example, the various shafts are selectively shimmed during initial assembly, and failure to keep track of these shims during an overhaul may result in hours of meticulous checking, possibly requiring professional help. Whatever the machine, **never** put all the various shims and washers together in one container.

In the case of plain bearing crankshafts, special semi-circular endfloat shims are often found at one or more of the bearing caps. Like the bearing inserts, these must be checked and a new shim selected from the available range if end play proves excessive.

In the case of the gearbox shafts, the position of one or more of the gear pinions will be controlled by plain round shims, rather like thin steel washers. Again, these can be obtained in varying thicknesses in most cases, allowing fine adjustment of endfloat of the individual pinions.

In the case of roller bearing big-ends, endfloat is allowed so that the connecting rod and piston can run true to the bore, a fairly generous allowance being made to compensate for machining tolerances. The play is controlled by thrust washers, and thus cannot be altered without pressing apart the big-end assembly.

On machines where pairs of bevel gears are used, typically those machines employing shaft final drive, shims are used to control the mesh depth of the gear teeth. This is a critical adjustment, and if too tight the gear teeth will wear rapidly, whilst excessive play will mean gear whine and backlash. Shims may take the form of shim washers, or may be in the form of specially-shaped shims fitted to gasket faces of bevel drive housings. Exact details of the arrangements used will be found in the appropriate Owners Workshop Manual.

In each of the above examples, measurement of endfloat is made using either a dial gauge or by inserting feeler gauges at a specified point, the exact method being described in the service manual for the machine concerned. Checking these clearances is an important stage of any overhaul, and can make a lot of difference to the smoothness and life of the rebuilt engine; do not overlook them.

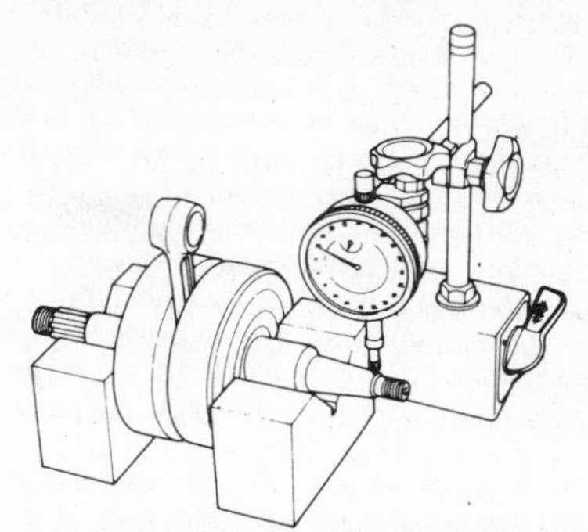

Fig. 4.18 Checking crankshaft runout

The shaft is supported on V-blocks and turned while the dial gauge probe rests on the shaft surface

Another significant application of shimming will be found in the shim adjustment used in the valve clearance setting of many dohc four-stroke fours. Each of the tubular cam followers houses a shim selected from a range produced by the manufacturer for this purpose. Clearance adjustment by this method is inclined to be more lasting than screw-and-locknut arrangements, though less simple to correct. The exact adjustment procedure will be found in the Haynes Owners Workshop Manual for the model concerned, but basically this requires each existing clearance to be measured and noted down. The existing shims are then removed and measured using a micrometer, and any new shim sizes calculated from this. Many dealers keep a large range of these shims in stock, and may agree to allow you to exchange unwanted shims for those of the correct size.

22 Exhaust system: repair

The exhaust system should be considered a semi-consumable item likely to last for around two years or so, depending upon the climate, the quality of the painted or plated finish of the system, road conditions and usage of the machine. During use, the system passes the exhaust gases, including a certain amount of acidic compounds, and numerous short journeys and cool weather will allow some of this to condense on the inside of the system as it cools down. Longer runs and hot temperatures reduce the amount of acidic deposits, and tend to make the system last longer. It is worth noting that many four-stroke systems will be found to have small drain holes at the lowest point of the system. These are meant to allow condensation to drain away, and it is worth checking and unblocking them regularly.

The exhaust gases from a two-stroke engine are invariably much more oily than those from a four-stroke because they contain a much higher percentage of burnt or partially burnt oil that is expelled from the exhaust ports, along with some unburnt fuel. Deposited on the inside of the exhaust pipe and silencer the oil will eventually turn to a sludge-like consistency, which eventually will impede the flow of the exhaust gases and perhaps block them off altogether. This is one of the most frequent causes of a mysterious loss of power or uneven running.

Most silencers have detachable baffles, often retained to the main body of the silencer by a bolt through its rearmost end. When this bolt is removed, it should be possible to withdraw the complete baffle assembly by gripping the bar across its end (just inside the silencer outlet) and drawing the assembly out with a pair of pliers. If it is stuck hard, a twisting motion will usually free it off. If all else fails, it may be necessary to apply heat with a blow lamp, taking care that none of the oil catches fire. Unfortunately this will often spoil the appearance if the silencer is chromium plated, so it is best to apply this drastic treatment only to silencers of the painted type, which can easily be re-sprayed afterwards with a heat-resistant finish.

A particularly effective way in which to clean out a silencer and exhaust pipe internally is to fill the entire system with a solution of caustic soda, after having blocked off one end. Here again this is somewhat drastic treatment as caustic soda is a particularly nasty chemical for which adequate safety precautions must be observed. Commence by mixing the caustic soda in the ratio of 3lbs to one gallon of fresh water, making up sufficient of the solution to fill the complete exhaust system after it has been removed from the machine and stood on end. Add the caustic soda to the water a little at a time, stirring until it has all dissolved. **Never** add it all at once or the water to the caustic soda, as this will cause a violent reaction accompanied by excessive heat. **Do not** plug the open end of the system – leave it in a well ventilated area, as it will give off noxious fumes whilst the solution is working. Leave it overnight, then carefully drain off the used liquid, flushing it away with copious quantities of water. At the same time, flush the entire exhaust system through until you are sure all traces have been removed.

The dangers in handling caustic soda cannot be overstressed. Always wear protective clothing, which includes eye protection and rubber gloves. If the solution comes into contact with the skin or the eyes, wash it away immediately wth clean running water and in the case of the eyes, seek immediate medical advice. Also make sure that the solution does not at any time come into contact with aluminium alloy. If it does it will react violently with aluminium, causing serious damage and generating an explosive gas. The mixing ratio suggested is the strongest that should be necessary for thoroughly cleaning the exhaust system.

Repairs to the exhaust system are difficult to carry out successfully. You will find that rusting tends to come through from the inside in many cases, so that once visible on the outside there will be very little metal left under the chrome or paint. More importantly, the internal baffles will have been affected too, and will eventually break up, making the system illegally noisy. As a general rule, the silencer or complete system should be renewed as soon as serious rusting is evident.

It is possible to weld back brackets or lugs which break off through vibration, though in the case of chrome-plated systems this will damage the finish. On painted systems, the welding can be disguised with a fresh coat of heat-resistant exhaust paint.

In an emergency, you can effect a temporary repair to a holed exhaust using a proprietary repair paste, or in the case of splits or larger holes, an exhaust bandage. The emphasis here is on temporary; use this method to get you by while a new system is on order.

Buying a replacement system is fraught with problems. The original equipment parts are probably the safest choice, though prices tend to be exorbitant. The cheaper 'pattern' systems may be good value, or can drop apart in a few months, and it is hard to guess which you are buying by visual inspection. In some cases, a non-original system may improve performance, though it can just as easily make it worse. In some cases, the new system may make an illegal amount of noise, though in the UK and in other countries noise legislation has been made more stringent to eliminate this factor during recent years. In the UK this legislation is complex, the overall noise output of any motorcycle being dependent on its capacity and also its year of manufacture. In the case of more recent models, it is illegal to use a non-approved replacement exhaust system, and in all cases, excessively noisy exhausts can do little to improve the motorcyclist's image.

The best advice here is to talk to reputable dealers, and to other owners of your model. Choose a system which is either original equipment, or if from a third-party manufacturer, one which is from an established and respected source, and which offers good value for money. If you intend to keep the machine for several years it may be worth looking for a system made in stainless steel in preference to chrome-plated mild steel. As long as you avoid dropping the machine, a stainless system will avoid regular expensive replacements.

23 Cooling system: problems and repairs

Most motorcycle engines are air-cooled, which means that, in this area, very little need be done by way of maintenance. In extreme instances, usually on off-road models, it is conceivable that the cooling fins on the cylinder barrel and head could become sufficiently clogged with dirt to cause cooling problems, but in practice this is a remote possibility.

With water-cooled engines the scope for cooling system difficulties is increased by a huge margin, due to the added complexity of the cooling system parts, and the relative vulnerability of items like the radiator when fitted to a motorcycle. It is argued that water cooling provides a more even cooling medium, which in turn allows finer tolerances and a higher output to be designed into a given engine. In addition, the water jacket blankets mechanical noise quite effectively, and with the introduction of increasingly stringent legislation limiting noise output, this factor alone may encourage the adoption of water-cooling in the future. Whether these factors justify the increased complexity is a subjective matter, however.

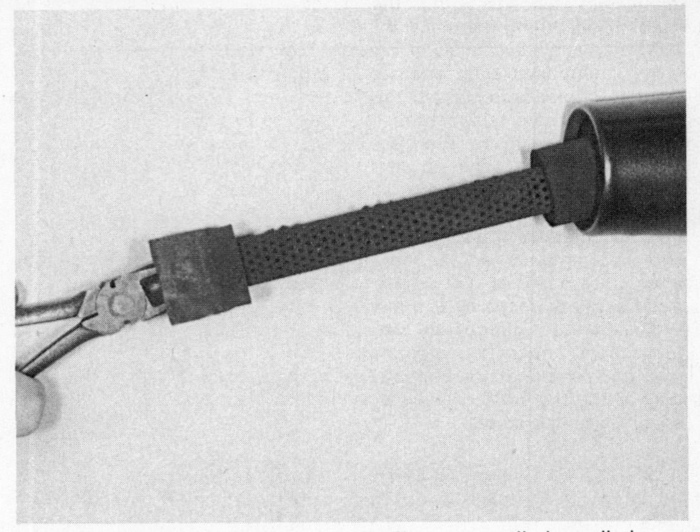

22.1 After removal of retaining bolt, baffle can usually be pulled clear with a slight twisting motion

Many water cooling system problems are at least partly due to the effects of corrosion within the system, though this can largely be avoided by using the prescribed mixture of **soft** water and inhibited glycol antifreeze. Tap water contains more impurities, and will allow corrosion to take place more readily. Clean, filtered, rain water is preferable, and best of all is distilled or demineralised water, as supplied for topping up batteries.

Other sources of leakage problems are the cooling system hoses, which will age and break in time. These should be checked regularly and renewed as soon as there are signs of perishing and cracking visible. Cooling system leakage at the cylinder head to barrel joint can mean compression pressure getting into the cooling system, or coolant getting into the cylinder, and neither condition is desirable. If renewing the head gasket and any O-rings fails to solve the problem, or if it is a recurrent one, check the head and barrel faces for warping or corrosion, and have these skimmed where necessary.

Some models can suffer water pump seal failures, and where the pump is internal, this can allow leakage of coolant into the transmission oil (two-strokes) or engine/transmission oil (four-strokes). The moral here is regular oil checks, and immediate investigation if you find traces of coolant. These may take the form of an oil/water emulsion; a thick white slimy deposit inside the crankcase and outer covers.

Radiator leaks are sometimes difficult to diagnose, and the real answer is to find a garage with cooling system pressure test facilities. This will allow the radiator to be put under pressure, so that cap opening pressure can be checked, and the radiator matrix checked to identify the source of a suspected leak. Before doing this, always find out the maximum test pressure allowed for the system on your particular machine, as well as the correct cap opening pressure. Radiator repairs are difficult to carry out, and unlikely to last very long, though if the alternative is an expensive replacement radiator it would be worth trying epoxy resin or putty if you can get to the source of the leak. Where the matrix itself is split or holed through corrosion, the source of the leak may show up as beads of coolant escaping as the machine reaches normal operating temperature. Leaks of this sort usually indicate that serious corrosion has taken place within the cooling system, and a new radiator will be the only satisfactory answer.

Overheating problems may be due to the general accumulations of dead insects or autumnal debris around the core of the radiator, and this should be cleaned out regularly. If the problem is caused by sludge or calcium deposits within the system, use a proprietary flushing solution, following the maker's instructions carefully.

If the system seems clean internally and externally, the likely culprit is the thermostat on machines so equipped. This is designed to regulate coolant circulation, the valve opening once the engine has reached a certain temperature. If it fails, the valve will tend to stick open, closed or in an intermediate position. Testing will require the behaviour of the valve to be checked whilst it is immersed in a container of water. The water is gradually heated, and the temperature monitored with a thermometer. The Haynes Owners Workshop Manual for your model will give details of the temperature at which the unit should commence to open, and that at which it should reach its fully open position. It is quite easy to assess whether the thermostat is functioning correctly by this method.

Temperature gauges are often fitted to water-cooled models, and the sender unit is a temperature-sensitive device which varies its resistance as the temperature rises. These can be checked in the same way as the thermostat, using a multimeter to check the resistance at specified temperatures.

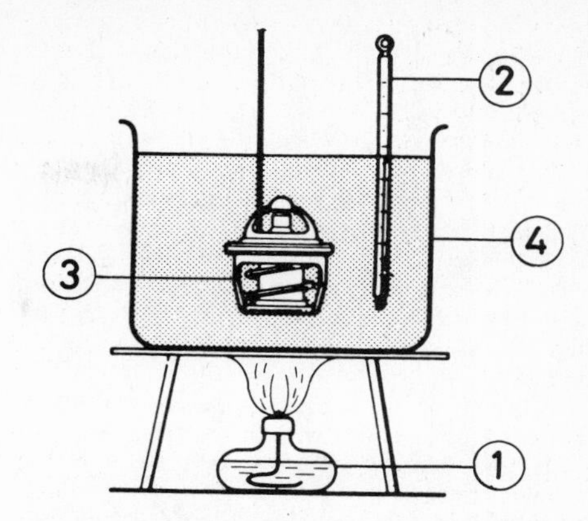

Fig. 4.19 Thermostat testing equipment

1 Heat source
2 Thermometer
3 Thermostat
4 Heat-proof container

Chapter 5 Frame and suspension related tasks

Contents

1 Introduction

In this Chapter we will be looking at the frame and suspension components, together with some of the more common operations relating to these areas. In many respects, the frame, wheels, suspension and brakes are the major safety-related areas of a motorcycle, and you will need to be particularly aware of this when dealing with the maintenance and repair of these parts; always remember that if you make a mistake during an engine rebuild, the engine might break down on you – a similar mistake with the brakes could kill you.

2 Front suspension: problems

Most machines employ oil-damped telescopic forks as the front suspension medium, often using air pressure to supplement the coil springing and to provide a means of adjusting the spring rate. Some models feature other refinements like adjustable damping rates and anti-dive systems linked to the front brake. Some smaller machines and mopeds use different systems, usually leading or trailing link forks.

The exact cause of suspension faults can be difficult to assess at times, and very often what feels like a fork problem may turn out to be play in the rear suspension. For this reason, whenever you are faced with some sort of steering or suspension fault it is best to check right through the front and rear suspension systems and also the steering head assembly.

Telescopic forks

Stiff fork operation can result from a number of faults, but most commonly this is due to the forks becoming twisted in the fork yokes. This can be rectified by slackening the wheel spindle nut and lower yoke pinch bolts, and then bouncing the forks repeatedly to allow them to settle in the correct position. The wheel spindle and pinch bolts should then be tightened to the specified torque figure.

Another reason for erratic fork action is damage to the stanchions as a result of impact, and this can cause the forks to stiffen or stick at a particular point. This may not be easily visible, and to check it you will have to dismantle the forks so that the stanchions can be tested for runout. This is best done with the stanchion supported on V-blocks, using a dial gauge. Failing this, a good indication can be given by rolling the stanchion across a dead flat surface like a sheet of glass.

Slight bending of the stanchion can be corrected using a press to straighten it. This method may work if damage is minimal, but carries the risk of invisible fatigue areas which could result in the stanchion breaking at a later date. Other drawbacks include cracking or flaking of the hard chrome surface. As a general rule it is always safer to fit new stanchions.

Poor damping can be caused by using the wrong grade or amount of oil, or by having different amounts in each leg. Check this by draining the old oil completely. (You may have to remove the fork legs to do this on some machines, see the appropriate Haynes Owners Workshop Manual for details.)

Measure out the correct amount and type of oil into a measuring cylinder or jug, then tip this into the leg, repeating the process on the remaining leg. Some models have a recommended oil level figure quoted, normally with the fork spring removed. Check this using a length of stiff wire or a steel rule as a dipstick, making sure that the level is similar on each leg.

Other causes of erratic damping include dirt or water in the damper valve(s), and which entail dismantling and cleaning the internal parts of each leg. There are a few exceptions to the above, specifically those Moto Guzzi models which employ separate damper inserts in the fork leg, and certain recent designs in which damping is carried out by one leg only. In these cases, reference to the appropriate Owners Workshop Manual is essential.

Weak fork springs will allow vague or imprecise fork action, and it is advisable to check the fork spring free length during oil checks or changes. Most manufacturers quote a standard length and also a service limit. If the spring(s) become permanently compressed to the service limit figure, they must be renewed as a pair.

Air leaks from air-assisted forks will prevent the forks operating normally, this will be masked to some extent by the fork springs. The most common cause is a worn or damaged seal. These seals have to withstand fork air pressure, in addition to retaining oil, so a little damage to a seal lip will inevitably cause loss of pressure. The seal renewal procedure will be found in the Haynes Owners Workshop Manual for that model. Note that most seals must be fitted facing in a particular direction, or they will not be able to withstand air pressure. Do not forget that a leaking air valve may be responsible for the problem; check this first because the valve is easier to renew than the seals.

Fork air pressure, on those models so equipped, should be checked regularly. This is of particular significance where the two fork legs are not interconnected, because imbalance between them can result in handling problems. You should use a syringe-type fork air pump, and also a suspension-type pressure gauge when carrying out checks and adjustments; those produced for use on tyres are not sufficiently accurate for this work. **Never** use a garage air-line to inflate the forks; you will probably destroy the seals in the attempt. Where the fork legs are not interconnected, take great care to make sure that the pressure in each leg is the same. This is of more importance than the exact pressure setting.

Other suspension systems

A good proportion of mopeds and scooter-type machines will be found to employ an alternative front suspension system to keep manufacturing cost down or because of limited space. With this sort of system, there is a mechanical linkage, controlled by damped or undamped suspension units. Maintenance is confined to checking for play in bushes and bearings, and regular lubrication of these pivot points. Most suspension problems will be traced to excessive play in the pivots, usually requiring the renewal of the various bushes or bearings, and possibly the pins upon which they bear.

3 Dismantling telescopic forks: releasing the damper bolt

Most current telescopic forks are of very similar construction, and a common problem during dismantling is releasing the damper rod bolt so that the stanchion and lower leg can be separated. What normally happens is that the damper rod rotates inside the stanchion, making removal of the bolt virtually impossible, though if you are lucky the pressure from the fork spring may exert enough friction to hold it still.

The manufacturer can usually supply a holding tool which is passed down inside the stanchion to engage with the damper rod head. This allows the rod to be immobilised while the bolt is released. For the home workshop, it is not worth purchasing the official tool in view of its infrequent use, so an alternative method must be found to prevent the rotation of the damper rod.

Most damper rods have a recess in the head, and it will be necessary to find something which will locate in this. The first problem is finding out the shape and size of the recess. It is difficult to get enough light into the stanchion to be able to see the damper rod, and it is helpful to be able to measure the recess when making up a holding tool. This can be done by sticking a piece of modelling clay to the end of a length of wooden dowel. If this is then pressed against the damper rod, you will be able to take an accurate impression of the recess.

On forks which have a hexagonal recess, you can make up a tool using a bolt with the appropriate size of hexagon attached to a length of steel tube. The bolt should be slid into the tube and either welded in place or drilled and pinned or riveted. At the upper end, drill a hole through the tube walls to allow a tommy bar to be used.

The solution is less easy where a round recess is used, and the majority of forks fall in this category. The most generally useful improvised tool that we have discovered to date is a piece of square bar, the ends of which have been ground to a blunt point. This can be pressed into the recess, the edges on the point providing enough 'bite' to hold the damper still. An alternative to this is to use a length of wooden dowelling, again with the end tapered.

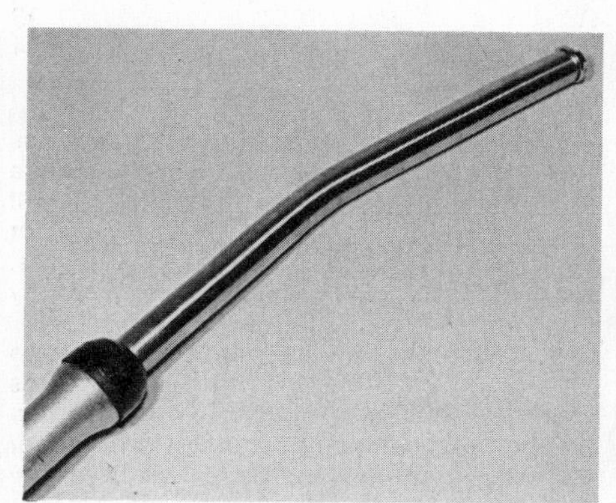
2.1 Impact damage to fork stanchions can cause problems even if it is not this obvious

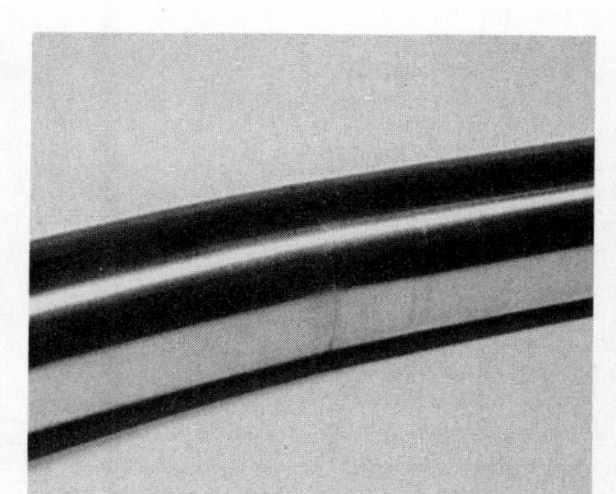
2.2 This stanchion is slightly creased at the bend and would not be safe if straightened

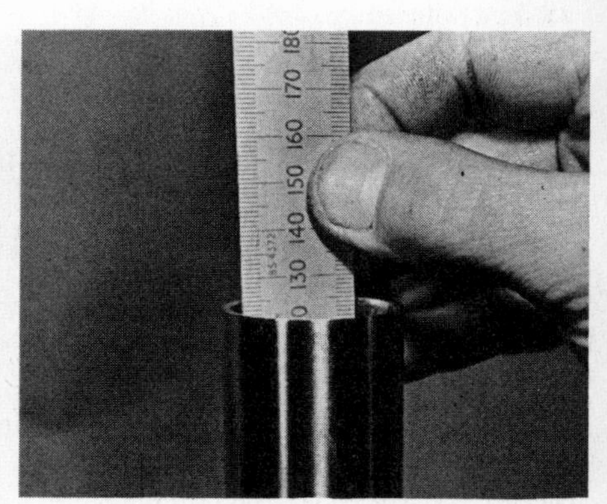

2.3 Either a steel rule or a length of stiff wire can be used to check fork oil level

3.1 A simple damper rod holding tool can be made up using a length of wooden dowel with a taper at the end

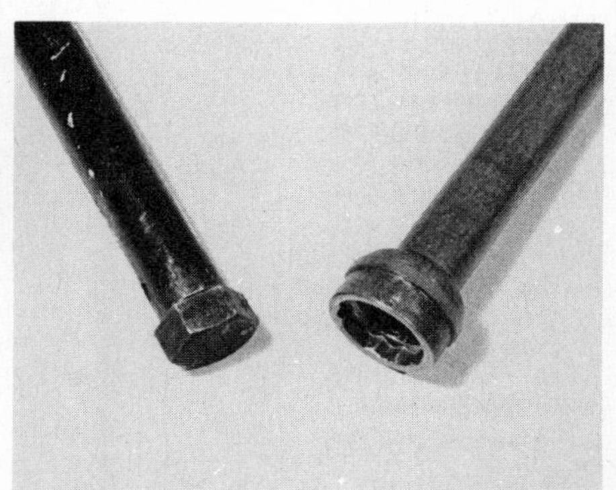
3.2 Where the damper rod has a hexagonal recess, try welding or pinning a bolt of suitable size to a length of tubing

3.3 The home-made tool can be passed down the stanchion until it locates in the damper rod head

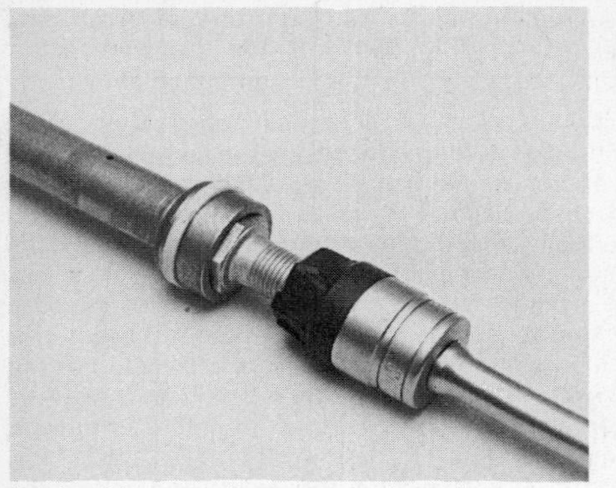

3.4 In this example, two nuts have been locked together at one end of the bolt, the assembly being taped to the socket for security

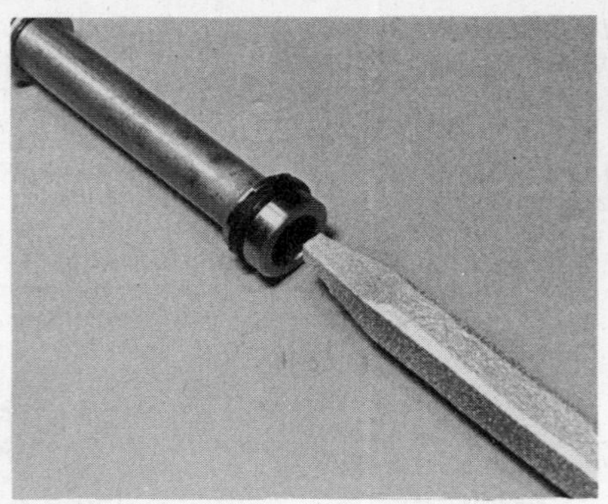

3.5 A length of square bar with the end ground to a blunt point has also proved to be a useful tool where a round recess is employed

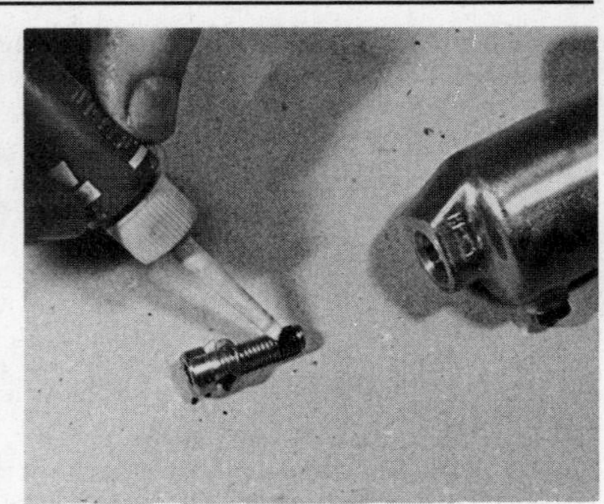

3.6 During installation, the damper rod bolt should be secured by applying a few drops of thread locking fluid

4 Fork stanchions: rust pitting

This is a fairly common problem in damp climates like that in the UK, and is not helped by the enduring fashion for exposed stanchions. These are coated with hard chrome to give the necessary smooth surface on which the oil seal runs. After a few years use, stone damage and the effects of road salt and water combine to cause rusting and flaking of the plated finish. This tends to be followed quite soon by oil leakage as the lips of the oil seals are torn by the sharp edges of the pitting. In some cases, normal fork action effectively pumps the damping oil out of the forks.

There is no satisfactory and cheap way of repairing this damage, and ultimately it will mean new stanchions. It is theoretically possible to have the chrome stripped and replated, and then ground back to size, but this is likely to cost almost as much as new parts.

In an emergency, you could try the well-known dodge of filling the rust pits with epoxy putty. The stanchion must first be degreased, preferably with contact cleaner, so that no trace of grease or oil remains. Apply a layer of epoxy putty to the pits, and allow it to harden. The putty can now be sanded back to the level of the surrounding plating, leaving as smooth a surface as possible. This repair method will certainly improve matters, and if the degreasing was thorough, should last for some time. It is doubtful whether it would be reliable in the case of air-assisted forks though.

Future damage of this type can be prevented by fitting fork gaiters over the exposed area of the stanchion. You should be able to find suitable gaiters at most motorcycle dealers, but if you have problems, try an off-road motorcycle specialist. Before you fit the gaiters, apply a film of multi-purpose grease to the stanchion surface to act as a barrier to condensation. Though the appearance of fork gaiters is not to everyone's taste, fitting them to undamaged stanchions will keep them that way.

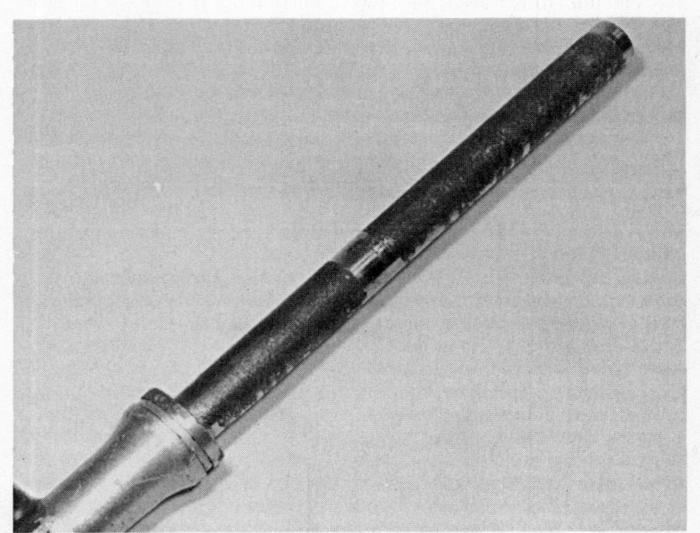

4.1 The inevitable result of the British climate, exposed stanchions and neglected maintenance mean that this stanchion is beyond repair

5 Rear suspension: problems

Most rear suspension faults result from wear in the bushes and linkages used, or from deterioration of the rear suspension unit(s). The wide variety of suspension designs make it impossible to generalise adequately on overhaul techniques, for which you will need the specific procedures described in the relevant Owners Workshop Manual.

It is worth making the point that regular cleaning and maintenance will pay off, especially well where any of the currently popular rising rate single-shock systems are concerned. Most of these employ large numbers of tiny needle roller and spherical bearings, most of which are positioned to take full advantage of the grit and water thrown off by the rear tyre. The tiny seals used to protect them seem to expire readily under this onslaught, and if they are not renewed immediately, the bearing will soon follow suit.

Rear suspension units are usually of sealed construction, and the only rebuilding possible is confined to fitting new springs. Note that if the springs are to be removed for renewal or checking of their free length, a spring compressor will be required to compress them safely;

4.2 Note how the rusted stanchion surface has destroyed the oil seal; a new seal would soon go the same way

these are discussed in Chapter 2. Many original equipment units are used because they are cheap, and this is one area where it is worth shopping around. Top quality units would probably be a waste of time and money on small commuter models and mopeds, but could be a good investment on larger machines. In some cases the handling characteristics can be transformed simply by fitting good quality rear suspension units. In the case of single-shock models, you may have a limited choice of units when the need for a replacement arises. Generally, though, this type of unit tends to be of higher quality as original equipment, so this need not be too much of a problem.

When buying replacement units other than replacement original equipment parts, it is worth buying from a reputable dealer; select the unit manufacturer who produces the best type for your model, then phone or write to him for the address of a local stockist. It is important that the spring and damper rate characteristics are suited to your bike, and that the mounting centres are correctly sized and spaced. You will have to rely on the vendor to ensure this, so avoid dubious or unknowledgeable sources who will try to sell you whatever they happen to have in stock regardless of its suitability.

For air-assisted units you should use a syringe-type air pump, and also a suspension-type pressure gauge when carrying out checks and adjustments; those produced for use on tyres are not sufficiently accurate for this work. **Never** use a garage air-line to inflate the units; you will probably destroy the seals in the attempt. Where the units are not linked, take great care to make sure that the pressure in each is the same.

6 Bearings, bushes and seals

The steering and suspension systems on all types of motorcycle are dependent on an assortment of bearing types, and wear or damage in these areas will invariably result in handling problems. In most cases the bearing or bush is lubricated during assembly, and the lubricant is sealed in place by a seal, which also serves to exclude abrasive dirt. Provided that the lubrication film is maintained, and the dirt excluded, the bearing surfaces will last for a very long time before wear allows play to develop. Once this occurs, the resulting movement will allow the seal to wear or be damaged, and the bearing or bush will then wear more rapidly.

Telescopic front forks

The stanchion and lower leg of a telescopic fork are machined to close tolerances during manufacture, the bearing surface being formed by the machined surface of the lower leg, or by renewable bushes. Lubrication is provided by the damping oil, and the assembly is closed by a seal pressed into the top of the lower leg. Where bushes are fitted, these are often coated with Teflon to reduce friction, and thus improve the action of the fork.

The oil seal in the top of the lower leg is in an exposed position, and vulnerable to damage from road dirt. To minimise this problem, a dust seal is usually fitted to road machines, the purpose of which is to prevent dirt from entering the seal and causing damage. In the case of off-road models, the dust seal is often replaced by fork gaiters, which effectively exclude all dirt from the stanchion surface. Once the oil seal lip becomes damaged by dirt, wear of the seal, stanchion and bearing surfaces is rapid, and it is worth checking these areas regularly for signs of deterioration; damage can be difficult and costly to rectify.

To overhaul the fork, you will need to refer to the relevant Haynes Owners Workshop Manual for your model. This will give details showing the dismantling and reassembly procedures, and also information on assessing wear. Oil seals are normally held in place with a wire circlip, and can sometimes be renewed without having to dismantle the fork, though it is preferable to do so to allow the rest of the fork to be checked and cleaned. With simple oil-damped coil-spring forks, a single-lip seal is usually fitted, but with air-assisted forks, a more complicated double-lipped seal is more common. This is because the seal must contain air pressure, as well as sealing in the damping oil.

Wear between the fork stanchion and lower leg will allow play between them, giving rise to poor fork action and juddering under braking. If the lower leg is unbushed, it is normally necessary to renew it (and possibly the stanchion) to resolve the problem, repair being impractical.

With bushed forks, the bushes can be renewed to restore fork action and precision. The upper bush, housed in the fork lower leg, can often be driven out by pulling the stanchion and lower leg apart sharply. On some models, though, a slide-hammer arrangement will be needed to displace the bush from its recess. The exact procedure will be described in the Owners Workshop Manual.

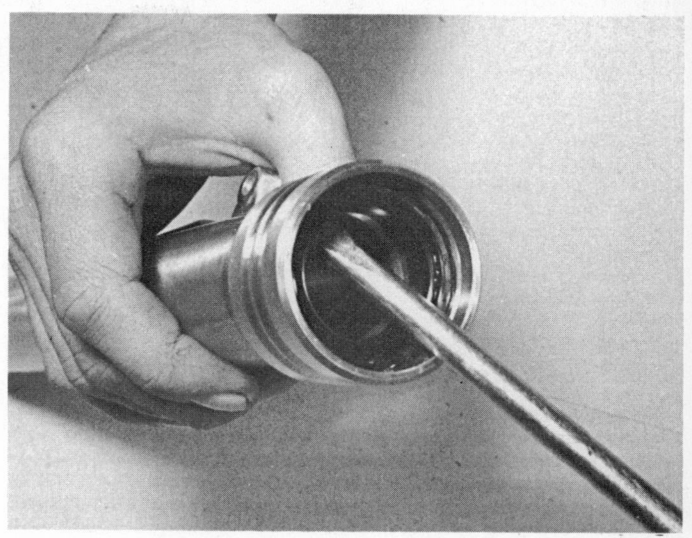

6.1 Oil seals can normally be removed by prising them out of the fork leg with a screwdriver. Note that the seal is normally retained by a circlip or wire retaining clip which must be removed first

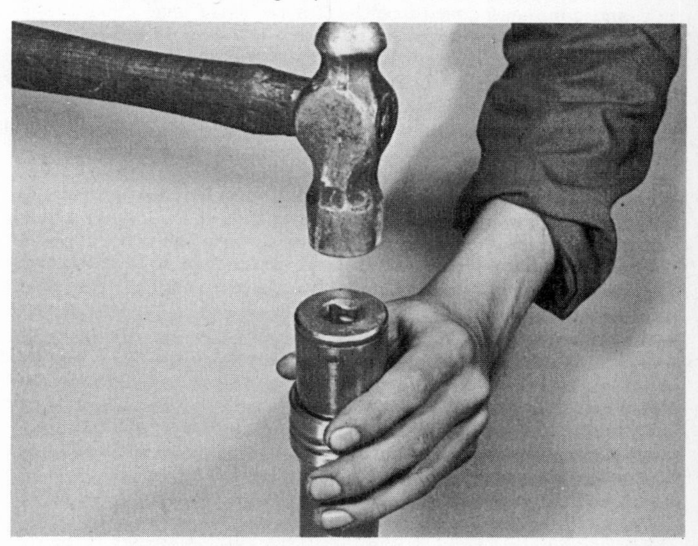

6.2 New seals can be installed using a large socket as a drift. Be sure to place the socket against the seal case, **not** the lip, which is easily damaged

Non-telescopic front suspension and rear suspension

In the case of the numerous variations of leading-link and trailing-link suspension used on many mopeds, scooters and commuter models, the suspension linkages often rely on plain bushes or small needle-roller bearing for their pivots. These are often closed by tiny grease seals to exclude dirt and water.

Rear suspension arrangements also vary widely. Simple swinging arm systems use metal or plastic bushes, bonded rubber bushes or needle roller or taper roller bearings to support the pivot shaft. With the more sophisticated rising-rate single-shock absorber systems, the complex suspension linkages usually rely on small needle-roller races, each closed by a grease seal.

These arrangements vary in their complexity and vulnerability to wear, but all will benefit from regular dismantling, cleaning and examination. The bushes or bearings, and their pivot shafts or pins should be examined closely for signs of wear or damage, and any suspect items renewed before the fault becomes more severe and extensive. The seals are best renewed as a precautionary measure each time the assembly is stripped down.

Wear in any of these pivots is best assessed by feel, and as a general rule, if play can be felt then the bush or bearing and its pin or shaft should be renewed. Severe wear will be easily visible, as will scoring, broken bushes or bearing rollers and other damage. It is not normally advisable to disturb any bush or bearing unless it is to be renewed; removal will usually destroy it.

The method employed for the removal and fitting of bushes and bearings will be governed by its location and accessibility. If access to both sides of its bore is unrestricted, the best approach is to employ a drawbolt arrangement, since this will allow firm, even pressure to be applied, minimising the risk of damage. An alternative is to use a socket or a length of bar or thick-walled tubing of slightly smaller diameter than that of the bush or bearing to drift it out of its recess.

Where a bush or bearing is fitted to a blind bore, some sort of extractor will be needed. These are available commercially from tool shops, and although expensive to buy have few viable home-made equivalents. Most have an internal puller section which can be locked in place, and a slide hammer-arrangement to permit removal.

One perennial problem is the removal of bonded rubber bushes. These invariably resist all attempts to remove them by civilised methods, unless a press can be used. The usual method is to cut away the rubber insert, leaving the outer steel tube in the bore. This can be weakened by hacksawing through its wall at one or more points. If the tube is then collapsed into the bore, the remains are easily extracted.

Steering head bearings

Most machines use either a cup-and-cone or a taper-roller bearing arrangement to support the steering head assembly in the frame. The cup-and-cone arrangement consists of hardened steel cups, pressed into the frame steering head tube which hold a number of uncaged balls. Completing the bearings are corresponding conical inner races, or cones. The taper-roller bearing consists of hardened steel outer races pressed into the steering head which correspond with tapered caged inner races on the steering stem. An adjuster and locknut arrangement allows the free play to be taken up without applying significant pressure to the bearing assemblies.

In time, and if periodic re-greasing of the bearing has been neglected, their condition will deteriorate, the races eventually becoming pitted or indented, making the steering feel stiff and notchy. This may not be immediately obvious in normal riding, but can cause the machine to feel vague and tend to wander. The remedy is to renew both bearing assemblies. The upper cone (inner race) is easily removed once the top yoke and adjuster nut have been released. The upper and lower cups (outer races) can be withdrawn using a universal slide-hammer type bearing extractor, or driven out of the head tube using a long drift. If using the drift method tap firmly and evenly around the race to ensure that it drives out squarely. It may prove advantageous to curve the end of the drift slightly to improve access. Note that extreme care should be taken with this method; if the race refuses to move the only alternative is to use a slide-hammer bearing extractor. The lower cone (inner race) will be pressed over the steering stem, and can usually be prised away from the lower yoke by tapping wedges below it; screwdriver blades are useful for levering it off the steering stem. When refitting the bearing lower cone (inner race) use a length of tubing with a slightly larger diameter than the steering stem to drive the bearing home. New cups (outer races) can be installed using a fabricated drawbolt tool as shown in Fig. 2.34. Full details of this process will be found in the appropriate Haynes Owners Workshop Manual, together with details of which parts need to be removed to gain access to the steering head assembly.

7 Brake: maintenance and repair

Regular maintenance and checks of the braking system are of obvious importance to the safety of the rider and other road users. Reference to the owner's handbook supplied with the machine will give details of the checks and adjustments which are needed on a particular model, whilst a factory service manual or the relevant Haynes Owners Workshop Manual will provide more detailed information on maintenance and overhaul of the system. Routine servicing of the

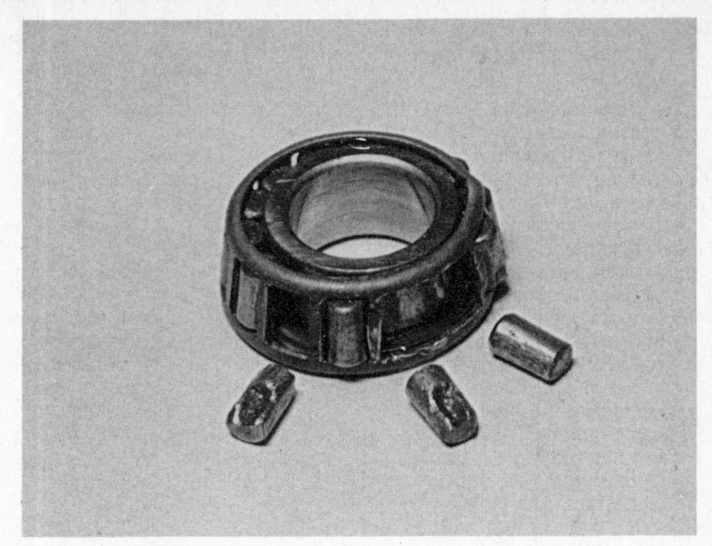

6.3 Bearing damage rarely reaches these extremes before the rider becomes aware of a problem

6.4 Steering head bearing cups and cones should be checked for wear or pitting. Once pitting occurs, as in this example, new races and balls are required

6.5 The old bearing cups can be removed by passing a long drift through the steering head and tapping them out

system, whether it is disc or drum based, will eliminate most failures and problems, but the additional points mentioned here may help in some cases.

Disc brake systems

Disc brake squeal is a common problem on many models and may be due to a number of factors. Always check the caliper carefully, removing the pads for close inspection and cleaning. Oil or grease contamination may provoke squealing, as can worn pads. In some cases, renewing the pads just before they reach the wear limit mark or line may help.

If new pads or a thorough cleaning of the caliper fail to solve the problem, the noise may be due to high frequency chattering between the pad(s) and the piston or fixed caliper half. Try removing the pad and cleaning off the backing metal. Note where the pad backing metal is marked where it is in contact with the caliper or piston. Apply a thin film of copper grease to this area. **On no account use other types of grease** – these will melt and contaminate the friction material when the pad gets hot. Assemble the pads, taking care not to get any copper grease on the friction surface.

Where the disc surface is drilled or slotted, it is not unknown for strange noises to be produced during braking. If the disc is in good condition, you may have to live with the noise, but if the disc is part-worn, you can try putting a slight chamfer on the edges of the holes or slots. If this fails, a new disc may be the only answer.

Disc brake judder can often be attributed to a warped disc, or one on which the surface has worn unevenly. Check the disc runout with the specified limits, using a dial gauge mounted on the fork leg. Measure the disc thickness, and check carefully for low spots, or scored areas. Where plain cast iron discs are used (*Brembo*-equipped machines, for example) the inevitable rust patch which marks the outline of the pads when the machine has been left unused for a time may cause the brake to grab at this point, but this should wear off as the disc is scrubbed clean by the pads and is thus only a temporary problem. Where drilled or slotted discs are fitted, check that the holes are unobstructed. If the disc is damaged in any of the ways mentioned, or worn below its service limit, fit a new one, together with new pads.

Disc brake fade can be due to pads of the wrong grade being fitted; cheap, unbranded far-eastern imports being notorious for this. Worse still, there have been spates of poor quality counterfeit pads sold in genuine-looking packaging. Although this problem is now less prevalent, it is so potentially dangerous that you should only buy replacement pads from reputable sources.

Even high-quality pads will cause fading when they are worn. This need not be because there is no friction material left; it is important to note that the friction material also functions as a thermal insulator between the disc and the caliper, and this is why the wear limit mark is reached while there is still a good thickness of material left. Ignore this reminder to renew the pads at your peril. Once the heat insulation becomes ineffective, heat can transfer to the caliper, and thus to the hydraulic system. This can boil the fluid under heavy braking, causing unexpected and serious fading.

Finally, always renew fluid at the specified intervals. Old fluid is dangerous because it is hygroscopic, absorbing water vapour from the air. Normal braking temperatures can then cause boiling, and the resulting air will make the brake fade in use. Regular changes of fluid will remove this moisture, which if left will cause corrosion of the system.

You might consider changing to a synthetic fluid, which will eliminate most of the above problems. The fluid is expensive, but of very high quality, and will need changing less often. It should also make the caliper and master cylinder last longer. If you do so, make sure that it is suitable for use in your system, and follow any special instructions regarding flushing the system and renewing caliper and master cylinder seals. It is a good idea to mark the reservoir cap permanently to indicate to yourself or any subsequent owner that synthetic fluid is in use in the system.

Poor braking effect or feel may be due to the type of system in use (some are better than others in this respect) or to general wear and tear. Points to look for are seized caliper pistons, or in the case of single piston calipers, seizure of the caliper pivot. If this is suspected, the caliper should be stripped and overhauled, following the detailed instructions given in the Haynes Owners Workshop Manual for your machine. On machines which employ an anti-dive system, remember that problems in this area can affect brake performance too.

7.1 This pad has reached the end of its life. Note that the groove through the centre of the friction material which indicates the wear limit has almost vanished

Bleeding – It will be of no surprise that air, even in small amounts, will seriously affect the operation of the brakes. If you suspect that air is present, the system must be bled to expel it. Whilst the procedure for bleeding brake systems varies from machine to machine, and is dependent on the exact arrangement used, the principle is to pump fluid from the master cylinder, through the pipes to the caliper. Here, the fluid is allowed to emerge through a bleed valve provided for this purpose and is caught in a suitable container. Special brake bleeding tubes can be purchased, and these help prevent air working back into the caliper by utilising a one-way valve.

The normal bleeding procedure for most models is as follows, but check the Owners Workshop Manual for any special procedures which might apply to your machine.

a) Clean the bleed valve and attach a length of tubing or a special bleed tube to it. Place the other end in the container which is to catch the fluid, submerging the end in hydraulic fluid
b) Remove the master cylinder reservoir and top up to the maximum level, using only the recommended type and grade of **new** hydraulic fluid
c) If the system is empty, open the bleed valve and pump the brake lever or pedal rapidly until fluid appears at the bleed tube, then check and top up the reservoir. Loosely refit the reservoir cap or cover to exclude dirt and to prevent spouts of fluid being expelled during the bleeding operation
d) With the bleed valve closed, pump the lever or pedal until resistance can be felt. Keep the pedal or lever depressed, then open the valve slightly until the flow of fluid slows to a near standstill. Close the valve, then release the pedal or lever
e) Check the level of fluid in the reservoir, and top it up where required
f) Repeat steps d) and e) until air bubbles no longer emerge with the expelled fluid, then tighten the valve fully and check brake operation
g) Remove the bleed tube and refit the bleed valve dust cap, wipe up any fluid spills, being particularly careful to remove fluid spots from the paintwork (see below). Dispose of the used fluid safely

Important note: *Hydraulic fluid will mark or damage paintwork and many plastic finishes. Keep the fluid well away from these areas, and protect vulnerable parts like the tank by covering it with clean cloth before working on the system. Never keep old fluid for re-use – it will be dirty and will also have absorbed moisture, lowering its safe working temperature.*

Brake lines – If an overhaul of a normally good system does not improve it, check the brake lines. These should be renewed at intervals specified by the manufacturer, irrespective of their apparent condition.

The inner reinforcing will gradually weaken in use, allowing the line to swell slightly under pressure. Even marginal swelling, probably imperceptible to the eye, can cause the brake to feel vague and spongy.

If you fit new brake lines, and intend keeping the bike for some time, consider fitting braided stainless Teflon lines instead of the standard type. These are thinner overall, but considerably tougher, which is why they are used for racing machines. They are undeniably more expensive, especially with multi-disc set ups, but a good investment if you are looking for optimum safety and performance. They also look good!

Caliper and master cylinder wear can cause all sorts of braking system problems. Its exposed position makes the caliper more vulnerable to the effects of water and road dirt, and it is often the caliper which gives problems first.

Single-sided calipers are prone to becoming seized on the caliper bracket pivot, and this will produce poor braking effort. If wear develops at the pivot, this may cause judder during braking. In both cases, it is normally necessary to renew the caliper assembly to resolve the problem. This sort of trouble can be avoided if the pivot seals are cleaned and inspected regularly. Keep the pivot well lubricated, and renew the seals if there is any sign of splitting.

Opposed-piston calipers do not pivot, and thus do not suffer the problems described above, but all calipers are prone to pistons sticking or seizing in their bores. Sometimes this is due to seal failure allowing moisture to get between the exposed end of the piston and the caliper bore. More commonly, neglected fluid changes will have allowed the fluid in the system to absorb moisture, and this in turn will have promoted corrosion between the caliper bore and piston.

Master cylinders are less likely to suffer problems than the caliper(s), but problems do occur. These can range from corrosion and blocked ports, to general wear and seal failure. Unlike calipers, seal failure is not always obvious; air can be drawn back into the system without external signs of leakage.

In the case of any suspected fault in the hydraulic system, it is vital that it should be investigated promptly, before the brake fails in service. The large variety of calipers and master cylinders precludes coverage here, so for specific information, refer to the Haynes Owners Workshop Manual for your model.

Drum brake systems

Other than regular maintenance and adjustment, details of which will be found in the workshop manual for the machine concerned, there is little to go wrong with a drum brake arrangement. Faults will occur as the machine gets older, however, and this will be more likely if maintenance has been neglected. The more common problems are described below.

Binding or dragging brakes may be due to corrosion in the operating mechanism causing seizure. With cable-operated systems, check that the cable works smoothly and freely after disconnecting it at the brake plate. If necessary, lubricate the cable to restore normal operation. Sometimes a cable will begin to fray at the upper end, this normally being caused by the nipple seizing in the handlebar lever or pedal. The frayed strands of the inner will tend to catch in the outer cable, holding the brake partially on. This is a dangerous situation, and the cable must be renewed at once.

Another likely spot for problems is the brake cam, which may become stiff in its bore due to old, dried out grease or corrosion. To prevent this, remove the brake shoes and operating arm and push out the cam for cleaning and lubrication. If this is attended to regularly the risk of seizure will be eliminated. Be careful not to apply an excessive amount of grease to the working faces of the cam, or the surplus grease may find its way onto the shoes.

Brake judder may be due to localised rusting or damage to the drum surface. Alternatively, check the drum surface for runout. This can sometimes be caused by the drum being pulled oval due to irregular spoke tension, where the wheel is of the wire spoked type. The best thing to do if you discover this problem is to have the wheel dealt with by a wheel builder. It may be possible to pull the drum back to the correct shape by adjusting the spoke tension correctly. Any remaining ovality can be corrected by skimming the drum, provided that this does not take the drum diameter beyond its service limit. To do this you will need to find someone with a lathe with a gap bed to allow the entire wheel to be set up intact. If the ovality is really bad, you may need to renew the wheel.

Weak or ineffective brakes can be caused by a number of problems, and you should check the obvious areas first, such as the condition of the shoes. If the lining material has become contaminated with oil or grease, new shoes should be fitted, having first degreased the brake parts and repaired the cause of the contamination. If there is no sign of grease on the linings, check for glazing of the friction surface. This can be caused by overheating the linings under heavy braking, or because of brake drag. The linings can be restored by roughening the surface with abrasive paper. Do this outside, and wear a particle mask to avoid inhalation of asbestos dust.

Temporary problems can be experienced if water gets into the brake, and this will tend to take some time to dry out naturally. To speed up this process, try riding for a while with the brake held lightly on; the friction will produce enough heat to help evaporate the water. Do not overdo this procedure; excessive heat will result in glazing (see above).

Mechanical problems include a loss of leverage at the operating arm. Ideally, the cable or operating rod should pull on the arm at an angle of approximately 90°. If the arm is positioned incorrectly, or if the shoes have worn down to the service limit, this angle may have altered, and less braking effort will be produced by a given amount of pressure at the lever or pedal. Check for shoe wear first (there is often a pointer on the brake cam which will show when the linings are worn to the point where renewal is advisable). If the brake arm angle still seems wrong, check that it is fitted in the correct position on the splines of the cam. On many brakes, a dot or line on the end of the cam should align with the split or dot in the arm.

A few machines may be encountered where twin leading shoes (tls) brakes are fitted. With this type of brake there is normally an adjuster in the link between the two cams, and it is important to make sure that this is set so that the two brake shoes operate in unison. The details of the adjustment procedure will be found in the Haynes Owners Workshop Manual for the machine concerned.

7.2 Bleeding operation is simplified by use of kit with one-way valve in tube

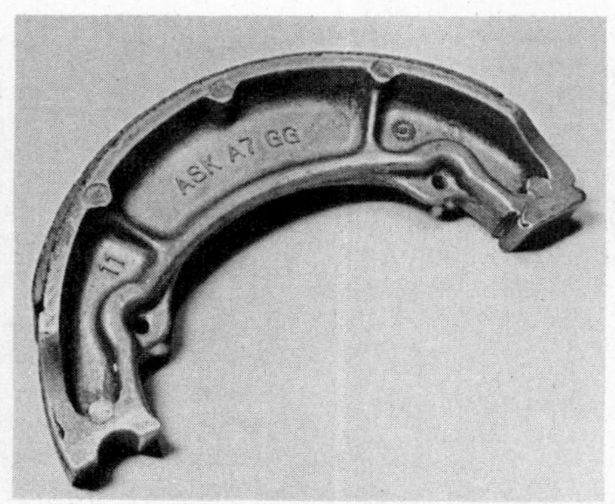

7.3 Drum brake shoes rarely wear evenly. As this example shows, wear occurs on the leading edge of the shoe

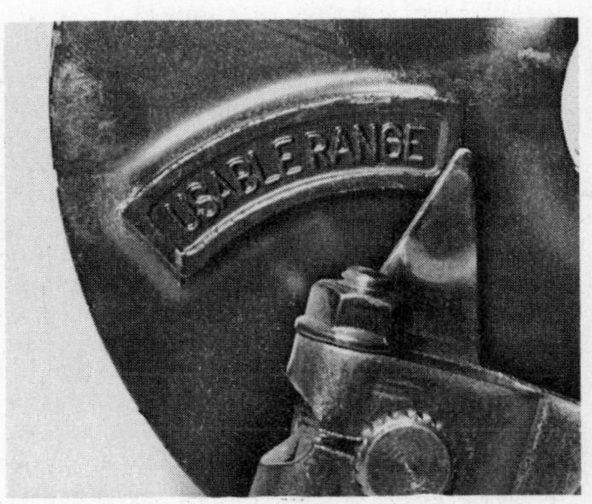

7.4 Majority of models incorporate a wear indicator plate on the brake cam

7.5 Typical brake arm and cam alignment marks

8 Chains and sprockets

Roller chains

The most familiar application of chains is in the final drive of most motorcycles. The exceptions to this are machines using shaft final drive, and a handful of models which employ toothed belt drives. The chain used for this type of application is a single-row roller chain, the size of which will be dependent on the weight and power output of the motorcycle. Roller chains are sometimes used in the engine to drive camshafts or balancer shafts, although this is more commonly done with Hy-vo or Morse chains (see below). A summary of the chain sizes and types in common use will be found in the reference section at the end of this manual.

Chains come in a wide variety of sizes, and may have their rollers sealed from the elements by O-rings. The Haynes Owners Workshop Manual for your machine will tell you what size and type of chain is fitted, and will also give full details about free play adjustment, lubrication methods and intervals, and how to remove and refit the chain.

As a general rule, smaller models are fitted with unsealed chains, and these can be removed by separating the joining link. This comprises a special link where one of the side plates can be detached after a spring clip has been prised off. When fitting this type of chain, always assemble the spring link so that the closed end of the clip faces in the direction of chain travel.

On larger machines an O-ring chain is often fitted. These have the advantage of long life; the O-rings seal out dirt and water from the rollers and bushes, and so there is less wear than with unsealed types. One disadvantage is that it is not normally possible to use a joining link. This means that the chain is made as a continuous loop to fit a particular model. One consequence of this is that it is normally necessary to remove the swinging arm assembly before the chain can be removed.

With all types of final drive chain, regular cleaning, inspection and lubrication is essential, and the recommended intervals for these checks will be given in the Owners handbook or Owners Workshop Manual for your machine. Note that in the case of O-ring chains, it is important never to use an aerosol lubricant unless it is marked as being suitable for this type of chain. Other types of lubricant may cause the O-rings to be damaged, and this will result in the rapid wear and failure of the chain.

Cleaning and lubrication is best carried out with the chain removed, though regular application of aerosol lubricant with the chain in position is preferable to infrequent full lubrication, and in the case of 'endless' chains, removal is likely to prove too time consuming to be carried out without a very good reason. Remove dirt and old lubricant with a degreasing solvent or paraffin (kerosene) and allow the chain to dry. Check the chain for wear in the manner outlined in the Haynes Owners Workshop Manual or owner's handbook. In addition, look for any signs of abnormal wear, such as wear or damage to the inner face of the side plates on one side. This can indicate mis-alignment of the sprockets, due to incorrectly positioned spacers, or more commonly, poor rear wheel alignment.

With a conventional (unsealed) chain, a good method of lubrication is to use a hot-immersion lubricant as shown in the accompanying photographs. This is supplied in a large metal container, and can be heated up over a gas burner or similar until it melts. The chain is placed in the hot lubricant, which penetrates the links and rollers. The chain is then removed and hung over the can until excess lubricant has drained off, and the remainder has set to a waxy consistency inside each bush. Lubrication by this method will extend the life of the chain considerably.

A quicker method, and one which is essential on O-ring chains, is to use either gear oil or a suitable aerosol lubricant applied to the chain while it is in position on the machine. This will give good protection to the external parts of the chain, but note that it will not penetrate unsealed bushes as well as a hot immersion type.

Sprockets should be checked along with the chain; if the sprocket teeth become worn or hooked, the chain will soon become damaged, and fitting a new chain to a part-worn sprocket should be avoided for this reason. As a rule, sprockets should always be renewed as a pair, never singly.

Sometimes it proves necessary either to shorten a chain or to add extra links to it, especially when sprocket sizes are being changed to increase or decrease the overall gear ratios. Separating a chain fitted with a spring link presents no problem but if the chain is of the endless type, it is then necessary to use a chain rivet extractor. This is a simple device that clamps on to a chain, gripping around one of the rollers. When the centre screw is tightened down on the centre of the rivet that passes through the centre of the roller, it draws off the side plate. When this is repeated on the second, adjoining roller, the side plate can be detached and the chain separated.

To rejoin the chain after links have been removed or added, the side plate should be pressed back over the rivet ends and the ends of the rivets peened over after the side plate has been pressed fully home. Special connecting links are available for this purpose, having rivets with softened ends. Note that it is considered bad practice to have more than one spring link in any chain run, and with many of today's high performance machines it is customary to have an endless chain, with no spring link at all.

When fitting a spring link, it is essential that the closed end of the spring clip faces the direction of travel of the chain. It is just as important that the ends of the clip, and the nose of its inner end locate correctly with the grooves cut into the ends of the rivets in the joining plate.

Where it is necessary to remove or add only one roller, a special cranked link must be used, which is connected to a double roller link, making a total of three rollers in all. Here again, it is inadvisable to use a second spring link as a joining medium; it must be securely riveted into position before fitting the joining link.

Do not join old and new lengths of chain, no matter how small the addition. Chains rarely wear evenly under normal circumstances, but when old and new parts are mixed, it will prove virtually impossible to achieve correct chain tension at all points along the chain run.

This information applies only to conventional chains and not those of the O-ring type for which special information should be sought.

Hy-Vo or Morse chains

As has been mentioned, this type of chain is commonly found inside the engine, where it is used to transmit drive to the camshafts, primary drive and balancer shafts. The multi-element links in this type of chain are shaped so that they fit the gear-type teeth of the driving and driven sprockets. These chains work in a near-ideal environment, and can be expected to last for a long time. When the time comes for renewal, as indicated by the inability of the tensioner mechanism (where fitted) to take up any more free play, you will have to strip the engine to carry out the work. The design of the chain means that joining links cannot be used, the chain being effectively endless.

8.1 Where a joining link is used, remove and refit the spring clip using pliers as shown. Take care not to distort it

8.2 Note that the closed end of the spring clip must face in the direction of normal chain travel

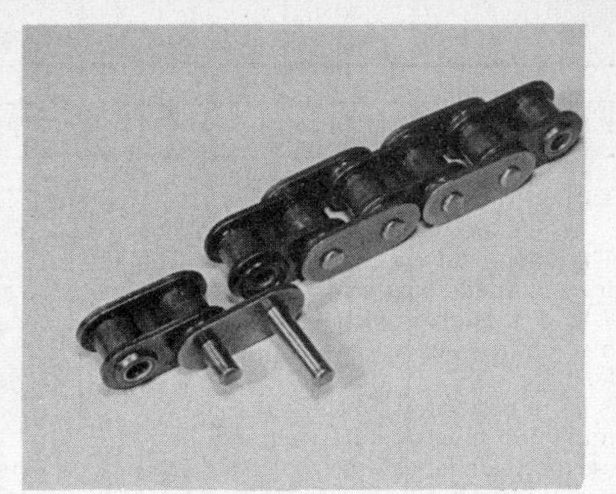

8.3 A section of O-ring chain showing the construction. Note the shouldered inner plates which carry the O-ring seals

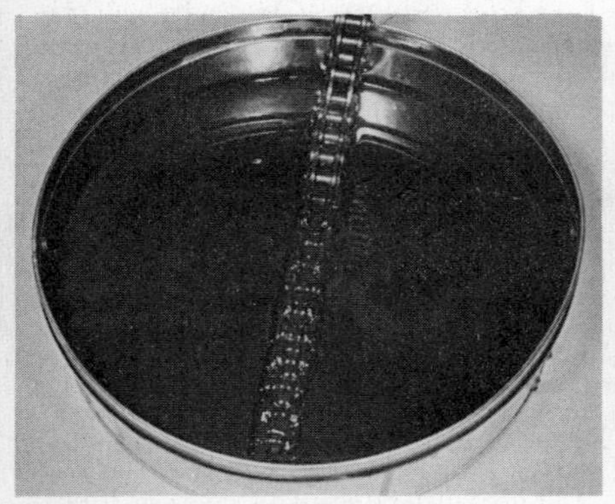

8.4 The best method of lubricating non O-ring chains is to use a hot-immersion lubricant

8.5 For intermediate lubrication or for use on O-ring chains, aerosol lubricants are very useful

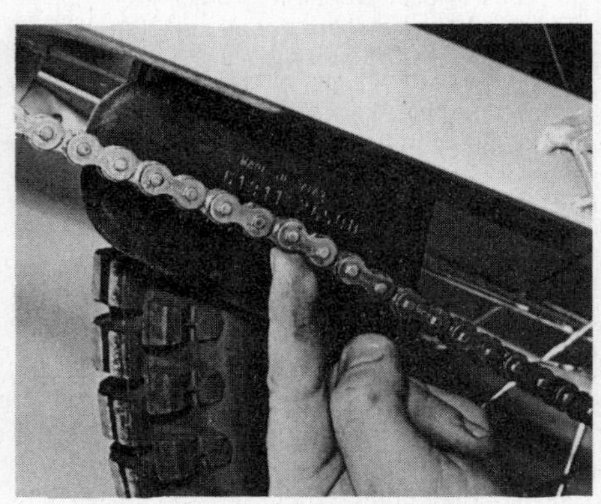

8.6 Correct chain tension is important in obtaining good chain and sprocket life – check regularly

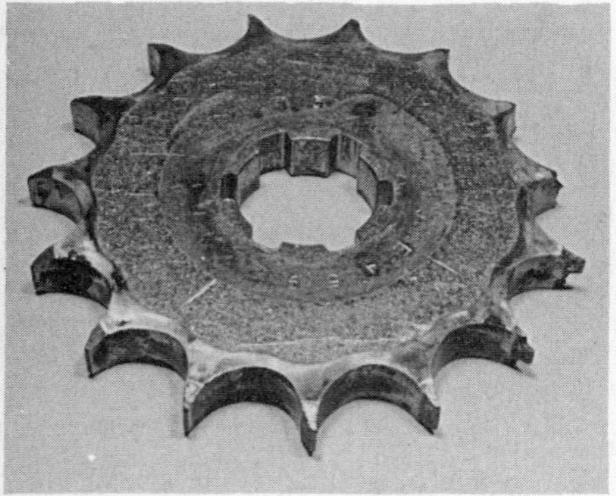

8.7 Always renew sprockets if they become worn. In this example the teeth have begun to break up at the ends

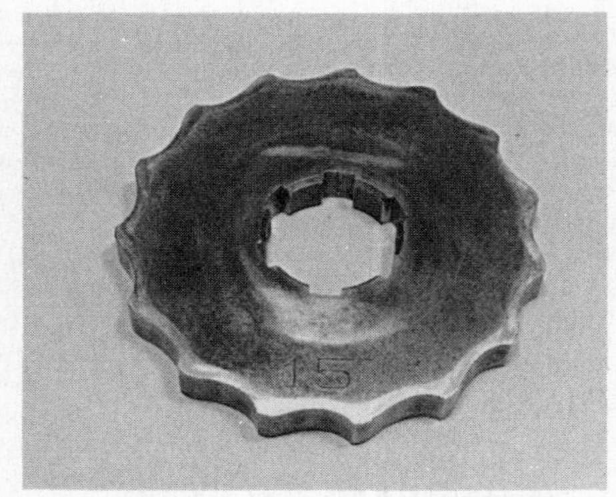

8.8 It is hard to imagine how this sprocket could have driven the machine!

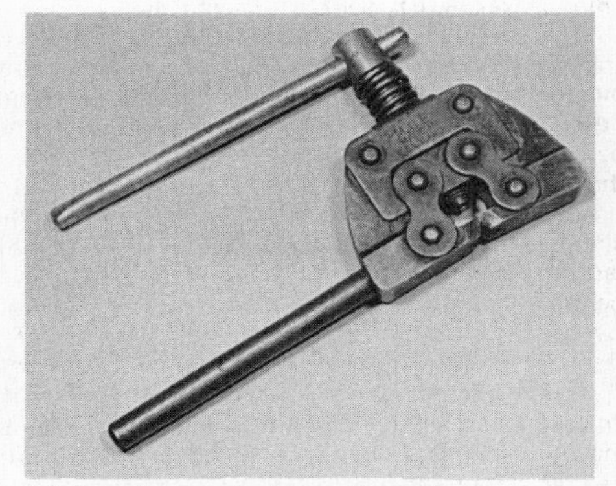

8.9 For altering chain length, a chain splitting tool will be required

8.10 The tool is used as shown to drive out the chain rivets, allowing the chain to be separated at the required point

8.11 Where an odd number of links is necessary, you may need to use a cranked link (half link)

9 Wheels: check and overhaul

Types of wheel

There are three basic wheel types in common use on motorcycles. The traditional wire spoked version, once used on virtually everything from mopeds upwards, has largely become displaced by the cast wheel, though ironically some recent 'factory custom' models have made wire spoked wheels a design feature, such are the vagaries of fashion.

The cast alloy wheel, itself once a popular sales feature, will be found on most middle-weight to large capacity machines. The cast construction lends itself well to mass production, the machining of the basic casting being a far less complex process than is wire wheel building. The technique allows new castings with various spoke designs to be produced as a styling feature of new models, and on some types of cast wheel, it is possible to use tubeless tyres.

The third type of construction, using an extruded aluminium alloy or steel rim riveted to pressed steel or aluminium spokes and a cast and machined centre, results in the composite wheel, a mass produced version of which enjoyed brief popularity with some buyers. It is still used in special applications, and on some mopeds and similar models.

Checking and servicing wire spoked wheels

Wire spoked wheels are unique in that they can be taken apart and rebuilt should the need arise, the big catch being that to do so requires time, patience and most of all, skill. The advantage is that you can repair minor damage, fit new parts where needed, and even alter the rim size should you find a need to do so, though the more complex jobs are best left to a professional wheel builder.

Routine checks on wire spoked wheels comprise spoke tension, rim runout and bearing free play. A quick way of identifying loose spokes is to tap each one in turn with a screwdriver. A sound spoke will be under tension, and will produce a fairly musical note as it is struck. A loose spoke will make a dull sound and may rattle or buzz. It is important to keep the spokes under tension; leaving them loose for any length of time will result in breakage.

The spokes are tightened by turning the threaded nipple using a special spoke key. At a push you could use a small self-locking wrench, but the correct tool is preferable. In almost every case, the nipple is fitted through holes in the rim (though a few wire spoked wheels may be found which use tubeless tyres. In these cases a solid rim is used, the spokes being inverted with the nipple at the hub end).

As the nipple is turned, the spoke is tightened, and in normal circumstances you need only bring the loose spoke up to a similar tension to the others. Be sure to remove any dirt from the spoke end before adjusting it, and apply a little oil or maintenance spray to help it turn easily. It is important that the nipple is turned just enough to restore tension; excessive tightening will pull the rim out of true.

If the spoke is really loose, requiring several turns to tension it correctly, you will need to remove the tyre and the rim tape so that the inside of the nipple can be checked. Any protruding spoke ends must be filed smooth and flush with the nipple head so that there is no risk of their puncturing the inner tube. You may also need to remove the tyre in situations where the nipple is corroded in place. Apply penetrating fluid to the threads, and use the screwdriver slot in the nipple head to turn it.

Rim runout is checked by arranging a wire pointer close to the rim edge and turning the wheel. As a general rule, about 1 mm or so of runout is acceptable, but check the exact figure in the Haynes Owners Workshop Manual for your model. Runout may be either radial (up-and-down) or axial (side-to-side). During this check, try rocking the rim from side to side. Any obvious free play here indicates worn bearings, and this must not be confused with runout. If you detect bearing play, the bearing(s) must be renewed, or the wheel will never run true. Bearing renewal is discussed later in this section.

If you find radial runout, the wheel can be trued by slackening several of the spokes on the 'high' side of the wheel, and then tightening the opposing spokes by a similar amount to pull the rim into shape. Slacken the spokes at the highest point by two or three turns, then slacken the adjacent spokes by progressively less turns to ensure that the rim is not pulled out of shape. Care must be taken to ensure that spokes from each side of the hub are slackened or tightened by a similar amount, or you will find that you have introduced axial play to add to your problems!

Axial play, or side-to-side wobble, will require that two or three spokes on the tight side of the hub are slackened, and the opposing spokes tightened to pull the rim sideways and into line. As you will appreciate, anything other than minor adjustments will tend to become complicated, with one alteration tending to cause distortion elsewhere. This is where the experience of the professional wheel builder will pay off, since he will be able to allow for this effect during adjustment. Most wheel builders will undertake trueing at reasonable cost, and in practice it may be quicker and easier to have this work done for you. This is all the more true if the distortion is significant. This usually indicates impact damage, and in such cases it is preferable to have the wheel examined properly in case there are signs of splits or deformation in the rim.

As has been mentioned, if you ignore loose spokes they will eventually break, leading to excessive stress and subsequent breakage of others. It is vital that any broken, bent or corroded spokes are renewed promptly, or you will end up having to have a complete wheel rebuild. Respoking requires the removal of the tyre, inner tube and rim tape to gain access to the nipple heads. You can order a new spoke and nipple from the local dealer for your model, though this may take time to arrange.

A local wheel builder, on the other hand, should be able to make up a spoke of the correct length to pattern, and this is probably a better way of getting a replacement. If just one or two spokes are to be renewed, you can do this at home, threading the new spoke into position, fitting the nipple and trueing the wheel as described above. Where a number of spokes are broken, bent or rusted, it will be quicker to have the wheel checked and the work done by an expert. It may be that he will advise a complete new set of spokes, and this should not cost too much. When the work is complete, check that all spoke ends are ground or filed smooth, and do not omit to fit the rim tape or band before the tyre and tube are refitted.

Significant damage which has resulted in a buckled rim, or just general old age and corrosion, will mean that the wheel will need complete rebuilding. This is definitely a job for the expert; you can try it at home if you wish, but be warned that you will spend many hours trying to get the rim true, possibly without success.

Take the complete wheel to the builder so that the offset (the position of the rim centre in relation to that of the hub) can be measured before the old spokes are cut away. Take the wheel spindle with you so that it can be used to support the wheel during the trueing operation. You should also check the wheel bearings and renew these if worn, or it may prove difficult to true the wheel. You may wish to get the hub blasted and polished or painted while the spokes are out of the way, in which case tell the wheel builder of your intentions. The bearings must be removed from the hub during any refinishing work. Bear in mind that you are not limited to the original rim type. Where a chrome plated steel rim was fitted, you may prefer to specify a replacement in alloy, which is a little lighter and easier to maintain.

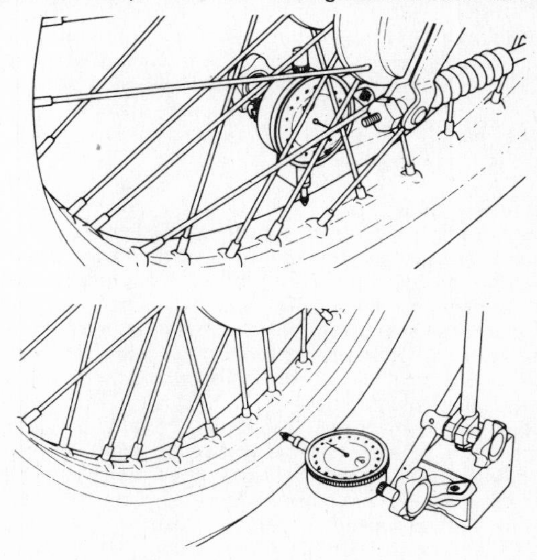

Fig. 5.1 A dial gauge is the most accurate way to check wheel runout – using a simple wire pointer is less precise but will usually be adequate

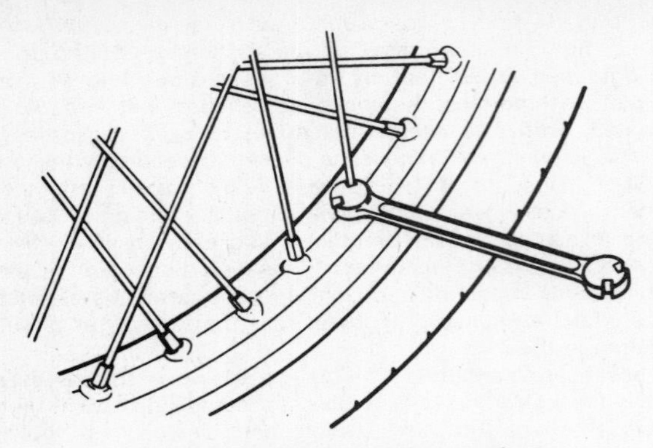

Fig. 5.2 Adjusting spoke tension

Checking cast alloy wheels

It is comparatively easy to check cast wheels, simply because there is little that can be done to correct any discrepancy found. Start by checking for bearing free play and rim runout as described above for wire spoked wheels. Worn bearings should, of course, be renewed (see below). Rim runout will be a problem to deal with. If this exceeds the limits set by the manufacturer, or if there is a distinct wobble caused by a deformed rim, the complete wheel must be renewed. There is no satisfactory way to straighten a twisted or bent wheel, and no attempt should be made to do so. Remember that the forces which caused the damage may well have set up potential fracture sites in the casting.

Corrosion of cast wheels is a common problem, but not a serious one unless it becomes extensive. Small chips or patches of damaged lacquer should be repaired by sanding and polishing the corroded area to match the original, and then touching in the lacquer coat to protect the metal. More extensive damage is best dealt with by removing the tyre, and also the tube, where fitted. Remove the wheel bearings and have the whole wheel cleaned up by bead blasting. The surface finish can be either left matt, or can be buffed up to a bright finish. Protect the cleaned metal by coating it with a proprietary clear wheel lacquer, which can be bought in aerosol cans from most car accessory shops.

Checking composite wheels

At first sight, a composite wheel would appear to be a good prospect for rebuilding, but in reality few of them are, at least as far as the manufacturer is concerned. In practice, you should make the usual checks for runout, damage and wheel bearing free play as described elsewhere in this section. In the case of wheels using steel spokes and alloy rims, check carefully for signs of corrosion where the rim meets the spokes; corrosion here is likely to be caused by electrolytic action between the two metals, and can seriously weaken the wheel structure.

The only alternative to buying a new wheel is to check the dealers and the motorcycle press for companies willing to undertake reconditioning work. There are a few of these around, though without direct experience it is difficult to make recommendations. Remember that from the manufactuer's point of view, such repairs are unauthorised, and thus will invalidate any warranty.

Renewing wheel bearings

A common arrangement for the wheel bearings on many machines is two ball or roller bearings pressed into the central bore of the wheel hub. The bearings are located internally by either a tubular spacer or by shoulders machined into the hub. The assembly is generally protected by one or more removable grease seals. In the case of the rear wheel, there is often a third bearing carried in a detachable sprocket carrier and cush drive unit, and a few machines use an additional needle roller race in the rear brake plate.

The above arrangement will be found on a large number of Japanese machines, where wheel bearing arrangements have become almost standardised, and in some cases, the same wheels are fitted to a number of similar models in a range. There are, of course, exceptions, and in the case of European models, a greater variety of bearing arrangements will be found. Common exceptions to the above arrangement include some mopeds, where bicycle-type cup-and-cone bearings may be employed. On many of the scooter-based mopeds, where the rear wheel attaches directly to an extension of the transmission housing, the wheel will be secured to the transmission output shaft by splines and a single retaining nut, the bearings being part of the transmission system. Most shaft drive models use a similar method of attaching the rear wheel, the shaft and bearings forming part of the bevel drive assembly. It is not possible to list all the various systems in detail in this book, and it follows that the appropriate Haynes Owners Workshop Manual should be consulted to see which arrangement applies to your model.

Where non-adjustable ball or roller bearings are employed, any discernible free play will require the renewal of the bearings. The precise method and sequence of removal and refitting will be described in the Owners Workshop Manual for the model concerned. As a rule, the wheel is removed and any separate grease seal removed from the hub. Occasionally, you may encounter a threaded bearing retainer at this stage. These require a special pin spanner to remove them, though in the absence of the correct tool you can often make up an improvised version from steel strip, drilling holes to correspond with those in the retainer and fitting small bolts to act as the locating pins. The accompanying line drawing shows an example of this type of tool. Always check the threads on the retainer before attempting removal; those on the right-hand side are often left-handed.

In a typical hub, such as that described above, the first bearing can be driven out after levering the spacer to one side, using a thin drift with a parallel tip, and working around the bearing inner race. This will drive out the bearing, and where one is fitted, the lipped seal on that side. The spacer can then be tipped out of the hub and the remaining bearing drifted out of its recess. Note that the removal of the bearings will place great stress on the balls or rollers and the races; it is not normally advisable to refit bearings once removed from the hub, so do not remove them without good cause.

Where taper roller bearings are fitted, it is generally possible to adjust them to remove free play, but only if the bearings are undamaged. Again, the workshop manual will give details, the normal procedure being to apply a very slight preload to the bearings. Cup-and-cone bearings are adjusted in a similar fashion, and there is normally an adjuster locked by a thin nut to facilitate this.

If new bearings are to be fitted, these can be ordered from the relevant dealer, though if you take the old bearings to a motor factor it may be possible to obtain standard replacements at a lower cost. The bearings can usually be matched from the part number stamped or etched on them, or by direct measurement. For further information on bearing types, see the reference section at the end of the book.

The new bearings should be packed with high-melting point grease and then tapped into position using a tubular drift of slightly smaller diameter than that of the outer race. Special tools are available for this purpose, but a socket of the correct size makes a useful alternative.

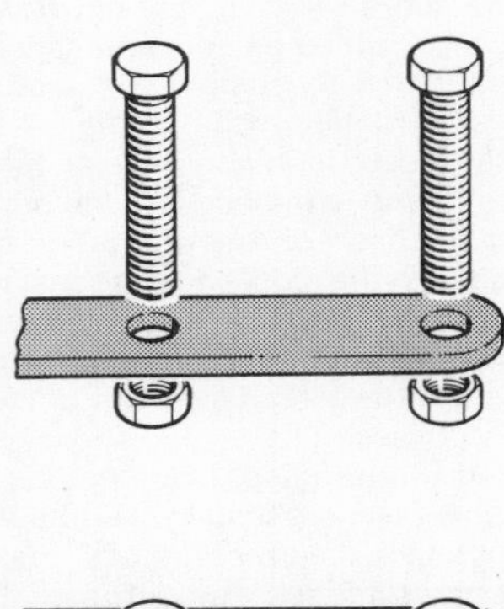

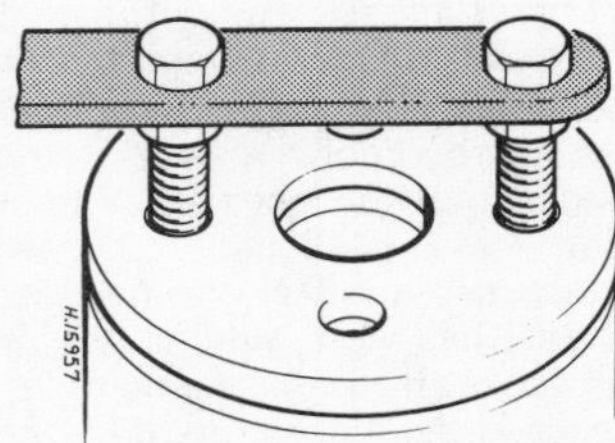

Fig. 5.3 Threaded retainers can be removed using the improvised pin spanner shown

9.1 Wheel bearings are generally protected by renewable seals which can be removed as shown

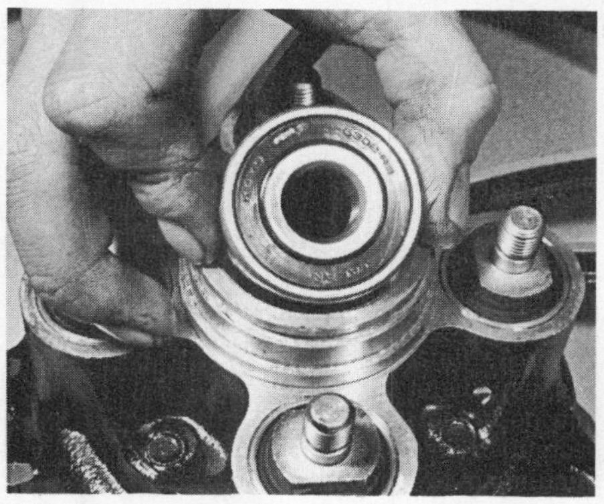

9.2 Many bearings have an integral grease seal on one side. The sealed face should always be fitted facing outwards

9.3 Where bearings are secured by a threaded retainer, remember to stake the retainer in position

9.4 The most common method for removal is to use a long drift passed through the hub as shown

9.5 Drive new bearings home squarely and fully. Select a drift which will locate on the bearing outer race

9.6 In the case of cup-and-cone bearings slacken and remove one of the threaded adjustable cones to permit dismantling

9.7 In this example, the bearing balls are held in place by a thin steel ring which should be prised out carefully

9.8 When assembling the bearings, retain the balls using grease during assembly. Adjust the cones until free play is *just* taken up

10 Wheel balancing

The front wheel should be statically balanced, complete with tyre and dust cap on valve. An out of balance wheel can produce dangerous wobbling at high speed.

Some tyres have a balance mark on the sidewall. This must be positioned adjacent to the valve. Even so, the wheel still requires balancing.

With the front wheel clear of the ground, spin the wheel several times. Each time, it will probably come to rest in the same position. Balance weights should be attached diametrically opposite the heavy spot, until the wheel will not come to rest in any set position, when spun.

Various types of weight are fitted to wheels. Those with wire spokes often use cylindrical weights with a slot to allow them to be fitted around a spoke. The weight is then slid down over the nipple to locate it. In the case of cast alloy wheels, the weights may be clamped to the rim edge, or stuck to the rim flange. In some cases, weights must be affixed in certain positions in relation to the spokes. These points should be checked at your local dealer or in the relevant Haynes Owners Workshop Manual before attempting to balance the wheel.

It is possible to have a wheel dynamically balanced at some dealers. This requires its removal.

Although the rear wheel is more tolerant to out-of-balance forces than is the front wheel, ideally this too should be balanced if a new tyre is fitted. Because of the drag of the final drive components the wheel must be removed from the machine and placed on a suitable free-running spindle before balancing takes place. Balancing can then be carried out as for the front wheel.

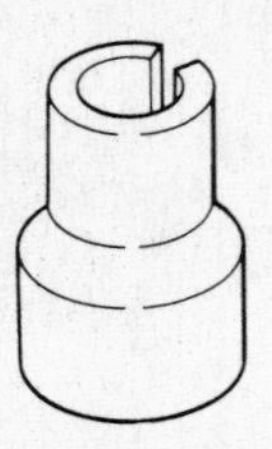

Spoked wheel type *Cast wheel type*

Fig. 5.4 Balance weight types

11.1 Most machines have alignment marks on the swinging arm end. Check that the adjusters are positioned in the same place on each side. If in doubt about the accuracy of the marks, check the wheel alignment using a straightedge (see text)

11 Wheel alignment

It is often forgotten that incorrect alignment of the rear wheel can have a significant effect on the handling and road holding of a motorcycle, with the tyre continually scrubbing at a slight angle to the axis of the machine. With a few exceptions, it is possible to position the wheel several degrees out of true, and there is a risk of this happening every time the rear wheel spindle nuts are disturbed. There are, as ever, exceptions. Machines which use shaft final drive have no requirement for adjustment of the rear wheel position, and so the problem of alignment never arises. Similarly, a few chain drive models, eg some Ducati models, employ a form of chain adjustment using an eccentric swinging arm pivot. In either case, mis-alignment can only occur as a result of physical damage.

Most machines have a set of alignment marks on the ends of the swinging arm, and these provide a useful check of the accuracy of the wheel position. For an infallible test of the alignment, however, it is best to employ the traditional method of placing a straightedge in the form of a length of wood or similar, against the side of the rear wheel. Position the straightedge so that it touches the wheel at two points, as near to the hub as possible. With the front wheel pointing straight ahead, make sure that the rear wheel lies parallel to the frame axis; the far end should not converge with or diverge from the front wheel, though an allowance will have to be made where the tyres are of different widths.

An alternative approach is to use a length of string stretched between the two wheels. This method, as well as the straightedge, is best explained in the accompanying illustration. Once the alignment is checked in this way, you can also check the accuracy of the alignment marks on the swinging arm ends. If these prove to be inaccurate, either make corrections to the existing marks, or use the straightedge each time the wheel is disturbed or the chain adjusted.

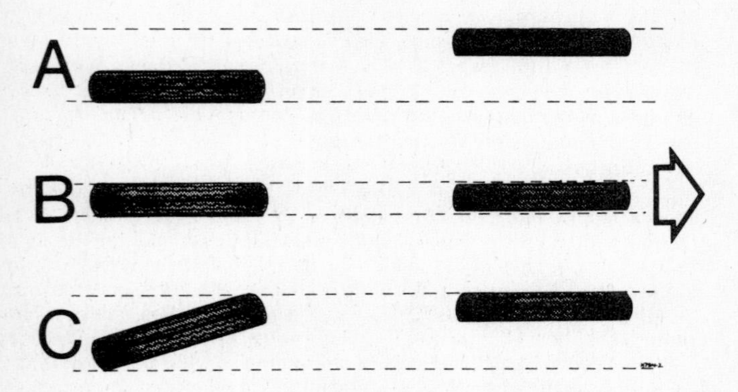

Fig. 5.5 Wheel alignment

(A) shows the rear wheel offset in relation to the front, and usually indicates that spacers have been positioned incorrectly
(B) shows the front and rear wheels correctly aligned
(C) depicts the rear wheel out of alignment in the swinging arm

12 Tubed tyres: removal, fitting and puncture repairs

At some time or other the need will arise to remove and replace the tyres, either as a result of a puncture or because replacements are necessary to offset wear. To the inexperienced, tyre changing represents a formidable task, yet if a few simple rules are observed and the technique learned the whole operation is surprisingly simple.

To remove the tyre from either wheel first detach the wheel from the machine. Deflate the tyre by removing the valve insert and when it is fully deflated, push the bead from the tyre away from the wheel rim on both sides so that the bead enters the centre well of the rim. Remove the locking cap and push the tyre valve into the tyre itself.

Insert a tyre lever close to the valve and lever the edge of the tyre over the outside of the wheel rim. Very little force should be necessary; if resistance is encountered it is probably due to the fact that the tyre beads have not entered the well of the wheel rim all the way round the tyre. Note that where the machine is fitted with cast alloy wheels, the risk of damage to the wheel rim can be minimised by the use of proprietary plastic rim protectors placed over the rim flange at the point where the tyre levers are inserted. Suitable rim protectors may be fabricated very easily from short lengths (4 – 6 inches) of thick-walled nylon petrol pipe which have been split down one side using a sharp knife. The use of rim protectors should be adopted whenever levers are used and, therefore, when the risk of damage is likely.

Once the tyre has been edged over the wheel rim, it is easy to work around the wheel rim so that the tyre is completely free to one side. At this stage, the inner tube can be removed.

Working from the other side of the wheel, ease the other edge of the tyre over the outside of the wheel rim that is furthest away. Continue to work around the rim until the tyre is free completely from the rim.

If a puncture has necessitated the removal of the tyre, reinflate the inner tube and immerse it in a bowl of water to trace the source of the leak. Mark its position and deflate the tube. Dry the tube and clean the area around the puncture with a petrol soaked rag. When the surface has dried, apply rubber solution and allow this to dry before removing the backing from the patch and applying the patch to the surface.

It is best to use a patch of self-vulcanising type, which will form a very permanent repair. Note that it may be necessary to remove a protective covering from the top surface of the patch, after it has sealed into position. Inner tubes made from synthetic rubber may require a special type of patch and adhesive, if a satisfactory bond is to be achieved.

Before refitting the tyre, check the inside to make sure that the object which caused the puncture is not trapped. Check the outside of the tyre, particularly the tread area, to make sure nothing is trapped that may cause a further puncture.

If the inner tube has been patched on a number of past occasions, or if there is a tear or large hole, it is preferable to discard it and fit a new one. Sudden deflation may cause an accident, particularly if it occurs with the front wheel.

To fit the tyre, inflate the inner tube sufficiently for it to assume a circular shape but only just. Then push it into the tyre so that it is enclosed completely. Lay the tyre on the wheel at an angle and insert the valve through the rim tape and the hole in the wheel rim. Attach the locking cap on the first few threads, sufficient to hold the valve captive in its correct location.

Starting at the point furthest from the valve, push the tyre bead over the edge of the wheel rim until it is located in the central well. Continue to work around the tyre in this fashion until the whole of one side of the tyre is on the rim. It may be necessary to use a tyre lever during the final stages.

Make sure that there is no pull on the tyre valve and again commencing with the area furthest from the valve, ease the other bead of the tyre over the edge of the rim. Finish with the area close to the valve, pushing the valve up into the tyre until the locking cap touches the rim. This will ensure the inner tube is not trapped when the last section of the bead is edged over the rim with a tyre lever.

Check that the inner tube is not trapped at any point. Reinflate the inner tube, and check that the tyre is seating correctly around the wheel rim. There should be a thin rib moulded around the wall of the tyre on both sides, which should be equidistant from the wheel rim at all points. If the tyre is unevenly located on the rim, try bouncing the wheel when the tyre is at the recommended pressure. It is probable that one of the beads has not pulled clear of the centre well.

Adjust the tyre pressure to that specified by the manufacturer.

Tyre replacement is aided by dusting the side walls particularly in the vicinity of the beads, with a liberal coating of french chalk. Washing-up liquid can also be used to good effect, but this has the disadvantage (where steel rims are used) of causing the inner surfaces of the wheel rim to corrode. Do not be over generous in the application of lubricant or tyre creep may occur.

On machines equipped with wire-spoked wheels, never fit the inner tube and tyre without the rim tape in position. If this procedure is overlooked there is good chance of the ends of the spoke nipples chafing the inner tube and causing a crop of punctures.

Never fit a tyre that has a damaged tread or side walls. Apart from the legal aspects, there is a very great risk of a blowout, which can have serious consequences on any two-wheel vehicle.

Tyre valves rarely give trouble, but it is always advisable to check whether the valve itself is leaking before removing the tyre. Do not forget to fit the dust cap, which forms an effective second seal.

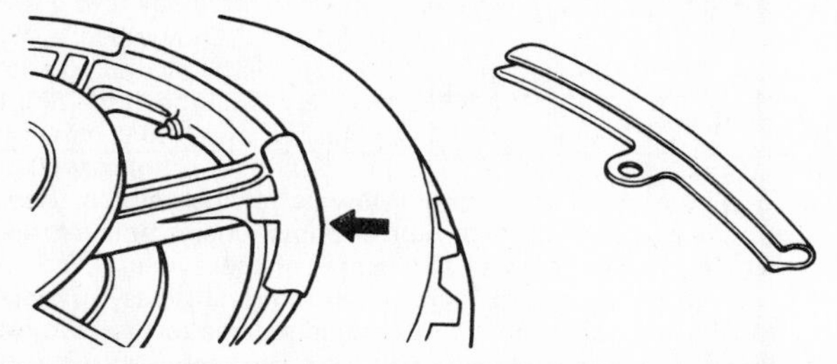

Fig. 5.6 Plastic rim protectors can be easily fabricated and prevent damage to cast alloy wheel rims

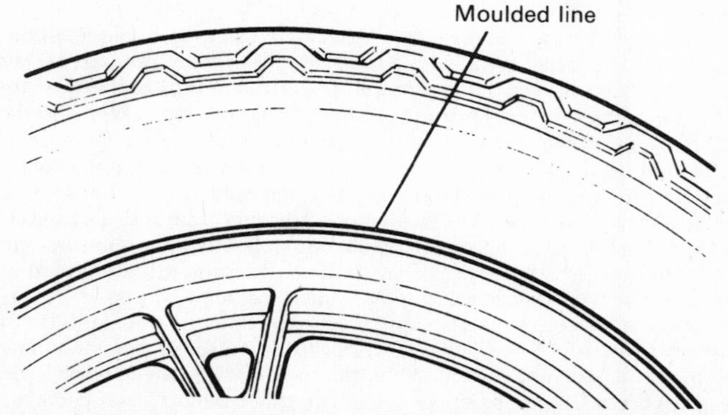

Fig. 5.7 Tyre moulded line should be equidistant from the wheel rim at all points

12.1 Tyres in this condition are potentially lethal, as well as illegal. You also stand a much greater chance of getting a puncture

13 Tubeless tyres: removal and fitting

It is strongly recommended that should a repair to a tubeless tyre be necessary, the wheel is removed from the machine and taken to a tyre fitting specialist who is willing to do the job or taken to an official dealer. This is because the force required to break the seal between the wheel rim and tyre bead is considerable and considered to be beyond the capabilities of an individual working with normal tyre removing tools. Any abortive attempt to break the rim to bead seal may also cause damage to the wheel rim, resulting in an expensive wheel replacement. If, however, a suitable bead releasing tool is available, and experience has already been gained in its use, tyre removal and refitting can be accomplished as follows.

Remove the wheel from the machine. Deflate the tyre by removing the valve insert and when it is fully deflated, push the bead of the tyre away from the wheel rim on both sides so that the bead enters the centre well of the rim. As noted, this operation will almost certainly require the use of a bead releasing tool.

Insert a tyre lever close to the valve and lever the edge of the tyre over the outside of the wheel rim. Very little force should be necessary; if resistance is encountered it is probably due to the fact that the tyre beads have not entered the well of the wheel rim all the way round the tyre. Should the initial problem persist, lubrication of the tyre bead and the inside edge and lip of the rim will facilitate removal. Use a recommended lubricant, a diluted solution of washing-up liquid or french chalk. Lubrication is usually recommended as an aid to tyre fitting but its use is equally desirable during removal. The risk of lever damage to wheel rims can be minimised by the use of proprietary plastic rim protectors placed over the rim flange at the point where the tyre levers are inserted. Suitable rim protectors may be fabricated very easily from short lengths (4 – 6 inches) of thick-walled nylon petrol pipe which have been split down one side using a sharp knife. The use of rim protectors should be adopted whenever levers are used and, therefore, when the risk of damage is likely.

Once the tyre has been edged over the wheel rim, it is easy to work around the wheel rim so that the tyre is completely free on one side.

Working from the other side of the wheel, ease the other edge of the tyre over the outside of the wheel rim, which is furthest away. Continue to work around the rim until the tyre is free completely from the rim.

Refer to the following Sections for details relating to puncture repair, renewal of tyres and tyre valves.

Refitting of the tyre is virtually a reversal of the removal procedure. If the tyre has a balance mark (usually a spot of coloured paint), this must be positioned alongside the valve. Similiarly, any arrow indicating direction of rotation must face the right way.

Starting at the point furthest from the valve, push the tyre bead over the edge of the wheel rim until it is located in the central well. Continue to work around the tyre in this fashion until the whole of one side of the tyre is on the rim. It may be necessary to use a tyre lever during the final stages. Here again, the use of a lubricant will aid fitting. It is recommended strongly that when refitting the tyre only a recommended lubricant is used because such lubricants also have sealing properties. Do not be over generous in the application of lubricant or tyre creep may occur.

Fitting the upper bead is similar to fitting the lower bead. Start by pushing the bead over the rim and into the well at a point diametrically opposite the tyre valve. Continue working round the tyre, each side of the starting point, ensuring that the bead opposite the working area is always in the well. Apply lubricant as necessary. Avoid using tyre levers unless absolutely essential, to help reduce damage to the soft wheel rim. The use of the levers should be required only when the final portion of bead is to be pushed over the rim.

Lubricate the tyre beads again prior to inflating the tyre, and check that the wheel rim is evenly positioned in relation to the tyre beads. Inflation of the tyre may well prove impossible without the use of a high pressure air hose. The tyre will retain air completely only when the beads are firmly against the rim edges at all points and it may be found when using a foot pump that air escapes at the same rate as it is pumped in. This problem may also be encountered when using an air hose on new tyres which have been compressed in storage and by virtue of their profile hold the beads away from the rim edges. To overcome this difficulty, a tourniquet may be placed around the circumference of the tyre, over the central area of the tread. The compression of the tread in this area will cause the beads to be pushed outwards in the desired direction. The type of tourniquet most widely used consists of a length of hose closed at both ends with a suitable clamp fitted to enable both ends to be connected. An ordinary tyre valve is fitted at one end of the tube so that after the hose has been secured around the tyre it may be inflated, giving a constricting effect. Another possible method of seating beads to obtain initial inflation is to press the tyre into the angle between a wall and the floor. With the airline attached to the valve additional pressure is then applied to the tyre by the hand and shin, as shown in the accompanying illustration. The application of pressure at four points around the tyre's circumference whilst simulaneously applying the airhose will often effect an initial seal between the tyre beads and wheel rim, thus allowing inflation to occur.

Having successfully accomplished inflation, increase the pressure to 40 psi and check that the tyre is evenly disposed on the wheel rim. This may be judged by checking that the thin positioning line found on each tyre wall is equidistant from the rim around the total circumference of the tyre. If this is not the case, deflate the tyre, apply additional lubrication and reinflate. Minor adjustments to the tyre position may be made by bouncing the wheel on the ground.

Always run the tyre at the recommended pressures and never under or over-inflate. The correct pressures for various weights and configurations are given in the owner's handbook or Owners Workshop Manual.

Tyre valves rarely give trouble, but it is always advisable to check whether the valve itself is leaking before removing the tyre. Do not forget to fit the dust cap, which forms an effective second seal.

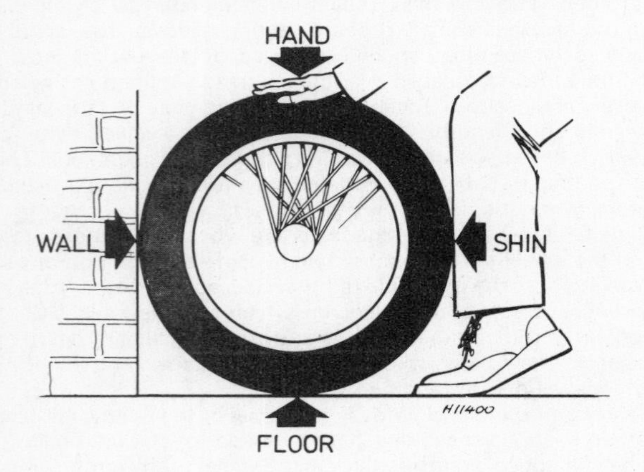

Fig. 5.8 Method of seating the beads on tubeless tyres

14 Tubeless tyres: puncture repair

The primary advantage of the tubeless tyre is its ability to accept penetration by sharp objects such as nails etc without loss of air. Even if loss of air is experienced, because there is no inner tube to rupture, in normal conditions a sudden blow-out is avoided.

If a puncture of the tyre occurs, the tyres should be removed for inspection for damage before any attempt is made at remedial action. The temporary repair of a punctured tyre by inserting a plug from the outside should not be attempted. Although this type of temporary repair is used widely on cars, it is strongly recommended that no such repair is carried out on a motorcycle tyre. Not only does the tyre have a thinner carcass, which does not give sufficient support to the plug, the consequences of a sudden deflation are often sufficiently serious that the risk of such an occurrence should be avoided at all costs.

The tyre should be inspected both inside and out for damage to the carcass. Unfortunately the inner lining of the tyre – which takes the place of the inner tube – may easily obscure any damage and some experience is required in making a correct assessment of the tyre condition.

There are two main types of tyre repair which are considered safe for adoption in repairing tubeless motorcycle tyres. The first type of repair consists of inserting a mushroom-headed plug into the hole from the inside of the tyre. The hole is prepared for insertion of the plug by reaming and the application of an adhesive. The second repair is carried out by buffing the inner lining in the damaged area and applying a cold or vulcanised patch. Because both inspection and repair, if they are to be carried out safely, require experience in this type of work, it is recommended that the tyre be placed in the hands of a repairer with the necessary skills, rather than repaired in the home workshop.

In the event of an emergency, the only recommended 'get-you-home' repair is to fit a standard inner tube of the correct size. If this course of action is adopted, care should be taken to ensure that the cause of the puncture has been removed before the inner tube is fitted. It will be found that the valve in the rim is considerably larger than the diameter of the inner tube valve stem. To prevent the ingress of road dirt, and to help support the valve, a spacer should be fitted over the valve.

In the event of the unavailability of tubeless tyres, ordinary tubed tyres can often be fitted. Use tyres of an equivalent type and grade to ensure their suitability. It is recommended that the advice of the tyre manufacturer or a reputable supplier is sought to ensure that a compatible replacement tyre is fitted.

15 Accident damage

Repairing an accident damaged machine at home can be a complicated business, and can also be dangerous unless you are very painstaking in your assessment of the extent of the damage. On the other hand, provided that the damage is not too great, it can be well worth doing. Where it is of a minor or cosmetic nature, like that caused if the machine is dropped while stationary, for example, you may not even need to bother with an insurance claim. In many instances you would have to pay an excess which would make this an expensive option anyway.

Insurance companies will often write off a machine with seemingly superficial damage, simply because the cost of the necessary replacement parts is so high. Another factor taken into account is the cost of labour involved when a damaged frame has to be renewed. In both areas the private owner is at an advantage.

Firstly, you may be prepared to accept some of the cosmetic damage which would normally have to be put right under an insurance claim, and you may be able to locate used parts from a local breaker rather than buying new parts at full price. Secondly, you have the advantage of not having to pay labour charges, which makes a labour-intensive job like frame replacement viable.

If you are unfortunate enough to be involved in an accident, try to keep in close touch with the bike during any subsequent insurance claim assessment. In all probability, the bike will be taken to the local dealer who will prepare an estimate for the repairs. Often a claims adjuster will visit the dealer to verify the extent of damage and to see if the estimate is realistic. If the insurance company decides that the machine is not worth repairing and writes it off, you should consider negotiating to purchase the damaged machine back from them and organising the repair work yourself. Before you do this be absolutely certain of the extent of any damage, and talk to the dealer about this.

Where you need to buy secondhand parts, check that they will fit your machine before you buy them. Remember that detail changes are often made at random throughout production runs. If possible take patterns with you so you can make comparisons.

16 Front-end damage

This is the most common area of accident damage where another vehicle is involved, and the usual result is bent forks, a damaged front mudguard, wheel and tyre and probably damage to the headlamp, turn signals and instrument panel. This will not be cheap to repair, and since most of the machines in the local breaker will have been scrapped after similar accidents, the chances of finding a good set of forks and a front wheel are slim.

If you can find these parts secondhand, all well and good, but be absolutely certain that they are undamaged. If the only option is new parts, check the prices very carefully before undertaking the repair. Another significant risk is that the frame will have been damaged. This will usually be evident around the steering head region, but may not be easily visible. **Do not take chances with frame damage;** have the frame checked on a jig by a suitably equipped dealer. Light damage may be repairable, again requiring jig work. Heavier damage means a new frame.

17 Handlebars, controls and switchgear damage

Handlebars are easily damaged in an accident, even where the machine is simply dropped at a standstill or at low speed. Normal chrome-plated tubular steel bars are not worth straightening, either from the point of view of cost or the risk of subsequent failure of a damaged set. Buy some new bars, and transfer the switches and controls to them.

If you are considering fitting non-original bars, make sure that they do not foul on the tank, or fairing (where fitted). It is also important that the existing cables, wiring and hydraulic hoses will reach; if these need to be renewed to suit the new handlebar, it could become an expensive and complicated prospect.

On machines fitted with separate steel or cast alloy handlebars, the situation is different; it is unlikely that you will have much choice in replacement assemblies, and so you will have to settle for the original type on most models.

If you can find them, complete control lever assemblies together with the associated switches, are a good prospect for secondhand purchase. If the only damage is to the lever blade, however, it makes sense to buy this new. Switch clusters in particular are expensive new, and a safe secondhand purchase.

Where the lever assembly includes a brake or clutch master cylinder, by all means look for a secondhand replacement, but be wary of acquiring a worn-out unit, or one which looks similar, but has a different master cylinder bore or stroke dimension.

18 Footrests and pedals damage

Depending on the arrangement used, it may be possible to straighten footrests, and also brake and gearchange pedals. If the footrests are of the folding type commonly used as pillion footrests, and sometimes for the rider's footrests, it is unlikely that repair would be cost effective, and new ones should be fitted. On machines with fixed footrest bars, as in the case of many mopeds and commuter models, these can be straightened unless very badly deformed.

The assembly should be removed from the machine for straightening. Heat the footrest using a blowlamp, or by pushing it into hot coals in a fire or boiler, until it glows a cherry red colour. This will soften the metal, allowing it to be straightened without risk of fracturing. Taking care to avoid burns, remove the footrest from the heat source and hammer it back into shape, using an anvil or a similar large piece of steel and a hammer.

Once the footrest has been allowed to cool, rub it down with progressively finer grades of abrasive paper until the surface is bright and smooth. Prime the bare metal as soon as possible to prevent corrosion, and then refinish to match the original paint coat. On items like footrests it is often easier and more economical to use a brushing enamel in preference to aerosol paints.

A similar technique can be used to straighten control pedals, but note that where these are chromium plated, the surface finish will be destroyed. You will either have to accept a painted finish on the repaired lever, or buy a new replacement. Note also that it is almost impossible to straighten aluminium alloy components.

19 Fuel tank, seat and side panel damage

These are all worth buying secondhand, the main problem being that they will be amongst the first things to be sold when a bike gets to the breakers. Tanks and side panels may need to be refinished to match the paint scheme on your machine, so budget for this, plus the cost of any decals. Despite this, a secondhand replacement is usually a more economical solution than repairing a dented tank. Removing dents from fuel tanks is a difficult process, often requiring the bottom of the tank to be cut out to allow access. Shallow dents and similar damage can be dealt with by using body filler, but if the damaged area was chromium plated, professional attention is essential; it is not possible to plate body filler. Remember that you must also budget for the cost of stripping and re-plating the tank.

Seats can be refurbished by fitting a new cover if required. You can either buy a cover and fit it yourself, or send the seat for recovering to any one of a number of specialists in this work. Alternatively, a local coach trimmer or upholsterer may be able to do the work for you.

20 Exhaust system damage

Secondhand exhausts are fine if they are in nearly new condition, otherwise it is better to buy new. One problem is that once a system has been used on a machine, it may prove reluctant to line up with the mounting points on another. Where the silencer is separate from the exhaust pipes, a secondhand replacement silencer could be considered, otherwise it is generally preferable to fit a new system, particularly on multi-cylinder models.

21 Frame and forks damage

Again, these are good value secondhand, but only if you can be sure that they are not damaged. It is worth having the frame alignment checked professionally before you use it. On no account buy a frame which has had its serial number defaced or altered – it is probably stolen and will create difficulties with re-registration even if the question of theft has been resolved.

Fork assemblies are also worth looking for, but the problem of accident damage applies here too. Only if you can be certain that no front-end damage has been suffered should you consider secondhand forks. Check that the forks move smoothly and evenly, and reject them if there are any signs of damage or if tight spots can be felt.

Frame and fork repairs are beyond the scope of most home workshops, and really require specialist alignment jigs to check properly. Slight damage can be repaired, but this will probably mean professional help. In some cases it may prove impossible to correct the damage economically, and a reputable dealer will be able to assess with some accuracy whether repair or replacement is the best choice.

If a fairing is fitted, it will almost certainly sustain damage in the event of an accident. According to the extent and location of the damage it may be possible to carry out a repair rather than resort to replacement, see Chapter 3.

22 Engine/transmission unit damage

These are well worth buying if your own engine is extensively damaged. You will be taking a chance on its condition, and you may have trouble obtaining a unit for a popular model. As with frames, defaced engine numbers almost always mean a stolen bike. Despite these reservations, this is very often an economical way to get the machine on the road again.

Where accident damage has been sustained, you may be able to replace the damaged parts, usually the crankcase outer covers, with new or used items. In such instances, remember to check first that more serious but less obvious damage has not occurred; there would be little point in buying a new outer cover if it later transpires that the crankshaft is bent or the crankcases cracked.

The cast alloy parts of engine/transmission units are easily damaged, and a common problem is a footrest being pushed through the outer cover if the machine is dropped. In some cases, the resulting cracks can be repaired by aluminium welding, or by home repairs using *Lumiweld*. These techniques are discussed in Chapter 3. The success or failure of such repairs is dependent on the extent of the damage, the type of alloy, and the skill of the repairer, but good results can often be obtained on small cracks or holes. If it is not possible to achieve a good cosmetic finish to the repair, consider disguising the area by spraying the cover with heat-resistant matt black paint, after filing and sanding the weld flat.

Welding the more important castings, such as the crankcases, is a job best left to an expert. In many instances, the problem of bearing surface alignment and the risk of distortion may render the proposed repair impractical, and if this is the case it will be safer to buy a new pair of crankcases. Similar problems may be encountered when attempting to weld cylinder heads; it may be possible to repair a crack, but the resulting distortion may make the head unusable. For this reason, always get professional advice.

23 Electrical and ancillary component damage

These are almost always worth buying secondhand, there being little risk to your safety if the replacement later proves to be defective. Lamp units, instrument panels and the various electrical parts like turn signal relays, CDI units and other ignition parts are expensive to buy new, and likely to be usable if taken from a scrapped machine. If possible, ride your own machine to the breakers, and fit the part there and then to check that it works.

Repairs to a cracked turn signal or tail lamp lens can be made by sticking the broken pieces back together, but extensive damage will require renewal of the lens.

24 Buying complete machines

Buying a complete accident-damaged, vandalised or stolen-recovered machine from a breaker can be a good basis for an accident rebuild, or even for building up a bike from scratch. Stolen/recovered machines or those which have been vandalised are obviously preferable to accident damaged ones. Most of these machines reach the breaker having been written off by the insurance companies as uneconomic to repair for various reasons. Most breakers will take considerable trouble to avoid stolen machines. Even so, always check all documentation, and ask the police to make sure that the machine is not reported stolen before you part with money. To this end, remember to note down engine and frame numbers when checking the machine.

Buying parts this way is usually more cost-effective than buying piecemeal as the need arises, provided that you have the space and time to dismantle the bike yourself. You may even be able to offset some of the cost by selling off surplus parts.

Chapter 6 Electrical and ignition related tasks

Contents

1 Introduction and general requirements for electrical work

This Chapter relates to the electrical components used on most motorcycles, including the ignition system, describing the more generally-used repair and test methods. For obvious reasons, it is not model specific, and like the rest of this book is designed to complement a workshop manual, not replace it.

The electrical system seems to cause more owners problems than any other aspect of their machine, and this is probably mostly a result of the generally intangible nature of electricity; you simply cannot see how it works. For much of the time you can deal with the normal type of electrical faults without any great understanding of electrical theory, and this Chapter was written with this approach in mind. For those wishing to investigate the theoretical aspects in greater depth it is suggested that you invest in a book on this aspect. A companion volume to this book, the *Haynes Motorcycle Electrical Manual*, would be a good starting point.

Electrical work is largely a self-contained area within the general subject of workshop practice, and as such requires a few specialised tools and safety considerations. Before we look at examples of actual workshop jobs, the notes below should be considered.

Safety considerations

There is generally more risk of damage to the electrical system than to you, should you make a mistake while working on the machine. Nevertheless there are a few simple precautions which can virtually eliminate this possibility:

a) When carrying out any electrical work, unless the system actually needs to be live, disconnect and remove the battery. The earth (ground) lead should always be disconnected first, and re-connected last. On most machines a negative (–) earth system is employed, so as a rule of thumb, disconnect the negative lead first. No matter how careful you are, it is easy to accidentally short circuit the system when connecting or disconnecting components and leads. Such shorts will not normally harm you, though the voltages produced in the ignition system can be dangerous. The biggest risk is instant and irreparable damage to an expensive electronic unit.

b) Always handle the battery with respect. It contains sulphuric acid, which apart from being useful as an electrolyte is corrosive. It will ruin clothing and fabrics, causes burns to the skin, and is very dangerous indeed if it gets into the eyes. If an accident does occur, wash the area with copious amounts of water immediately. The acid can be neutralised by an alkaline solution, and sodium bicarbonate added to water will help prevent damage to clothing if you act quickly enough.

In cases of splashes on the skin, again use plenty of water to wash away and dilute the acid. If the skin is damaged, seek medical advice. To eliminate the risk of eye injury, always wear goggles when handling batteries. If acid enters the eyes, irrigate them immediately with plenty of clean water, then arrange prompt medical attention.

Beware of fire risk when dealing with batteries. The gases produced in the battery during charging are potentially explosive, so on no account smoke near a battery under charge, or carry out charging in the vicinity of a heater or other source of flame. To reduce the risk of a potentially explosive build-up of gases, carry out the charging operation in a well-ventilated area, or leave the workshop door and windows open to allow ventilation. Also ensure that the battery vent is clear and that cell caps are loosened. Each year many people are seriously injured by batteries which explode because these precautions are not observed.

Special tools and equipment

The basic tools outlined earlier in Chapter 2 will be required for electrical work, as well as mechanical work. In addition, there are a few extra pieces of equipment which you will need. To summarise, the following items would make up a good basic electrical toolkit:

A multimeter should be considered an essential purchase, simply because of its general usefulness when making checks of the system. You can get away with a simple battery-and-hub test circuit for continuity checks, but a meter will also allow detailed checks of the components to be made where the manufacturer has specified precise limits or tolerances for some components. It should be noted that in the case of some electronic assemblies, the polarity of the test meter is of significance. In such cases, the manufacturer will recommend a particular make and model of meter, and this is mentioned in the Haynes Owners Workshop manual for that machine. On many current models, very little electrical fault diagnosis can be done without a meter of some sort.

A soldering iron will allow you to repair connections and broken electrical leads properly, as well as being useful for other jobs around the machine. For electrical work only, a 25 watt iron will suffice, but if you intend to use it for soldering heavier items like control cables, get a higher wattage model. You will also need a reel of multi-cored solder.

A pair of side cutters will be needed for cutting electrical cable to length, and will also be handy for other cutting jobs around the machine.

Wire strippers are useful for removing the insulation from electrical cables. You can get by with a sharp knife, but an inexpensive wire stripper is a better choice. Some of these tools also allow crimp-on terminals to be fitted, and these can be bought as kits together with a selection of terminals.

A selection of terminals will be needed to replace any which are broken or corroded. These can be bought in kits (see above) or as required from vehicle electrical specialists.

An assortment of colour coded insulated wire will be needed for repairs and alterations to the system. It is obviously preferable that the new wires match the colour coding of the original. You can either buy this new, in reels or cut to length, or as a cheap alternative, try a local motorcycle breaker for the wiring harness from a scrapped machine.

Whether you obtain new or used wire, it should be noted that its current-carrying capacity is determined by the effective diameter of the conductor. The conductor area of any replacement wire must be at least equal to that of the original. If it is smaller, the circuit concerned may not receive sufficient power, and the risk of an electrical fire is increased due to the conductor becoming hot in use.

PVC insulating tape is useful for repairing breaks in the wiring insulation and covering repairs. You can also use it to tape extra wires neatly to the existing harness, and in emergencies only, to tape up bare wire ends or wires which have been twisted together to make a temporary connection. Keep a roll in the toolkit carried on the machine.

Nylon cable ties are invaluable for tidying up the harness and strapping wires (or control cables) to the frame. You will find them a useful replacement for the original metal ties when these break off. Again, it is worth carrying a few of these on the machine, taped to the frame below the seat, for emergency repairs of all sorts.

Consumable materials include such things as aerosol contact cleaner, WD40, abrasive paper and fine files, small nuts and bolts, fibre or nylon washers and silicone grease.

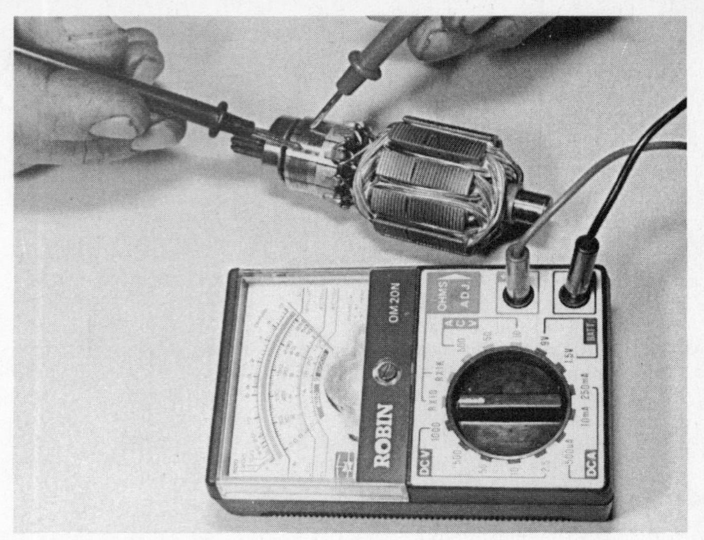

1.1 A simple inexpensive multimeter will allow basic resistance checks to be made, in addition to voltage tests. Note that the current (DCA) range is limited to milliamps

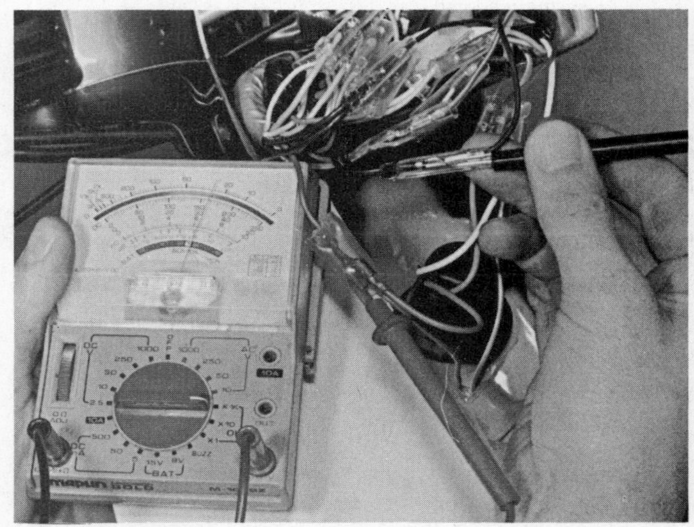

1.2 This is a slightly more expensive meter, giving improved accuracy and a more easily-read scale. The current range allows measurement up to 10A, and is useful for checking low-current circuits

2 Circuit testing equipment and methods

In the event of most electrical faults, the first problem to be faced is tracking down the cause. In many instances the fault will be due to a simple mechanical failure such as a broken wire, or a loose or corroded connection. To pinpoint the source of the trouble you will need the wiring diagram for the machine, and some sort of test equipment. In its simplest form, a continuity tester can be made up using a dry battery, a torch bulb and a couple of pieces of insulated wire. An example of this arrangement is shown in the accompanying illustration, and you can add refinements like home-made test probes or an on/off switch as required.

A better alternative which will give much greater scope during testing is to use a multimeter. Set on the resistance range, this can be used in exactly the same way as a simple continuity tester, and some meters provide a special continuity test setting where an internal buzzer sounds; invaluable where you cannot easily look at the meter while making the test.

The resistance scale will often show up problems which might get missed with a simple continuity test. For example, an erratic lighting system fault such as an abnormally dim headlight could be caused by a corroded connection causing high resistance. This might dim a test lamp imperceptibly, but you should be able to get a definitive reading from a multimeter. Set on the dc volts scale, and with the battery connected, you can check whether battery voltage is present on a particular terminal or connector.

Basic testing procedures

For the purposes of continuity checks or resistance measurements, disconnect and remove the machine's battery to avoid the risk of battery voltage damaging the tester or meter. Using the wiring diagram supplied in the machine's handbook or the appropriate Owners Workshop manual, locate the suspect wiring run, using the wiring colour code to guide you. Check the continuity between each end of a wiring run, moving through the circuit until the break is located.

Where you need to check for voltage on a circuit, leave the battery connected, and set the meter to the appropriate volts range (this will normally be 0 – 10 volts dc in the case of 6 volt machines, and 0 – 20 volts on 12 volt systems). Without disconnecting any of the wiring, connect the correct meter probe to a sound earth (ground) point on the frame. On negative (–) earth machines this means the black meter probe, and on positive (+) earth models the red probe. Connect the remaining probe to the connector terminal to be tested, pushing it into the back of the connector. Switch on the ignition, and where necessary, the circuit under test, and note whether battery voltage is shown. If you do not get a reading, make similar checks back along the wiring harness until the break is located.

On mopeds, and similar machines where direct lighting is used, the power to the headlamp is fed direct from the flywheel generator, rather than from the battery. It follows that you will need to run the engine to check for power on the affected circuits. Note also that direct lighting current is unrectified, or ac. This means that you must set the meter to the **ac volts** range for this type of check. This is also the case on larger machines where voltage checks are made on the alternator output; this too is ac before it passes through the rectifier.

When you need to make detailed checks of specified components, particularly any electronic units, you will need the details of the test procedure and the table of test result figures. These will be found in the Haynes Owners Workshop Manual for the particular model. Note that in some cases, accurate tests can only be made using the test meter specified by the manufacturer; using another make or type of meter may give misleading results. This will usually be made clear in the workshop manual.

In the case of ignition system components, especially those used in CDI systems, very high voltages are present. These can be unpleasant or even dangerous if care is not taken to avoid shocks, and no testing is normally done with these systems under power. (An exception to this rule is the use of specially designed CDI test equipment which is a little outside our scope.) Always read through test procedure carefully before starting work. You will normally have to remove the unit from the machine for testing, and the battery must be disconnected before you do this. In some cases the unit terminal must be shorted out before testing starts, this is to discharge any residual charge in the capacitors in the unit. The manual will advise where this is necessary; failure to do so may damage the unit or the test meter.

What to look for

Although it is impossible to list all the areas where faults will occur, there are certain areas which are more likely than others to be causing problems, and it is always best to check these first. You should also refer to any fault diagnosis table or flow chart applicable to the specific model.

Loose or corroded connections are the single most common cause of electrical faults, and may be responsible for a number of symptoms. Common problems are intermittent or erratic operation of lighting circuits, and turn signal systems are especially prone to this sort of fault. The cure is to work through each connection in the circuit, separating all connectors and checking for looseness, corrosion or the presence of water. Clean the connection carefully, or if badly corroded, renew it and pack the connector with silicone grease to prevent water getting into it in the future.

This applies equally to earth (ground) connections, which can disrupt the circuit just as easily. Areas especially prone to earthing faults include turn signal and stop/tail lamps, where it often causes erratic operation and can result in frequently blown bulbs.

Bulb failure may be due to old age, vibration, loose or damaged wiring (see above) or fluctuations in the supply. If the bulb envelope is clean and a section of the filament is damaged, vibration is the most likely cause. Check the mounting of the lamp and where necessary remount it using rubber washers to damp the vibration.

Blown bulbs can be due to current surge or other supply problems, in which case the bulb envelope will normally have become blackened, or will have a silvery appearance. If the fault keeps recurring, check the charging system voltage – you could have a faulty voltage regulator or rectifier. On direct lighting machines, make sure that the bulbs are all of the correct wattage. If one bulb is of the wrong rating, excess current may blow it or other bulbs. On simple machines with direct lighting systems, a resistor unit is often used to shunt excess current to earth. If this fails, or if it becomes partially or fully disconnected from the circuit, repeated bulb failure is likely.

Overcharging or undercharging of the battery may be due to a failure of the voltage regulator or rectifier unit, and this should be removed and tested using the test details in the Owners Workshop Manual. Other factors can include a defective battery, where a dead cell may provoke this sort of problem. In this case, either get the battery tested, or try a new one to see if this resolves the fault.

Turn signal system faults are usually caused by a blown or out-of-circuit bulb. If this happens, the remaining lamp on that side will flash rapidly and dimly. Check and renew or refit the bulb as required. If this does not solve the problem, check the lamp wiring next. If the turn signal relay fails, the fault will apply to both sides of the machine. If one side of the circuit is dead, check the wiring to the turn signal switch, the switch terminals, and the wiring from the switch to the lamps.

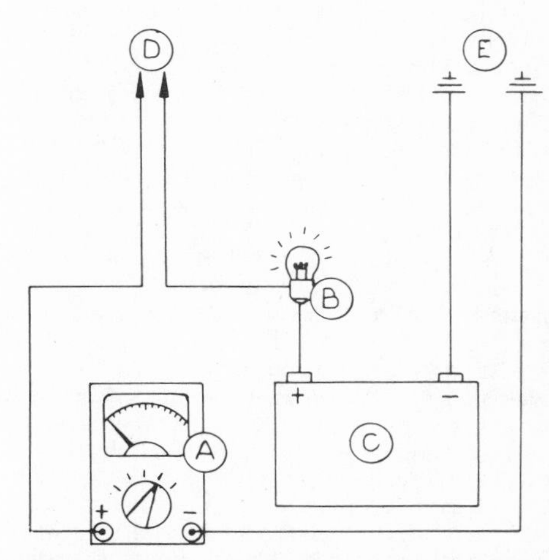

Fig. 6.1 A battery-and-bulb test circuit or a multimeter can be used for simple continuity checks

In the example shown, (D) and (E) are the positive (+) and negative (–) probes respectively. (A) is a multimeter set on the resistance range, (B) and (C) denote the torch bulb and dry battery

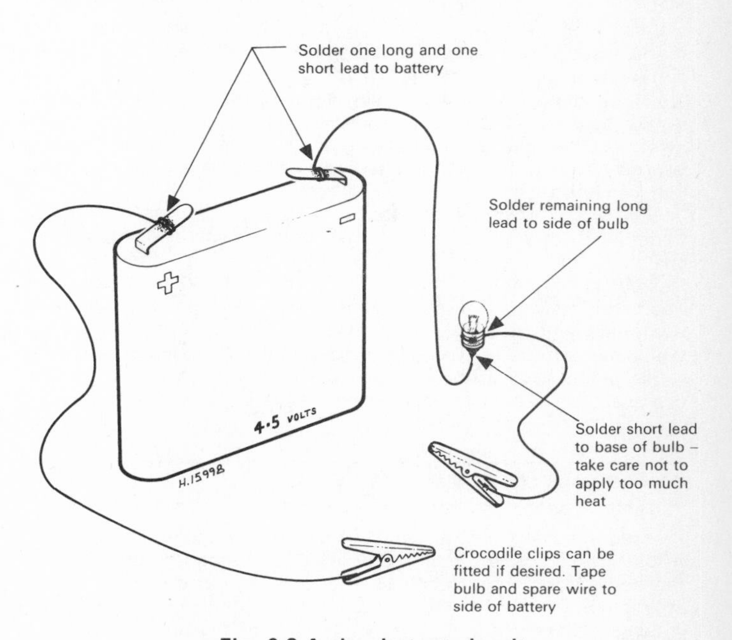

Fig. 6.2 A simple test circuit

A basic continuity tester can be made up as shown using a dry battery, a torch bulb, some lengths of wire and two crocodile clips

3 Wiring harness: repair and renewal

At some stage you may well need to make repairs to parts of the wiring harness to replace damaged wires or connectors. This sort of job requires careful soldering of the new terminals, after cutting off the old ones with side cutters. The spade terminals used in multi-pin connectors are held in place by a small tang, and the terminal can usually be freed after this has been displaced with a small screwdriver blade. Remember to fit the wire through the connector housing before crimping or soldering the new terminal to it.

If a section of wire has become damaged, cut through the insulation at the damaged point and strip back a few millimeters of insulation on each end. Tin the exposed wire, then overlap the ends and solder them together. For details of soldering techniques, refer to section 6 of Chapter 3. Wrap the repair after it has cooled off, using PVC tape.

In some instances, especially on older machines, it may be necessary to renew a whole section of wiring. If you can, try to get a spare wiring harness from a similar model at a breaker, and just swap sections of wiring as required. If you have to make up your own section of harness from scratch, try to match wiring colours as far as possible, and be sure that the conductor gauge of the new wire is at least equal to that of the old one. Group wires together and run them through PVC sleeving for protection, remembering to fit this before any connections are soldered into place.

If the damage is extensive or general to the whole system, it may be worth fitting a new harness. This is a long involved process, but not beyond the capabilities of most owners if approached systematically. Again, get a good used harness if you can; it will be cheaper and easier than buying it new or making up your own.

Points to watch out for are extra wires, which will be found on most models. These are for circuits applicable to other markets, and are present simply because it is easier for the manufacturer to make a single universal harness than many variations of it. Occasionally, these 'spare' wires can be used to replace a damaged one in the harness, but check its connections and diameter carefully first. Another interesting problem is colour changes in wires; this happens sometimes, a wire being joined to one of another colour inside the harness sleeving. As might be expected, this can be very interesting, if not downright confusing.

One really interesting point on some machines, usually the small-capacity Japanese models, is the use of small diodes *inside* the harness. These are used to stop current backfeeding between circuits, and can be very difficult to deal with. They are sometimes not mentioned in the wiring diagram, which confuses matters still further. If you suspect the presence of a diode within a wiring run, check for continuity in **both** directions. The diode should pass current in one direction only, acting as a one-way valve. You will have real problems if the diode has failed, for obvious reasons. If the diode is external, it is an easy matter to plug in a new unit, but in the case of 'hidden' diodes, you may have to bypass the affected area of wiring with an external link, carrying the new diode.

Finally, a word about Italian electrical systems. These are traditionally of dubious design, use poor quality connections and switches, and it is uncommon to find that the wiring diagram supplied resembles that of the machine in question. They are a source of unending frustration to owners, and wiring work on such systems is not for the faint-hearted. In many cases it is less confusing to throw the whole harness away and set about designing and making up your own. If this, or any other wiring job, seems too daunting, remember that you can always take the bike to a vehicle electrical specialist....

4 Additions to the electrical system

Before adding any electrical accessory to the system on your machine, you should be sure that it is capable of handling the extra demands. Most larger machines with alternators and batteries will be able to cope with small extra loads with little difficulty, but beware of making additions to mopeds or small trail bikes, where the system is balanced against a specific load. In such cases, the system is load-dependent to the extent that even a change of bulb wattage may cause problems. There is no spare capacity in the system, and so no additions should be made.

On slightly larger machines where a small battery is fitted to provide power for turn signals and a brake light, note that this is again limited in capacity. Adding accessories is unlikely to be successful, and may only result in a flat battery.

Additions to the electrical system on larger machines should not exceed the overall output of the alternator. Where the alternator wattage is given you can calculate spare capacity by adding up the existing load and subtracting this from the rated output. When totalling the load on the system, you need only consider those items which impose a constant load, namely the headlamp, tail lamp, instrument lamps and the ignition system. The ignition system will pose something of a problem, since the power requirement for this is rarely quoted, but as a rough guide, allow about 20W to cover the ignition system plus intermittent loads like the brake lamp, turn signals and horn.

Most electrical accessories will be supplied with a fitting kit, including the necessary wiring, connectors and switch. The feed for the accessory circuit should be taken from a fused supply point, having taken the precaution of disconnecting the battery before work commences. Some machines are provided with a fused accessory supply, and this will be indicated on the wiring diagram. On other models, take the supply from a suitable point in the wiring system, making sure that where necessary it is controlled by the ignition switch (note that a few accessories, such as some anti-theft alarms, may need to be connected to the battery even when the ignition is switched off). You may find that there is a spare fuse connection which can be used, otherwise fit an in-line fuse holder, complete with a fuse of the correct rating.

Determining the fuse rating for an accessory circuit can be accomplished by dividing the wattage of the accessory by the voltage of the electrical system. In the case of an auxiliary lamp rated at 60W to be used in a 12 volt system, this would require a 5A fuse (60 divided by 12). Note that a lamp of similar wattage used in a 6 volt system will require a 10A fuse, and that correspondingly heavier wiring should be used. Wiring sizes and their corresponding load-carrying capacity will be found in the reference section at the end of this manual.

Fitting a switch can prove something of a problem on many machines. You may be able to fit an add-on handlebar switch such as the universal dip/horn switches sold by most dealers. A neater method would be to find a handlebar switch cluster from another machine, and having extra switch positions. This is a rather complicated approach, and it will necessitate searching the breaker's stock of switchgear, but can produce a neat and convenient solution. Where ready access is less essential, consider making up a switch panel from steel or aluminium sheet and attaching it to a convenient location point below the tank, to the top yoke, or to the fairing (where fitted).

When connecting up the new circuit, make sure that you use the correct gauge of wire, and that insulated connectors are used throughout. Waterproof the connectors with silicone grease during assembly. If accessory switches are to be used, remember that those intended for use in cars will not be waterproof; take care to site them well away from likely sources of water contamination, or take steps to seal them from the elements.

5 Batteries

Note: *Before attempting any work on the battery, refer to the Safety considerations discussed in Section 1*

On all but the smallest machines a lead-acid battery is used to store electricity to be fed into the system as the need arises. As has already been mentioned, the sulphuric acid used as the electrolyte is highly corrosive, and the appropriate precautions must be taken whenever the battery is to be worked on. It is well worth noting that many charging system faults can be attributed to a defective battery – always check and recharge the battery before looking elsewhere.

In most cases, motorcycle batteries can be expected to last for about three years. If you maintain the battery carefully, this life can be extended somewhat, and equally, neglect will tend to shorten its working life. Maintenance consists of keeping the outside of the battery clean and dry, and the terminals clean and corrosion-free. Remember that dirt on and around the battery will have a certain amount of acid in it, so take care. The terminals should be taken apart for cleaning from time to time, and coated in petroleum jelly to inhibit further corrosion.

The level of the electrolyte must be checked regularly, and if necessary topped up, using distilled or demineralised water. You should not need to add acid to the battery once it has been filled initially; the only fluid loss is water which will evaporate. In cases where frequent topping up is required, this is a good indication that the charging system output is too high. In most cases this will be due to a faulty voltage regulator/rectifier unit. This should be checked and rectified promptly, or the battery and even the alternator may be damaged.

As the battery ages, and under the effects of vibration, there will be a gradual accumulation of lead sulphate around the plates in the battery cells. This will tend to accumulate around the bottom of each cell, and where the battery has a clear or translucent case you may be able to see these deposits. As the sediment builds up, it will eventually cause the cell(s) affected to short out internally, and this indicates the imminent failure of the battery. As time goes on, the battery will be less able to hold a full charge, and will have to be renewed. On models with electric start, this point will come earlier, because an old battery which can cope with normal loads may have insufficient energy to operate the starter motor.

It is not easy to check the battery's ability to work under load at home; this requires a special tester. You can ask an electrical specialist to check the battery for you, however. About the best that can be done at home is to check the charge in each cell, using a hydrometer. This device has a calibrated float which measures the specific gravity of the electrolyte. If one cell remains consistently weak, it is a good indication that it has become sulphated. If you buy a hydrometer so that you can check your battery, bear in mind that the cells in motorcycle batteries are small; get a small sized hydrometer, or there may not be enough electrolyte to show a reading on the float.

Some machines are fitted with sealed batteries. These incorporate a special leak-proof venting system which will release gas pressure during charging, but will keep the acid safely inside. These batteries are considered to be maintenance-free, there being no requirement for topping up the electrolyte. At the same time, it is almost impossible to check the condition of this type of battery. If normal re-charging fails to resolve a battery problem, you will have to buy a new one.

Another feature you may come across is a small battery sensor lead attached to a small terminal on the top of the casing. This is a low-voltage tap into one of the cells, and is connected to the battery electrolyte level gauge or warning lamp in the instrument panel. If the electrolyte level drops below the end of the sensor, the rider is alerted to the need for topping up. Where the sensor is removable, it is a good idea to take it out for cleaning occasionally, otherwise no maintenance is required.

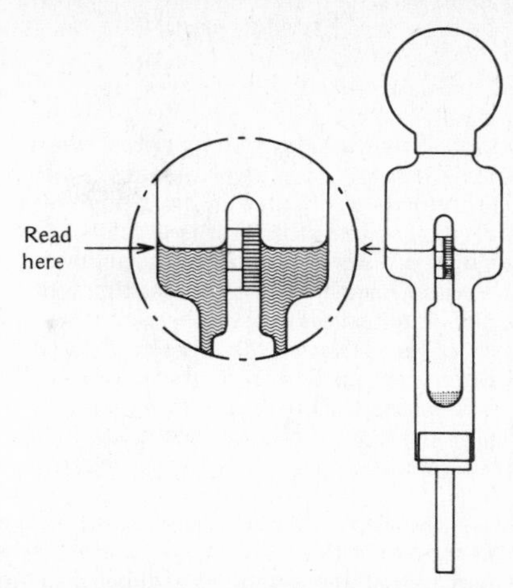

Fig. 6.3 Battery specific gravity should be checked with a hydrometer

5.1 Some motorcycle batteries contain a sensor to detect low electrolyte levels, this being connected to a warning lamp in the instrument panel

Battery charging

The battery on most machines will require occasional recharging off the machine, especially when a lot of short night-time trips are done, and the charging system cannot keep up with demand. If you need to do this, it must be remembered that the battery will be of limited capacity compared with those used on cars.

The rate at which the battery can safely be charged is dependent on its capacity, which is expressed in units of ampere-hours, or Ah, this figure being given in the specifications for the machine, or occasionally, printed on the battery case. As an example, a small two-stroke model with direct lighting will usually have a battery rated at about 6 volts, 4Ah. On larger models the battery capacity will be around 12 volts, 14Ah.

The maximum safe charging rate is 1/10 of the rated capacity, which means in the case of the two examples given, a maximum of 0.4 amps and 1.4 amps respectively. This poses something of a problem when using normal DIY battery chargers, because these are designed for use on the much heavier car batteries. These chargers are current controlled, which means in practice that they will charge at whatever rate the battery will accept, normally around 2 – 4 amps. This is much too high for motorcycle batteries and can seriously damage them. In the case of the smaller of the two given examples, the current from a car battery charger is probably enough to destory the battery, possibly causing it to explode.

The answer is to use a special charger designed for use on motorcycle batteries. These have a much lower output, and are often constant current types, producing about $^1/_2$ amp output. Whilst it would take some time to recharge a 14Ah battery at this low rate, it is safe to use on all motorcycle batteries. These chargers can be bought through motorcycle dealers, and you should either check that you get the correct output voltage for your machine (6 or 12 volt) or get a dual voltage version.

When using any type of charger, always check that the voltage setting is right for the battery, that the charger leads are connected round the right way, and that the battery vent pipe is clear. On batteries other than maintenance-free types, slacken the cell caps to ensure that there is no risk of excess pressure building up inside the battery.

6 Switches

With very few exceptions, electrical switches on motorcycles are sealed assemblies, and are thus difficult to repair in the event of failure. You should start by deciding what exactly the problem is; if it is a simple case of water contamination, you can try spraying WD40 or a similar silicone-based product into the switch mechanism. This will displace any water inside it, and will often solve the problem without the need for further dismantling. It follows that the better your access to the switch, the more likely you are to succeed, but even spraying the fluid through the extension nozzle supplied will usually reach the internal parts quite well.

Where the switch in question serves only a single function and is relatively inexpensive, it is rarely worth spending too much time on it; it will be quicker and probably more successful to renew items like brake light or stand switches. With more complex switch assemblies, like the handlebar clusters, the cost of a new assembly makes further investigation worthwhile. Start by isolating the battery, then trace back the switch wiring and disconnect it. The connections are usually hidden inside the headlamp shell.

You can check which contacts are faulty at this stage, using a multimeter. The wiring diagram normally includes smaller block diagrams of the switch contacts connected in each position, and you can check each of these in turn until the fault is located.

Further dismantling may or may not be possible, depending on the type and make of the switch used. In some cases, the switch halves serve to locate individual switches, and these are easily unscrewed for separate inspection. In other designs, the switch mechanisms are housed direct in the outer casing, and with these you will have to actually dismantle the mechanism. Only you can decide if it is worth risking this operation; there is a good chance that it will be difficult to get it all back together again, even if you can get it apart in the first place.

If the switch is useless as it stands, there is no real loss in attempting a repair, and the worst that can happen is that the attempt fails. The method of dismantling the switch is entirely dependent on the type concerned, and you will have to figure this out as you progress. It is a good idea to work on a clean surface so that any escaping parts can be recovered easily, and to make a sketch of the switch internals as a guide during assembly. Remember that you will not be able to buy new parts for the switches; any repairs are entirely down to your own ingenuity. If the repair attempt fails, it will have been an interesting undertaking anyway. As a last resort before you order a new assembly, check the local breakers to see if they have a suitable used replacement in stock.

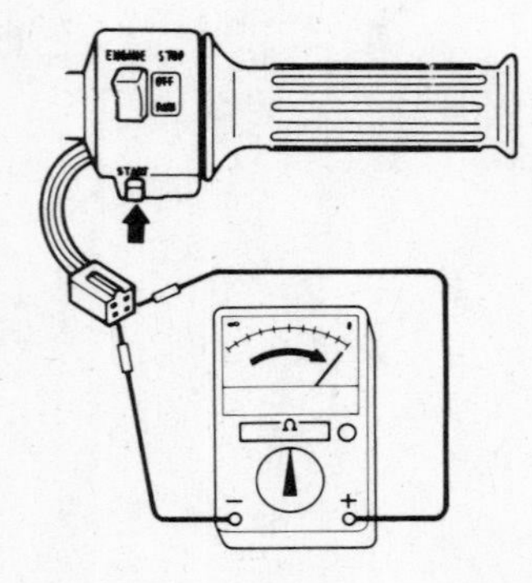

Fig. 6.4 Simple continuity checks can often save unnecessary dismantling of the switch

7 Generators

Most machines use ac (alternating current) generators of some description. These can take the form of simple flywheel generators such as those used on lightweight two-strokes, mopeds and similar models where direct lighting is used. Alternatively, the more sophisticated machines employ single or 3-phase alternators. There is great variety in the types and designs in current use, and if you develop generator problems it is essential to refer to the machine's workshop manual for the specific test procedures and specifications. Few manufacturers still use dc generators (dynamos) and on older models where these may be encountered it is difficult to get parts easily. If you encounter a dynamo-equipped model, the appropriate test details will be found in the Owners Workshop Manual for the machine. There are a few specialists who can offer reconditioning services for these devices, and who advertise regularly in the classic motorcycle press.

Many alternator and flywheel generator problems can be traced back to wiring faults or defects in rectifiers and voltage regulators, where these are fitted. Always check these areas first, and where a battery is fitted, make sure that it is in good condition and fully charged, or any test readings may be erroneous.

Simple resistance tests will reveal faults such as open circuits or shorted windings, and if this is the problem, the stator assembly (and/or rotor on 3-phase arrangements) may need to be renewed. An alternative is to phone a local auto electrical specialist and check

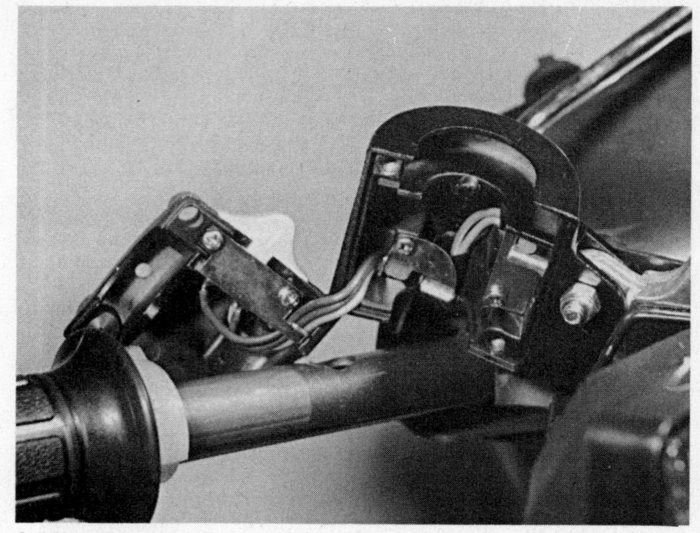

6.1 Handlebar switch assemblies are usually in two halves clamped around the handlebar

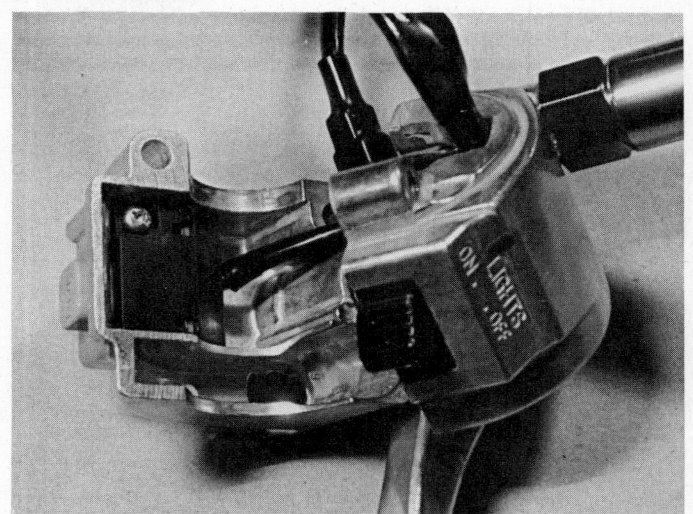

6.2 It is often possible to remove the switch internals, either as sub-assemblies or in component form. If this can be done, it may be possible to gain access to the switch contacts to permit cleaning

6.3 Dismantling to this level is normally safe enough, though if you intend to dismantle the component parts, do so on a clean bench and beware of detent balls and springs which may escape

whether they will be able to rewind the existing part for you. This may take a few days, but could be less expensive than buying a new assembly. Other alternative sources are motorcycle breakers, but take along your old unit to make sure that the proposed replacement will fit. Do not forget that manufacturers buy in electrical parts from various sources, so check that yours is the same type as the replacement unit offered.

On permanent magnet (single phase) alternators and flywheel generators, it is possible for the rotor to become demagnetised, and this will reduce the output of the generator significantly. Again, an auto-electrical specialist may be able to remagnetise the rotor for you, the other option being to buy a new one. Causes of demagnetisation include reversed wiring connections and impact to the rotor, so avoid this happening when working on machines so equipped. This is one reason why a rotor must never be struck to jar it free during removal.

In the case of three phase alternators, always check the brushes and the slip rings (where applicable) if poor output is a problem. If the brushes are worn or the slip rings dirty, the output will drop markedly over a period of time. Fit new brushes, and burnish the slip rings with fine abrasive paper or an ink eraser to give a bright finish. If the slip rings are very dirty, take the rotor to an auto-electrical specialist and have the surface skimmed smooth in a lathe.

Where the output drops suddenly on three phase units, the cause may be an open circuit in one of the windings. This may not be immediately obvious, but is worth checking if the performance of the electrical system has dropped off noticeably, and no other cause can be found. The windings can be tested using a multimeter.

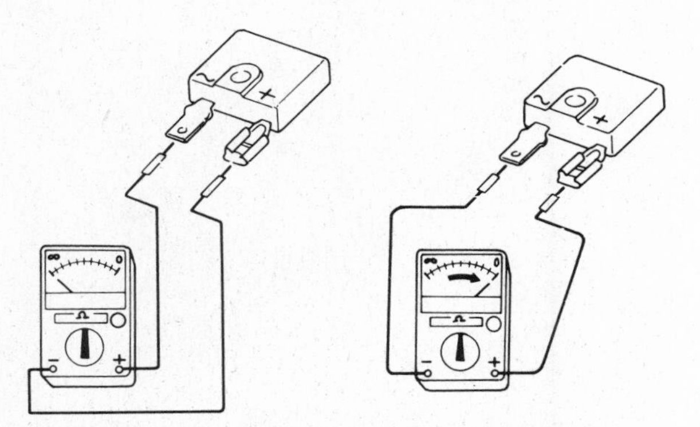

Fig. 6.5 Plug-in diode blocks can be checked with a multimeter

If the probe connections are reversed (left) infinite resistance should be shown. If connected correctly (right) a very low resistance reading should be indicated

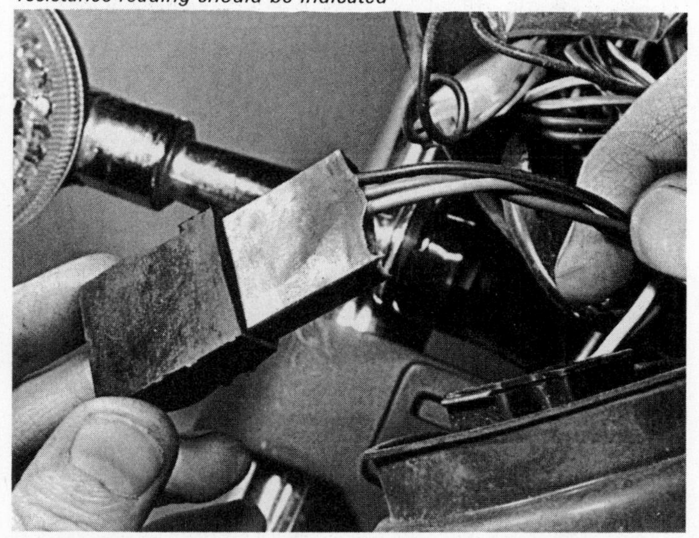

7.1 Many small machines use a simple half-wave rectifier like this to convert the flywheel generator output to dc. The rectifier is normally plugged into a connector, and the location may not be readily apparent

8 Starter motor and circuit

The starter motor circuit comprises the starter motor, a relay, a switch and the battery. On some models this system is interlocked with other switches to prevent the starter motor operating unless the transmission is in neutral, the clutch disengaged or the side stand retracted.

Initial system checks

The starter motor draws a very heavy current while it is turning the engine over, and this is why the motor is switched indirectly through a relay. If the motor was controlled by the handlebar switch contacts they would be burnt out in no time under the heavy current passed through the circuit. This should always be kept in mind if a starter fault develops. Many problems will eventually be traced to a partly discharged battery, or one with a weak cell. During normal riding the battery may be perfectly adequate for the loads imposed upon it, but when the starter operates, the current demands may just prove too much.

With the ignition switched on, check that the kill switch is set to 'RUN' and the gearbox is in neutral. Press the starter button and listen for a response from the relay; this should click into engagement. Repeat this check, this time with the headlamp switched on. Normally, the headlamp will dim as the starter motor turns. If the motor fails to run, but the headlamp goes out completely, it is a good indication of a discharged or faulty battery.

The easiest and quickest way to check this is to fit another battery and see if this cures the fault. This need not be a motorcycle battery; anything of the correct voltage will be fine, such that you could connect a car battery to the terminals of the motorcycle battery, using insulated jump leads. Make sure that you do not cause short circuits when doing this, and be sure to observe polarity. If the engine now starts, you know that the problem lies with the battery, not the starter circuit; remove and charge the battery, and if this fails to improve the situation, fit a new one.

Checking the starter relay

The starter relay is a heavy-duty switch which is controlled remotely from the handlebar switch. When this is pressed, power is fed to the relay, a low current being used to close the main contacts electromagnetically. When closed, these connect the heavy starter motor cable direct to the battery.

A quick check of the relay operation is to disconnect the heavy duty cable between it and the battery, making sure that the cable end is kept clear of any earth point. Connect a multimeter or continuity test circuit to the two large terminals on the relay. (One of these will be the battery terminal which was just disconnected, the other holds the heavy lead to the motor.)

Switch on the ignition and press the starter button. You should hear a click as the relay contacts close, and the meter should indicate zero resistance. When the button is released, the contacts should open again. If there is no response from the relay, make sure that there is battery voltage at the thin lead from the starter button to the solenoid when the button is depressed. If no voltage is shown here, the fault lies in the switch circuit. If you read battery voltage, the relay has failed and must be repaired or renewed. If you can get the unit apart, you can attempt a repair. Check that the solenoid plunger is free to move, and clean up the heavy-duty switch contacts with fine abrasive paper. Where the unit is of sealed construction, or the solenoid windings have failed, renewal will be your only option.

Sometimes the above test may show that the relay is working normally, even though the motor will not run. This can be caused by burnt contacts in the relay; they may handle the test current without difficulty, but be incapable if carrying the heavier motor current. It is difficult to devise a test for this, and if you cannot dismantle the unit for examination, the best option here is to substitute a sound relay and see whether this resolves the fault.

Overhauling the starter motor

If the above checks have indicated a motor fault, you should first make enquiries at a dealer to see whether you can get parts for the motor used on your model. In some cases, a good range of repair parts is available, though in others the motor must be renewed as a unit in the event of a fault.

The Haynes Owners Workshop Manual will give details of the procedure for removing the motor and on dismantling it for examination. The first area to check is the brushes and the copper-coloured commutator on which they bear. Make a note of the brush connections, then disconnect the brush leads and remove them. Worn or sticking brushes can cause the motor to fail or turn abnormally slowly. Check the brush lengths, where these are specified, and renew them if worn down to the service limit. If your motor is not supported by spare parts backup, take the worn brushes to an auto-electrical specialist and get hold of the nearest size and type you can. The new brushes can then be filed to shape to fit the brush holders.

Clean the brush holder unit and also the commutator using electrical switch cleaner, removing all accumulated dust and dirt. The surface of the commutator should be examined closely for scoring and general wear. If lightly scored, use a strip of fine glass paper to smooth the surface down, removing no more of the soft copper material than is essential to restore the surface. The copper segments should be polished smooth using a special fine abrasive paper produced for this purpose. In the UK, this is sold under the brand name of *Crokus;* an auto-electrical specialist will probably be able to advise on and supply alternative types.

Where damage is severe, it is best to have the problem dealt with professionally. The surface of the commutator is cleaned up by taking a fine cut across it while the armature is turning at high speed in a lathe. After any resurfacing work, check the depth of the slots between the commutator segments, and if necessary cut them back to the specified depth (usually around 0.5 – 1.0 mm, or 0.002 – 0.004 in).

A tool for carrying out this work can be made up from an old hacksaw blade. Grind the sides of the teeth until they are exactly the same width as the grooves, then bind the other end with PVC tape to form a handle. The grooves can then be cut back to the correct depth, making sure that the sides are parallel and the bottom of the groove is kept square.

If the above methods fail to get the motor to work, you will have to check the armature and field coil windings for open or short circuits, the procedure for this being given in the workshop manual. Look closely around the armature where the windings are soldered to the commutator segments; blobs of solder here indicate overloading of the windings, the melted solder having been flung off. This will produce 'dead' pairs of segments, which will result in the motor not operating if it previously came to rest with these segments in contact with the brushes, a common cause of intermittent motor failures. If you can get to the wire ends easily, you could try soldering them back into place. Where this is impossible, you will have to buy a new armature assembly, or get the old one rewound professionally.

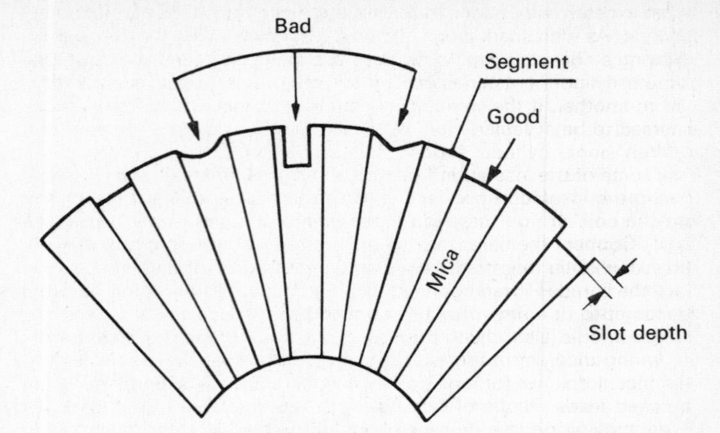

Fig. 6.6 Commutator slot condition

8.1 Examine commutator segments for wear or scoring. Use fine abrasive paper to clean up burnt or dirty segments. If damage is more severe, have the commutator skimmed in a lathe and then re-cut the gaps between each segment

9 Ignition system

The ignition systems used on most current machines are fairly reliable, many requiring no attention other than checking, cleaning or renewal of the spark plug(s). The method of checking the system in the event of a fault will depend upon the arrangement used, and you should refer to the workshop manual for detailed information.

In all cases, the first thing to check is whether there is a spark at the plug electrodes. Remove the plug or plugs from their threads in the cylinder head, refitting each one into its plug cap. Arrange the plugs so that the metal body is in contact with bare metal; the cylinder head fins being a suitable location. Note that in the case of CDI systems, it is **essential** that the plug body is earthed; if the HT circuit remains open, the CDI unit may be damaged.

Check that neutral is selected and that the engine kill switch is set to the 'RUN' position. Switch on the ignition and crank the engine using the kickstart lever or electric starter, observing the plug electrodes. Each plug should produce a fat and regular blue spark. If the spark looks weak or yellowish, this indicates partial failure of the system, whilst no spark at all speaks for itself. It is always worth trying new plugs at this stage; these are relatively inexpensive, and eliminating them as a source of trouble may save much unnecessary work. Note also that it is essential to ensure that the plug is of the correct type for your machine; use of a plug of the wrong heat range or reach will often cause ignition problems.

The plug cap may also be a source of trouble. Most caps incorporate a resistor, the purpose of which is to suppress radio and television interference. Occasionally these may break down, present-

9.1 No engine will run properly using a plug like the one shown on the left. The earth electrode has almost melted away, indicating that the wrong grade of plug had been fitted, or that there was another engine fault present. Note also the porcelain insulator nose round the centre electrode has split

ing an excessive resistance to the HT pulse and thus preventing normal sparking. As with spark plugs, the best way of checking for this fault is by fitting a new plug cap. In the case of a roadside breakdown, owners of multi-cylinder machines can try swapping plug caps from one HT lead to another; if the weak spark moves with the cap, then it can be assumed to be at fault.

With single cylinder models, try removing the cap and stripping back some of the insulation from the end of the HT lead (take care not to strip the insulation back too far, especially if the HT lead is sealed into the coil, or you may find that the lead will no longer reach the plug!) Connect the bared wire conductor directly to the plug terminal and see whether this effects a cure. If the plug sparks normally, you can leave the bared HT lead connected to the plug to get you home, but do not forget to fit a new plug cap as soon as possible.

You should also check through the ignition system wiring, using the wiring diagram in the workshop manual or owners handbook for reference. Look out for loose, corroded or waterlogged connections or damaged leads. Faults of this type must be checked and eliminated before moving on to a detailed check of the system components.

Flywheel generator systems

These devices normally employ an ignition source coil which powers the system. This forms part of the generator stator assembly, the remaining coils being used to power the lighting and charging circuits. The ignition timing is often fixed, and uses contact breakers housed inside the generator to control sparking. You will find systems of this type on most mopeds, and on some small trail bikes and commuter models. Note that the battery, where one is fitted, plays no part in powering the ignition circuit.

With this type of system, the condition of the contacts and their gap setting are of vital importance, and any burning or pitting of the contact surfaces or inaccuracy of the gap setting will affect the operation of the system to a marked extent. Always check and clean (or better still, renew) the contacts, and check and adjust the gap before looking at other aspects of the system. Badly burned contact faces indicate a faulty capacitor (condenser), which will allow excessive arcing across the contacts. The contact breaker assembly should be renewed if it is badly burnt, but you should also fit a new capacitor at the same time. These are not easy to test without elaborate equipment, but they are cheap enough to warrant renewal as a precautionary measure.

Light burning or pitting of the contact faces can be corrected by restoring the surfaces with a fine Swiss file or with fine abrasive paper. Try to leave the contact faces as smooth as possible and slightly convex in shape. Note that if too much material is removed, it may prove impossible to obtain the correct gap setting. In view of the need to remove the flywheel rotor to gain access to the contact breakers, it is normally preferable to fit a new contact breaker set, rather than spend time and effort trying to rectify damaged contact faces.

Other possible causes of ignition faults are damaged source coil windings, a demagnetised rotor, or a faulty ignition HT coil, HT lead or plug cap. These should be checked and eliminated systematically. On a few models, the ignition timing is variable by moving the position of the stator assembly, and details will be found in the workshop manual about checking and resetting the timing, where this is possible.

In many cases, no ignition timing adjustment is possible, but slight alterations can be made by varying the contact breaker gap within the specified range. Again, the Owners Workshop Manual will give details of this.

Note that under normal circumstances, the timing setting should not require attention, and this is unlikely to be the true cause of the problem. To check the timing you will need to refer to the relevant specifications, which in the case of most small two-strokes will give the setting as a piston position before top dead centre (BTDC). The firing position is normally shown as a mark on the edge of the flywheel rotor, and the contact breaker points should be at the point of separation when this mark aligns with a fixed index mark on the crankcase. It is not normally necessary to verify the accuracy of the timing marks, but if you have reason to suspect that they may be incorrect, note that this can only be checked accurately using a dial gauge and a spark plug thread adaptor (see Chapter 2).

Coil and contact breaker systems

These are similar to the arrangement described above, but derive their power from the battery.

In addition to the checks described above, make sure that battery

9.2 On two-stroke engines using contact breaker ignition, the contact gap is especially important and should be checked regularly

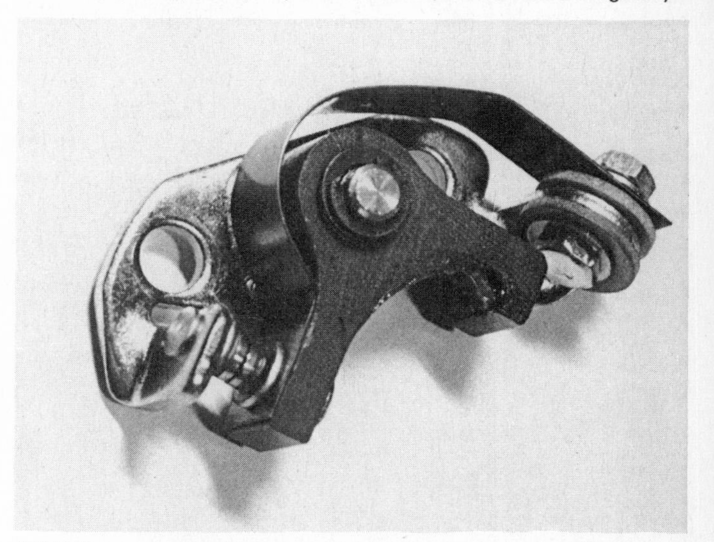

9.3 If the contact set is removed it is easier to examine and clean the contact faces

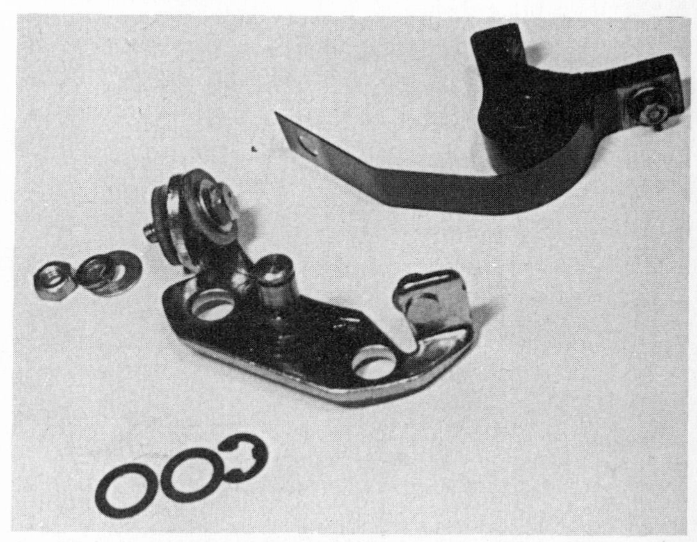

9.4 If dismantling the contact breaker assembly make sure that you note the correct position of any insulating washers or spacers

voltage is being supplied to the ignition coil(s), checking back through the wiring and engine kill switch to the ignition switch. The workshop manual will provide details on contact breaker gap and ignition timing procedures, and also resistance tests on the ignition coil(s).

With coil and contact breaker systems, the ignition timing is usually advanced automatically, using a mechanical automatic timing unit (ATU). This should be checked for wear or damage, referring to the Owners Workshop Manual for details. Make sure that the weights move smoothly and freely, without signs of sticking. Apply a drop of light machine oil to each pivot, and to the cam support spindle.

Capacitor discharge ignition (CDI) systems

Many current models employ CDI systems, ranging in complexity from simple replacements of the flywheel generator types to sophisticated units incorporating electronic advance circuits. When dealing with these systems it is essential that you follow the procedures outlined in the relevant workshop manual. Failure to do so may lead to damage to the CDI unit, which will be expensive to replace, and can even cause dangerous electric shocks to the unwary owner if mishandled.

It is worth mentioning that electronic systems can be damaged if the engine is cranked with the plug leads detached. The HT spark must always be able to earth out, and you should ensure that this is always possible by refitting the plug(s) in the plug cap(s) and ensuring that the plug body is earthed.

Testing electronic systems is often limited unless you have access to expensive test equipment, and the best way of tracking down a fault is to substitute each suspect part in turn until the defective one is located. The factory service manual or the relevant Haynes Owners Workshop Manual will describe the test procedures for each model in some detail, but it is noteworthy that many dealers prefer to test by substitution because it is quicker, easier and often more conclusive.

In the event of ignition failure there is little that can be done at the roadside, but you could usefully check simple mechanical breaks in the system wiring. One area that may often cause problems is the wiring between the ignition pickup and the CDI unit. This carries only a tiny trigger pulse, so if the wiring connections become detached, broken or contaminated by water, the system will fail to produce a spark. As with most electrical connections, packing with silicone grease will prevent this type of problem.

The ignition pickup coils can normally be checked with a multimeter. If possible, check the actual resistance figures, but if you do not have access to the precise specified readings, check that the coils are not open circuited (infinite resistance) or short circuited (zero resistance). Anything between these extremes is a fair indication that the coils are intact.

If the failure is complete, and the pickup coils are intact, this suggests a failure of the CDI unit itself. Where you can localise the fault, perhaps to cylinders 1 and 4 on a four-cylinder model for example, the CDI unit is unlikely to be responsible. Check instead the ignition HT coil which controls these two cylinders.

Appendix

Lubricants

Motor oils

Motorcycle manufacturers invariably list one or more grades of motor oil for use in the engine/transmission unit (four-strokes) or in the gearbox (two-strokes). Motor oils are classified according to their viscosity and performance classification.

Viscosity is denoted by the SAE number of the oil. This is based on a viscosity standard set by the Society of Automotive Engineers (SAE) and relates to the time taken for a given volume of oil to flow through a calibrated capillary tube at specified test temperatures. The higher the number, the greater the viscosity (thickness) of the oil.

Monograde oils are classified by a single number, with the SAE prefix. In the case of four of these grades, the SAE rating applies to viscosity at 100°C, the four grades being SAE 20, SAE 30, SAE 40 and SAE 50.

A further six grades of oil are rated according to tests at temperatures between -5°C and 30°C, and have a W suffix to denote this. The six grades are SAE OW, SAE 5W, SAE 10W, SAE 15W, SAE 20W and SAE 25W.

Few motorcycle manufacturers recommend a monograde oil, and this is because current engines are designed and built with modern multigrade types in mind. Monograde oils still have an application in older engine designs, particularly those where ball and roller bearings are used throughout, especially in the case of the big-end assembly. In these applications, an SAE 30 or SAE 40 grade is normally recommended for winter use, with a change to an SAE 50 grade for summer.

Multigrade oils, the type specified for use in most modern engines, are based on a monograde to which has been added polymers and other additives to allow it to meet a range of viscosity requirements. At low temperatures, the oil is relatively thin, providing good performance when the engine is first started. As engine temperature rises, the additives reduce thinning, allowing a higher viscosity rating to be achieved. These characteristics are particularly well suited to engines in which plain bearings are used extensively.

Many motorcycle manufacturers specify an SAE 10W/40 motor oil. In this example, the oil behaves as an SAE 10W oil at low temperatures, but remains sufficiently viscous at high temperatures to retain SAE 40 characteristics.

Additives are included in most motor oils to reduce oxidation and improve the load-carrying capabilities. In addition, dispersants are added to keep carbon, metal particles and water in suspension. These prevent the formation of sludge deposits, the contaminants being removed when the oil is changed.

Oil performance is indicated by a suffix, based on standards formulated by the American Petroleum Institute (API) and the Society of Automotive Engineers (SAE). For petrol engines, this suffix begins with an 'S', followed by a second letter, currently ranging from 'A' (lowest quality) through to 'F' (highest quality). Most motorcycle manufacturers specify an SE or SF quality oil, and lower grade products should be avoided to prevent premature engine wear. Bear in mind that motorcycle engines put great pressures on their lubricants; the use of inexpensive but low quality lubricants is a false economy.

Synthetics are becoming increasingly popular choices for motorcycle engine lubrication. Originally formulated for racing, these advanced-specification lubricants are now widely available for road use. Claimed advantages are higher performance than equivalent conventional mineral-based oils, and suitability for use in demanding situations such as turbocharged engines. The only disadvantage is the relatively high cost of these lubricants.

Castor-based oils are often used in racing engines, and have no real application for road use. These vegetable-based oils offer exceptional lubrication performance, but must be changed much more frequently than mineral oils. All castor-cased oils are more expensive than equivalent mineral oils, making them prohibitively expensive for road use.

Note: *Under no circumstances should castor and mineral oils be mixed, or a thick sludge will be formed in the engine oilways, possibly resulting in the failure of the lubrication system. Where an engine has been run on a castor-based oil, all traces of the old oil must be removed before switching to a mineral oil, ideally by stripping and cleaning all engine parts.*

Choosing an oil can prove less straightforward than it might seem. The motorcycle manufacturer will normally include in the rider's handbook a list of oil brands and grades which they endorse, and it is generally safe to assume that these are well suited to a particular model. The major oil companies all produce oils of equivalent viscosity and quality, and where the SAE number and API suffix is shown on the packaging, the choice of an alternative brand of the correct grade is easy enough. Unfortunately, there are one or two oil companies who omit to give viscosity information on the cans or bottles in which the oil is sold.

As a general rule, most modern road-going machines will require a high-API (SE or SF) multigrade mineral oil with a viscosity of around SAE 10W/40. Any reputable brand of oil meeting these requirements should be safe to use in your machine, but do check the manufacturer's recommendations before choosing an oil.

Gear oils

A few machines will require a gear oil for use in a separate gearbox, though the vast majority employ a shared motor oil supply for the engine and the transmission components. Some shaft-drive models require gear oil as a lubricant in the rear bevel housing.

Gear oils are traditionally mineral-based monograde oils, containing high-pressure additives to deal with the high loadings imposed by meshed gear teeth. More modern derivatives offer limited multigrade ranges. Hypoid gear oils are often specified for shaft drive applications.

The variety of gear oils is less easily summarised than equivalent motor oil grades, and the best advice here is to check carefully the exact specification stipulated by the manufacturer, and to use an oil conforming to this. If there is any doubt as to the requirements of your machine, seek the advice of a reputable dealer, who will be able to recommend a suitable product.

Two-stroke engine oils

Two-stroke engines are lubricated by a total-loss system in which oil is fed to the engine either mixed with the incoming fuel, or by direct injection. Any excess oil is burnt along with the fuel, and so the oil need not have the dispersal additives normally associated with four-stroke motor oils.

Most two-stroke oils are monograde types of around SAE 30 or SAE 40, and include additives to reduce carbon build-up and plug fouling. In some cases, separate types are produced for pump injection, tank-mixing and pre-mixing. The distinction between the last two types is significant; tank-mixing types can be added direct to the fuel tank in the recommended proportions, whilst pre-mix types require the fuel and oil to be mixed together *before* adding to the machine's fuel tank. With the exception of a few mopeds, almost all current road-going two-strokes use a pump fed arrangement, tank-mix and pre-mix systems being favoured for competition use where weight and complexity can be reduced by omitting the tank, pump and oil pipes.

Synthetic two-stroke oils are becoming increasingly popular, despite the higher cost. All-synthetic two-stroke oils must not be used with conventional mineral-based types, and generally require the mixture ratio to be reduced. They are primarily intended for racing use, and may not be suitable for use in pump-lubricated road engines.

Part-synthetics can normally be used on road machines, including pump-fed engines, without alterations to the mixture ratio. Like the all-synthetic types, part-synthetics claim to deliver higher lubrication performance and to reduce smoke emission from the exhaust.

Choosing a two-stroke oil – For most road-going two-strokes, a conventional good quality two-stroke oil is all that is required, and many of these can be used in pump or tank-mixed applications. Engine oil costs add appreciably to the cost of running a two-stroke machine, and there is little to be gained by using an expensive synthetic racing oil in the average restricted moped.

In the case of high-performance two-strokes, a part-synthetic oil may be justifiable, though the cost consideration will apply here as well. Only in the case of pure competition machines is a full synthetic oil worth consideration. Given the relatively low mileages covered by (most) competition bikes, the extra expense can be warranted by the greater protection and cleaner running offered by these oils.

Fork and suspension oils

Fork oils are monograde lubricants containing additives to make them better suited to this type of application. Amongst these are additives to reduce oxidation and to minimise corrosion of the suspension components. Most important of all are the anti-foaming additives. As the name suggests, these prevent the formation of air bubbles, as the damping oil is forced through the damping orifices at high speed and pressure. Where foaming is allowed to occur, cavitation of the oil, and consequently erratic damping performance, results.

Many of the larger oil companies produce 'fork oil', giving no indication of its viscosity. Whilst this is generally suitable for use in the majority of machines, it is worth noting that *Bell-Ray* and *Silkolene* each produce a wide range of viscosities, and this allows the damping effect to be tailored to personal preference. This is of obvious use in racing applications, where fine-tuning of the suspension may have some bearing on the competitiveness of the machine. It is also of great use for road machines, allowing the owner to change the damping performance of his machine to suit his own riding style.

Suspension fluids can be added to the range of fork oils, and are similar products designed for use in both telescopic front forks and in rebuildable or refillable rear suspension units. Like conventional fork oils, these are available in a range of viscosities, and these can be mixed to obtain the required damping effect. Mainly intended for racing machines, these products can be used for road-going machines where the damping oil in the rear unit(s) can be changed; for obvious reasons, they are of little use where the damper unit is of sealed construction.

Filter oils

These are specially-formulated oils designed for use with foam element air filters. This type of oil has additives to make it 'tacky' and is thus able to trap dust and moisture more effectively than conventional lubricating oils. Originally designed for off-road use, these products can usefully be used on roadgoing machines which use this type of filter.

Greases

Greases are an alternative to oil for lubrication purposes in applications where a liquid lubricant would be inappropriate. Most general-purpose greases combine a mineral oil with a lithium soap base to produce a general purpose high melting-point grease suitable for general lubrication of wheel and steering head bearings and suspension pivots.

Molybdenum disulphide grease is sometimes specified for assembly use, where it provides initial lubrication to protect engine bearing surfaces in the first few vital seconds of running before full oil circulation is restored. It is also useful for some general lubrication applications, providing a degree of dry lubrication where a conventional grease would fail.

Graphite grease is used in similar applications to molybdenum disulphide grease, and for the same reasons. The graphite content maintains lubrication after the grease has hardened with age or burnt off with heat.

Copper grease is a high-temperature product able to provide lubrication on engine and brake parts which reach very high temperatures. The copper content remains effective as a lubricant where any conventional grease would have burnt off, and is of particular value in preventing seizure of exhaust system fasteners, and as an anti-squeal agent for use with disc brake systems.

Silicone grease is used in electrical switchgear and connectors as a means of excluding moisture and preventing corrosion. It is also of use as a lubricant with many plastic parts.

Chain lubricants

Chain lubricants are supplied in either hot-immersion or aerosol forms, the comparative merits of the two types being discussed in detail in Chapter 1. Note that some types of aerosol lubricant are unsuitable for use on O-ring chains; check this before buying, or the O-rings which seal each roller assembly may be damaged.

Replacement bearings, bushes and seals

Bushes

Most machines use numerous bushes throughout the engine/transmission unit and also in various chassis applications. In the event of wear or damage the bush should, of course, be renewed. Replacement bushes are generally easily obtained through an authorised dealer for your model.

In a few instances, you may discover that a particular bush is not supplied as a separate part, typically in the case of bought-in assemblies like starter motors or similar items. If the cost of a new assembly is prohibitive and a second-hand assembly cannot be obtained, consult an engineering company who may be able to make up and fit a suitable bush for you.

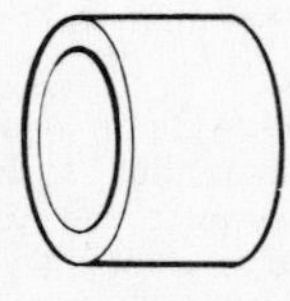

Plain bearings

Many machines use plain big-end and main bearings with renewable inserts, and similar arrangements are to be found on some camshafts. As a general rule, these are installed on a selective fit basis. The bearing journals are machined to a nominal size, which is then checked and given a sizing code number or letter. The precise method of sizing depends on the manufacturer concerned, but will usually take the form of a letter, number or colour mark placed near the journal. Using this marking in conjunction with a table of insert sizes produced by the manufacturer, the correct insert can be selected to obtain the specified oil clearance. It follows that you will need the bearing insert selection details, and this will be found in the relevant manufacturer's service manual or the Haynes Owners Workshop Manual for your model. Note that in the absence of the necessary selection information it is probably best to take the assembly to a dealer for the new bearing inserts to be ordered after checking and measurement, and the same action is advised if you are unsure about the condition of the journals.

If the exisiting shells are to be checked for wear with a view to reusing them, this is best carried out using Plastigage to check the oil clearance as described in Chapter 4, Section 8; the only alternative is to calculate the clearance after careful measurement of each component. If the clearance exceeds the manufacturer's specified limit the shells must be renewed.

A certain amount of general wear can be accommodated by selecting a slightly smaller diameter bearing insert, though it should be noted that if serious wear or damage has occurred, this is unlikely to prove successful. In case of severe wear, scoring or ovality, you are left with the choice between buying a new crankshaft or camshaft, or having the old one reconditioned by re-metalling and grinding back to the standard journal sizes. The latter work should be entrusted to a reputable engineering company. Note that unlike car engines, there is normally no provision to grind the journals undersize and fit appropriate inserts to suit.

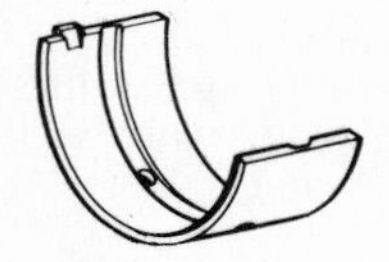

Ball and roller bearings

Where ball, roller or needle roller bearings are to be renewed, it is worth noting that these can sometimes be obtained through a local bearing stockist at significantly lower prices than the dealer for the machine would charge. This does not indicate excessive profit-making on the part of the dealer; you are paying extra for repackaging and handling charges through the manufacturer's spares network.

You can locate a bearing supplier through Yellow Pages, or by asking a local garage or motorcycle dealer. Remember that you are unlikely to be able to obtain specialist bearings off the shelf, though it may be possible to order them. The more common sizes, such as those used for wheel bearing and transmission bearing applications are relatively easy to obtain.

The size and type of bearing will be indicated by a number and/or letters stamped on the edge of the bearing outer race. As an example, a bearing marked '6003RS' will be found to have an inside diameter of 17 mm, an outside diameter of 35 mm, a width of 10 mm and a rubber seal on one side. Although each manufacturer uses their own system of identification, it is usually possible to cross-reference these where necessary. If this fails, it is normally possible to find a replacement of the correct size and type by measuring the old item. It is preferable to take the old bearing with you when ordering any replacement parts, as this will allow the bearing supplier to check the markings or dimensions against his catalogues. It goes without saying that if the bearing supplier does not have the required size or type listed in his catalogues, you will have to order a replacement through a motorcycle dealer for that make of machine.

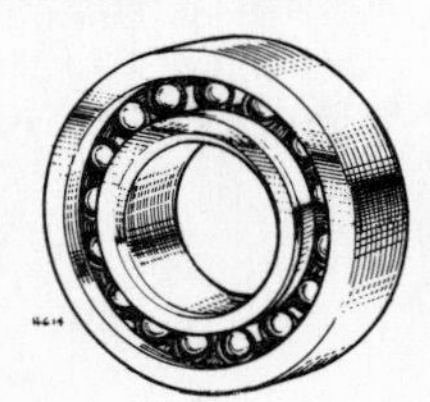

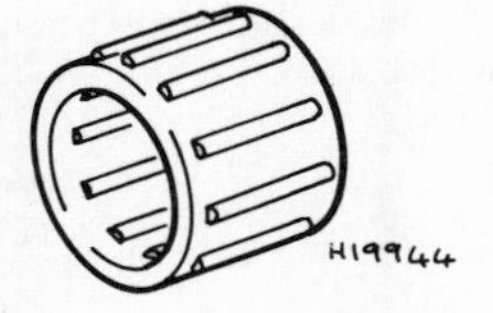

Seals

Like ball and roller bearings, oil, grease and some dust seals are often of a standard size, and are thus available commercially from bearing stockists. Once again, the seal dimensions and type are indicated by a reference number moulded into the seal case, and the old component should be taken along when ordering so that the correct type can be supplied. Some seals are for specialised applications or may be of an unusual size, and in such cases you will have to order the new parts from a motorcycle dealer for your model.

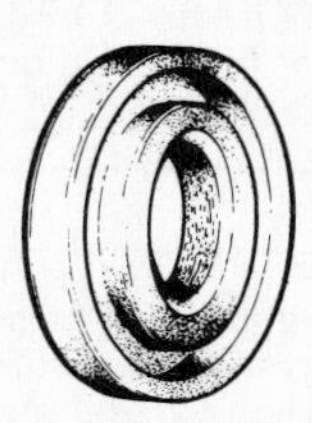

Fuels

Gasoline, which is normally called petrol or motor spirit in the UK, is a complex product. The subject of the production of gasoline and other products from simple crude oil could easily fill a book, and it is not proposed to delve too deeply into it here. Suffice it to say that the product sold as 'petrol' in the filling station can be described as a mixture of volatile hydrocarbons and various additives designed to be burnt as a fuel in a wide variety of internal combustion engines. In this section some of the main factors governing the suitability of different grades of pump petrol for use in motorcycle engines will be examined.

Octane rating

The octane rating of petrol describes its resistance to 'knocking', by which is meant the inclination of the petrol to explode violently, rather than to burn in a rapid but predictable manner. The higher the octane number, the greater the resistance to this undesirable behaviour. The standard method of describing octane rating is by the 'Research Octane Number' (RON), which describes the degree of knock resistance under normal driving conditions. A second standard, the 'Motor Octane Number' (MON) is being adopted as more representative of fuel behaviour at higher speeds, and both ratings are sometimes quoted. At the time of writing the octane numbers of pump petrol on sale in the UK should conform to the following ratings:

Leaded 4-star .... 97 RON minimum
Premium unleaded .... 95 RON minimum
Super unleaded .... 98 RON minimum

Leaded 2-star (90 RON) and 3-star (93 RON) are no longer available in the UK

Gasoline additives

The common gasoline additives used in pump petrol sold in the UK are summarised below. Note that the proportions of these additives vary, particularly in the case of anti-icing additives, which must be balanced against the proportion of high-volatility fractions in the fuel to suit seasonal weather conditions.

Additive	Function	Effect
Anti-oxidant	Stabilises fuel	Prevents gum formation
Lead	Boosts octane	Reduces knocking
Detergent	Reduces deposits	Keeps engine in tune and clean
Anti-icing	Lowers freezing point of moisture	Prevents carburettor icing in cold weather
Anti-corrosion	Reduces corrosion	Increases engine life

Unleaded petrol

There is a well-justified move to reduce the levels of lead in pump petrol sold in the UK, and the current leaded grades have had restricted lead levels (a maximum of 0.15 grammes per litre) since 1st January 1986, since when pump petrol has been termed 'low-lead'. Prior to this the accepted level was 0.40 grammes per litre.

Unleaded petrol is defined as containing less than 0.013 grammes of lead per litre, and the term 'unleaded' means that no lead is added to the fuel. Note that it is **not** lead-free. Premium unleaded and super unleaded are available in the UK, their octane ratings being shown above. Note that regular unleaded (91 RON) may be found in other countries.

It should be noted that unleaded petrol can be used only in engines designed to operate with this type of fuel, and used in unsuitable engines it will normally lead to problems of knocking and possible damage to the valves.

The situation regarding the use of unleaded fuels in motorcycles remains unclear at the time of writing. None of the models currently on sale in the UK need to run on unleaded petrol, though most Japanese models can do so. Many European models must either use leaded fuel, or may require engine modifications to do so. The safest course of action with *any* machine at present is to consult the dealer for that model to check which type of fuel you should use.

Standard torque settings

Motorcycle manufacturers generally specify torque setting values for the majority of fasteners on the machine, and in some cases also specify a tightening sequence where more than one fastener is concerned. Where no torque setting is given fasteners can be secured according to the table below.

Fastener type (thread diameter)	**kgf m**	**lbf ft**
5 mm bolt or nut	0.45 – 0.6	3.5 – 4.5
6 mm bolt or nut	0.8 – 1.2	6 – 9
8 mm bolt or nut	1.8 – 2.5	13 – 18
10 mm bolt or nut	3.0 – 4.0	22 – 29
12 mm bolt or nut	5.0 – 6.0	36 – 43
5 mm screw	0.35 – 0.5	2.5 – 3.6
6 mm screw	0.7 – 1.1	5 – 8
6 mm flange bolt	1.0 – 1.4	7 – 10
8 mm flange bolt	2.4 – 3.0	17 – 22
10 mm flange bolt	3.0 – 4.0	22 – 29

Fasteners

This section describes the most common types of fasteners, washers, pins, rivets and retainers likely to be encountered on most recent models, together with an illustration to facilitate identification. Some additional types are listed where these are likely to be of interest in the home workshop. The list is by no means exhaustive, but relates to the most common types.

Note that in the case of general-purpose fasteners used throughout the machine, ordinary mild steel or high tensile steel is normally used, with a bright zinc or cadmium coating to resist corrosion. If you purchase replacement fasteners of this type, it is more economical to buy them in quantity. Most motorcycle magazines will carry advertisements for suitable suppliers, many of whom will supply selections of popular sizes, together with suitable plain washers. Note that if a selection of '100 nuts, bolts and washers' is advertised, this normally means 25 nuts, 25 bolts and 50 plain washers, not 100 of each.

Assortments of fasteners will usually include a selection of bolts and setscrews in a variety of lengths and diameters. Bear in mind that these may have to be shortened to suit a particular application, and unless stainless steel is specified, cut ends of steel bolts will rust.

In some more specialised areas, such as stand pivot bolts, special materials or hardening processes may have been used to resist wear or breakage. These fasteners often have non-standard head sizes or shank diameters. If you need to renew any fastener in a highly-stressed application, be sure to replace it with a suitable equivalent to avoid subsequent failure, *not* with an ordinary mild steel item, or it may fail in use. The best way to avoid this risk is to purchase official replacement parts from a motorcycle dealer.

Hexagon-headed bolts will be found extensively in many applications on most motorcycles. The parts to be joined are located on the plain shank area, the threaded section screwing into a thread in the frame or engine, or secured by a nut.

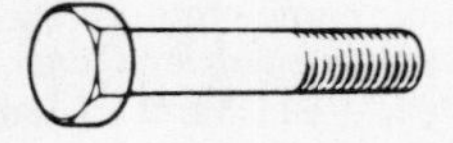

Hexagon-headed setscrews are similar in appearance to bolts, but note that the threaded area extends up to the underside of the head. Setscrews may be used where a thin fitting or bracket is attached to a frame lug or similar.

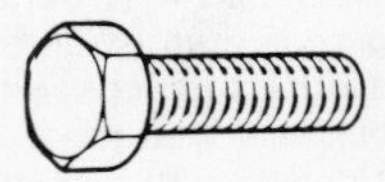

Flanged bolts or setscrews are often used in place of Phillips-headed casing screws, and in this application are less vulnerable to damage during removal and installation. Larger sizes are often used in place of conventional bolts or setscrews, where the need for a separate washer is avoided.

Socket screws, also called *Allen* screws may be found as standard fitments on some models, and are worthwile replacements for Phillips-headed casing screws. Driven by hexagon keys of appropriate size, the screw heads are highly resistant to damage. In addition to the normal cheese head type shown, socket screws are also available in countersunk and button head forms, as well as grub screws.

Torx screws will not be encountered on many Japanese models, other than in special applications. Recent BMW models, however, make extensive use of this type of driving recess, for which special Torx keys will be required.

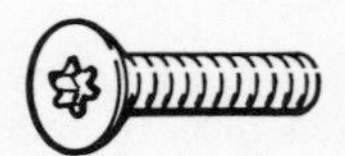

Casing screws used on most Japanese models are of the pan head Phillips pattern shown. This type of screw head is very easy to damage during removal, and it is vital that the correct shape and size of screwdriver is used. It is preferable to slacken and tighten these screws using an impact driver, or to replace them with socket screws.

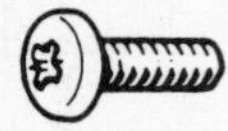

Self-tapping screws come in a wide variety of designs and sizes, and are used to retain some brackets and covers, particularly on thin steel panels. These screws have hardened threads which are capable of cutting their own corresponding thread in most thin materials, once a pilot hole has been punched or drilled to provide a starting point.

Plain nuts will be found in many areas of all machines, and are used to secure the ends of bolts, setscrews and studs. Like the bolts and setscrews, these should be bright plated to avoid corrosion, or better still, stainless steel.

Thin nuts, as the name suggests, are of unusually thin section, and are designed for use as locknuts, where they are tightened against a plain nut to secure it, or in special applications where space is restricted.

Washer faced, or flanged nuts will be encountered regularly on Japanese models, where they are used in preference to a plain nut and washer.

Domed nuts are used on the end of shafts or threaded studs and fasteners both to retain them and to provide a decorative finish. They will also exclude dirt and water from the threads. Note that the projecting thread end must be of the correct length, or the nut will become bottomed out before the fitting is secure. Pack with plain washers where necessary.

Nylon-insert self-locking nuts are available in a variety of designs, and are used where engine or road vibration might otherwise cause loosening. The nylon collar grips the thread to which the nut is fitted, resisting accidental slackening in use. Note that nylon-insert self-locking nuts are intended to be renewed each time they are removed; they will not grip securely once the nylon collar has been deformed by the thread.

Stiff nuts are also used for their self-locking properties, and come in a variety of types. That illustrated has a raised collar inside which the thread has been deliberately split and deformed so that it grips the thread to which it is fitted. Variations include special nuts which have been crushed so that the threaded hole becomes slightly oval. Like other self-locking nuts, they should be renewed once removed.

Castellated nuts are used in conjunction with a split pin (cotter pin) to secure items like wheel spindles. The nut is torque-tightened, and a new split pin fitted through the spindle end, which is drilled for this purpose. The nut itself can be re-used indefinitely, but the split pin must be renewed each time the nut is refitted. Where an R-pin is used in place of a split pin, this too can be re-used.

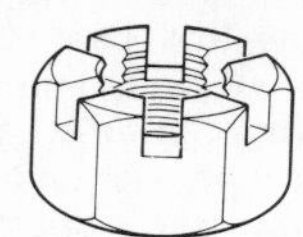

Plain washers spread the load applied to items to be secured and allow the fastener to turn without damage to the fitting. Extra large diameter plain washers, called 'Penny washers' are used for even better load-spreading capabilities on soft or fragile materials.

Spring washers are designed to prevent a plain nut from working loose once tightened, and are available in square or rectangular section.

Double-coil spring washers are a variation on the plain single-coil spring washer shown above.

Toothed, or serrated, lockwashers are available in internal or external patterns and are used to lock plain nuts in place. Note that they can also be useful where the teeth cut through a paint film to provide a sound earth connection.

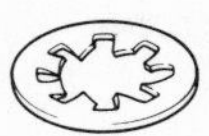

Wave washers in varying designs are sometimes used to take up unwanted free play, to avoid rattles or unwanted movement. These are often found in this role on items like kickstart levers, where a wave washer is used to steady the pivot and hold it in place.

Split pins (also known as cotter pins) are used in conjunction with a washer or a castellated nut (qv) as a method of retention. Available in a wide range of sizes, the pin is cut to the required length after fitting. Split pins should not be reused.

R-pins are more durable replacements for the humble split pin in some applications. They are especially useful in places where repeated removal is required, as in the case of the wheel spindle nuts.

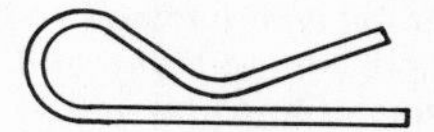

Roll pins (also called spring tension pins) are thin tubular steel pins with a longitudinal split. They are used in place of hollow dowel pins in some applications, and can also be used to retain a component on a shaft. They should be fitted and removed using a stepped parallel punch of the correct diameter, to prevent distortion.

Taper pins were traditionally used to secure items like pulleys to machinery shafts. In the workshop, small taper pins can be useful when pinning together a kickstart and its shaft to overcome stripped splines.

Cotter pins are a variation of the taper pin, having a thread and a nut at one end to allow the pin to be drawn into position. They are used to secure components to shaft ends, a common application being the retention of pedal cranks to mopeds and bicycles.

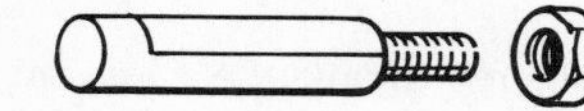

Plain rivets are the traditional method of riveting together two parts, provided that access to both sides of the workpiece is available. Materials include steel, brass and aluminium alloys, and head forms include domed and countersunk. Rivets are normally solid, though hollow copper rivets are used in some applications such as securing brake linings to shoes, on older machines.

Pop rivets, or blind rivets, allow pieces to be fastened from one side only. These rivets are available in a range of diameters and lengths, and head materials vary from steel to aluminium and Monel. Head designs include specialist shapes used to secure trim or upholstery. The rivets are set (fitted) using a special tool.

Circlips are a simple method of retaining a component to a shaft end, or to secure a shaft or pin in a bore. Typical applications are the securing of piston gudgeon pins, and holding gears on shafts. There are two basic types, Seeger, or flat-section circlips, and plain wire circlips. Note that the two must never be interchanged – the retaining groove for the Seeger type must be parallel sided, and if this type of clip is fitted to a rounded groove it may work out in service. Circlips are available in a wide range of sizes, and as internal or external types.

H.20465

Drive chains

With the exception of shaft drive and belt drive machines, roller chain is the normal method of transmitting drive from the transmission output shaft to the rear wheel. Despite the chain's undeniable complexity, and its vulnerability to damage and wear as a result of poor lubrication and the effects of road dirt and water, the roller chain remains an effective means of transmission.

Chain sizes

The size of final drive chain used is determined by the motorcycle manufacturer and is selected according to the power output of the machine, plus other factors such as the likely use of the machine. As an example, a trail bike is likely to use a heavier chain than a road-going commuter machine of similar engine size, to allow for the more punishing conditions likely to be experienced by the off-road model.

As a general rule, it should not be necessary to depart from the size or type of chain recommended by the motorcycle manufacturer, and in many cases to do so would lead to problems with running clearances or chain life. The size of chain can be checked in the owners handbook or in the appropriate Haynes Owners Workshop Manual. Alternatively, check the size markings on the chain itself. With most chains, the size is indicated by a three digit number, often prefixed by letters indicating the chain type. Examples of the chain sizing numbers are given below.

Reference number	Size (pitch x width)
415	$1/2 \times 3/16$ in
420	$1/2 \times 1/4$ in
428	$1/2 \times 5/16$ in
520	$5/8 \times 1/4$ in
525	$5/8 \times 5/16$ in
530	$5/8 \times 3/8$ in
630	$3/4 \times 3/8$ in

Chain types

Most chain manufacturers produce a range of chain types in various sizes. As mentioned above, it is normally preferable to fit replacement chains of the same type as originally specified; fitting a heavier duty chain may cause the chain to come into contact with the chain guard, frame or engine casings due to the increased width. The same is true of O-ring chains, which are unavoidably wider than an equivalent conventional type. If you decide to fit an uprated type of chain, always check carefully that clearance problems will not result from the change.

Standard chains are used on most small-to-medium capacity models, and normally have fairly lightweight, waisted sideplates. The chain is usually joined with a conventional spring link.

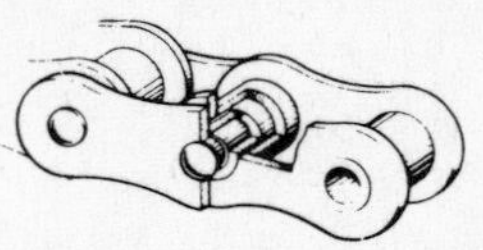

Heavy duty chains are similar in dimension to the standard types, other than in the thickness of the sideplates which are thicker to give improved strength. This type of chain is often used on off-road models, and can usefully replace the standard type on all models where the increased overall width does not cause clearance problems.

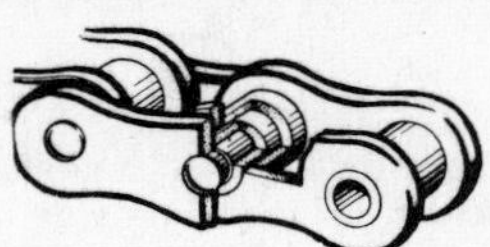

Self-lubricating chains incorporate oil-impregnated sintered metal bushes inside the rollers, and these are designed to gradually release lubricating oil in service.

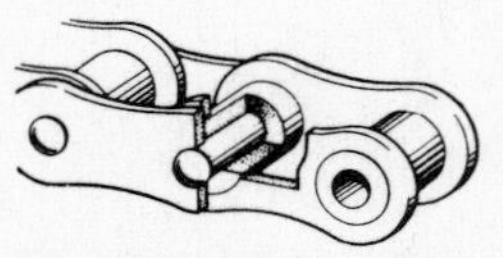

O-ring chains are often used on larger machines, and feature a semi-sealed construction. Lubricant is kept in place within the roller assemblies by small O-rings, a feature which also excludes road dirt and thus extends chain life considerably. O-ring chains are normally joined by riveted soft links because the O-ring construction precludes the use of a conventional spring link. Note that it is essential that only lubricants designed for use with O-ring chains are used; unsuitable lubricants may damage the O-rings.

High-performance chains are frequently specified for use on high performance models where great strength is required. The heavy-duty sideplates are normally parallel-sided rather than being waisted, and the chain may be of conventional or O-ring design. The chain ends are invariably joined by a riveted soft link in the interests of security.

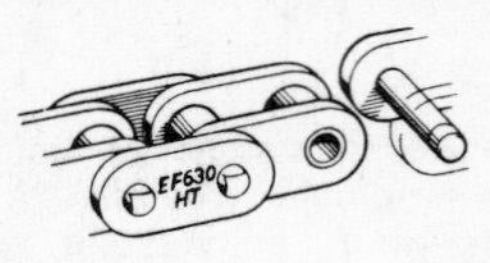

Chain accessories

Spring links are the traditional method of joining the chain ends once the chain has been fitted around the sprockets. The spring link comprises a special sideplate and bearing pin assembly which is passed through the bushes at each end of the chain. A second sideplate is then fitted over the pin ends, and this is secured with a spring clip fitted over grooves in the pin ends. Note that the closed end of the clip must always face in the direction of chain travel.

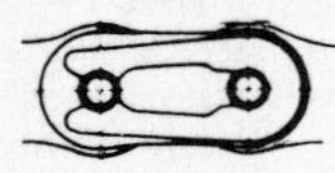

Soft links are used in applications where a spring link is judged to be insufficiently secure, notably where O-ring or heavy-duty types of chain are employed. Instead of the spring clip used on spring links, the pin ends are unhardened, and this allows them to be riveted over to secure the outer sideplate. Note that this type of link must be renewed every time the chain is removed and refitted, and the link should be checked periodically for security. It is a good idea to apply a paint mark to the soft link to make future identification easier.

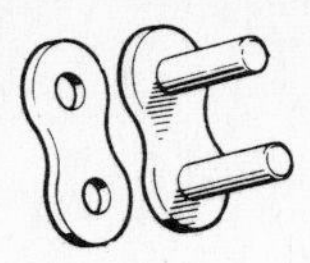

Cranked links, or 1/2 links are used where the machine needs an odd number of links in its chain, and are rarely encountered on production machines. It should be noted that the use of a 1/2 link weakens the chain overall, and their use should be avoided where possible. They are not safe for use on high-performance machines.

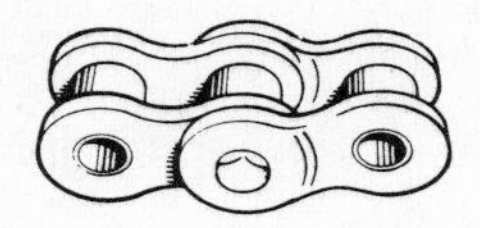

Tyre sizes and markings

The manufacturer will specify the size and type of tyre to be used on a particular model, this information being given in the owner's handbook, and also in the appropriate Haynes Owners Workshop Manual. Any replacement tyre should be of the same type and size as the original, or an alternative type and size where this is authorised by the manufacturer. In this section, the tyre sidewall markings are explained to make it easier to understand what this information means in practice. Unless stated otherwise, the following information relates to conventional diagonal ply (crossply) motorcycle tyres.

Tyre sizes

Amongst the information included on the tyre sidewall is a section relating to the tyre size, speed rating and load index. This is indicated in the accompanying line drawing. In the example shown, this part of the sidewall marking shows: 130/90H17 68H, which can be interpreted as follows.

The overall width of the tyre is shown first, and in this example is given in millimetres. This is followed by a slash mark, then a second figure indicating the aspect ratio of the tyre section. In the example this is 90, and indicates that the height of the tyre section is 90% of the width.

The letter H which follows is a speed rating indication, H denoting that this tyre is suitable for use on a machine capable of up to 130 mph (210 kph). Next comes the rim diameter, shown in inches. The final section of the sidewall marking gives a load index number, followed by the tyre speed rating again.

On some tyres, the size is given in inches, rather than in millimetres, the latter sizing relating to modern low-profile tyres. A list of the older imperial sizings, together with the nearest metric equivalents is shown below.

Old sizes,	Low-profile equivalents	
Imperial	Imperial	Metric
2.50/2.75	3.10	80/90
3.00/3.25	3.60	90/90
3.50	4.10	100/90
4.00	4.25/85	110/90
4.25	4.70	120/90
4.50/5.00	5.10	130/90

Speed ratings

The table shown below relates to the system of speed rating markings adopted by the European Tyre Manufacturers, though in practice most of the markings will be found on tyres from outside Europe. Note that the symbols 'L', 'M' and 'P' are recent additions, and on older tyres still in use, these markings will not be found. Note that the speed rating denotes the theoretical top speed of the machine, not the speed at which it is normally used.

The speed rating system relates more to the level of stress likely to be imposed by a class of motorcycle over a broad range of speeds, rather than to its actual top speed. This is why a motorcycle manufactuer may stipulate a V-rated tyre for a machine which is perhaps limited to a top speed of 115 mph. On no account be tempted to fit an S-rated or H-rated tyre to save money; you will be placing yourself and others at risk, and may be breaking the law by doing so.

Construction	Speed symbol	Maximum speed
Diagonal	L	75 mph (120 kph)
Diagonal	M	81 mph (130 kph)
Diagonal	P	93 mph (150 kph)
Diagonal	S	113 mph (180 kph)
Diagonal	T	118 mph (190 kph)
Diagonal	H	130 mph (210 kph)
Diagonal	V	speeds above 130 mph (210 kph) at reduced loadings*
Bias belted	VB	speeds above 130 mph (210 kph) at reduced loadings*
Radial	VR	speeds above 130 mph (210 kph) at reduced loadings*
Radial	ZR	speeds above 149 mph (240 kph) at reduced loadings

* *Note – maximum approved speed may be marked on tyre, eg V260*

Load Index

The load index number denotes the maximum load that a tyre can carry at the speed indicated by its Speed Symbol, and under the service conditions specified by the manufacturers. The numbers and their load capacities are as follows.

LI	Kg	LI	Kg	LI	Kg
32	112	48	180	64	280
33	115	49	185	65	290
34	118	50	190	66	300
35	121	51	195	67	307
36	125	52	200	68	315
37	128	53	206	69	325
38	132	54	212	70	335
39	136	55	218	71	345
40	140	56	224	72	355
41	145	57	230	73	365
42	150	58	236	74	375
43	155	59	243	75	387
44	160	60	250	76	400
45	165	61	257	77	412
46	170	62	265	78	425
47	175	63	272	79	437
				80	450

Radial and bias belted tyres

Recent innovations in tyre development include radial and bias belted tyre construction, in addition to the traditional diagonal (cross-ply) tyres. These terms relate to the way in which the casing plies are configured, those of the radial and bias belted versions offering a lighter and more flexible construction. This in turn ensures that the tyre tread is kept in better contact with the road surface, offering improved grip and reduced wear rates, albeit at greater initial cost.

Not all machines are suitable for use with radial or bias belted tyres; the wheels must be able to accept tubeless tyres, and the degree of success depends to some extent on suspension design. Unless the machine was fitted with this type of tyre as original equipment, always consult the machine manufacturer or importer before making the change.

Note that the tyre construction is indicated by the speed symbol. In the case of cross-ply tyres, a single letter is used, whilst on radial or bias belted tyres, this is indicated by a second digit (see above).

Tyre safety checklist

Your safety, and that of other road users, depends to a great extent on tyre condition and suitability. The checklist which follows describes the most important checks applicable to your tyres.

1) Make sure that the tyres fitted to your machine are of the type and size recommended by the motorcycle manufacturer, paying particular attention to speed ratings. The use of the wrong type or size of tyre is dangerous, and may be illegal.

Tubeless tyres may only be used on wheels designed for this type of

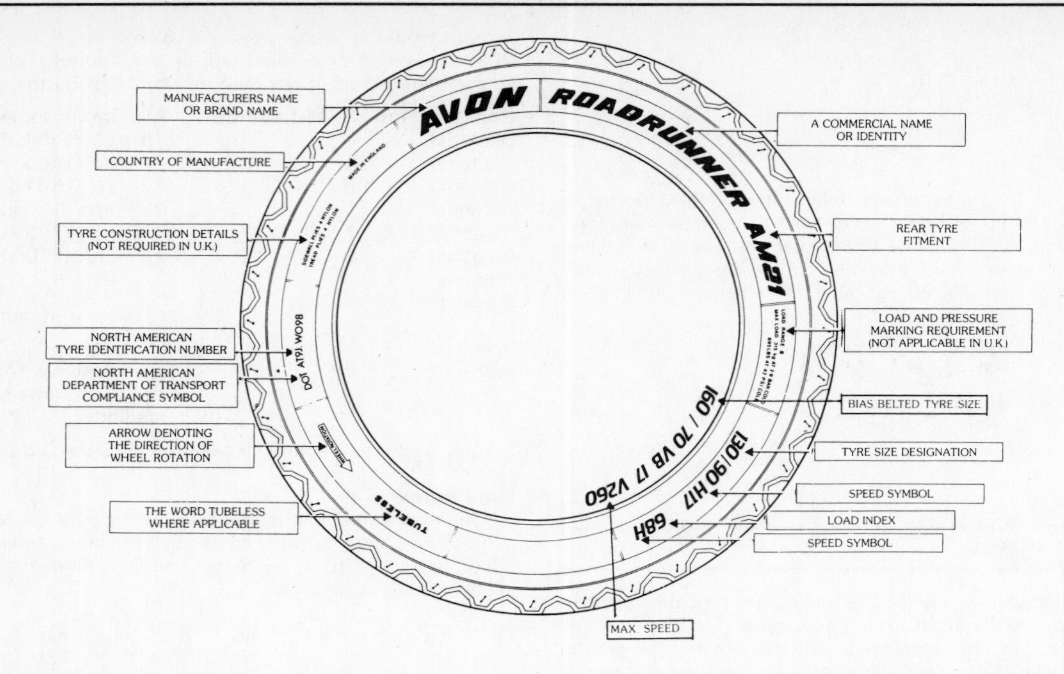

tyre. If you wish to use tubeless tyres on other wheels, check whether this is permissible with the manufacturer or importer. Note that tubeless tyres cannot normally be used on wire-spoked wheels, and that some cast alloy wheels may be unsuitable.

2) Check tyre condition regularly, looking for wear or damage. In the UK, it is illegal to use a tyre on which the tread depth is less than 1 mm throughout a continuous circumferential band measuring at least 3/4 of the breadth of the tyre. In practice, many machines will suffer a significant loss of adhesion and cornering precision well before this legal minimum is reached, and 2 mm is a more realistic figure to work to.

In the case of damage, it is illegal to use a tyre which has a cut deep enough to expose the casing plies or cords, or one which shows signs of bulges or other indications of the separation of the sidewall or tread from the underlying casing.

Remove stones or pieces of metal which have become embedded in the treads, and check that the underlying casing has not been damaged. Note that foreign objects left in the treads can gradually work in deeper, causing punctures.

Remove any oil or grease from the tyres. Apart from the obvious risk of skidding, oil or grease anywhere on the tyre will attack the rubber if left for any length of time. Remove using a petrol-moistened rag.

3) Check tyre pressures regularly, preferably each day before the machine is ridden, and at least once a week. Riding a machine with incorrect tyre pressures is illegal and dangerous, and will shorten tyre life considerably.

Pressures must be checked when the tyre is **cold** – never after the machine has been ridden. Tyres warm up in use, and the air pressure will rise accordingly. The specified pressures apply to a tyre at atmospheric temperature.

Use an accurate pocket gauge to check pressures; do not rely on filling station forecourt gauges, which may be inaccurate.

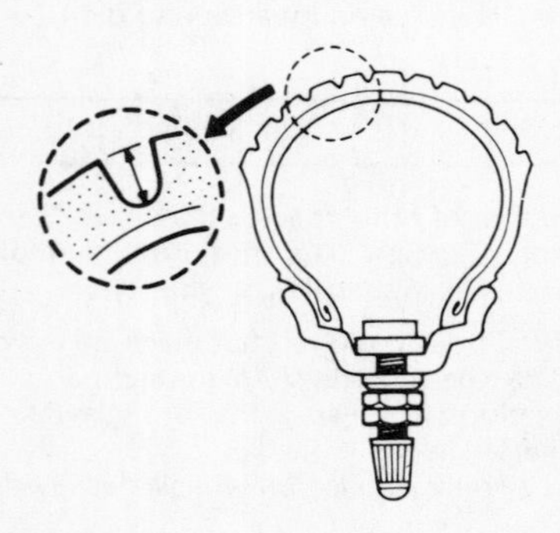

4) Inner tubes should be renewed each time a new tyre is fitted, and whenever there have been more than two or three puncture repairs. In the case of tubeless tyres, fit a new valve assembly each time the tyre is renewed. In all cases, keep the valve clean, and always refit the dust cap to exclude dirt and to provide a secondary seal in case of valve failure.

5) Check wheel condition regularly, and repair or renew the wheel if defects are noted. Rim runout must not exceed the specified limit (about 1 mm for most models), and wheel balance should be checked each time a new tyre is fitted, or whenever imbalance is felt. Check rear wheel alignment when adjusting the final drive chain tension.

6) In the event of punctures, note that in the case of tubeless tyres the tyre must be removed from the rim and a headed plug fitted from the **inside**. Repairs made using a plain plug fitted from the outside are illegal. Tubeless tyres should be renewed if a second puncture occurs.

Tubed tyres should be repaired using a good-quality self-vulcanising patch. If several previous repairs are found, or if the tube has started to deteriorate, fit a new tube of the correct size. Always locate and remove the object which first caused the puncture, and check that any spoke ends are ground flat and that the rim band is in good condition.

Electrical cable sizes and ratings

The size of an electrical cable is expressed in terms of the area of the copper conductor, and this indicates its current-carrying capacity and its voltage drop (loss due to electrical resistance) over a given length. The conductor is almost invariably made from copper, and is composed of a number of thin strands rather than a single thick one, to allow the cable to bend easily without risk of the conductor breaking. The cable size is shown as the number of strands and the diameter of each one, eg; 14/0.30 would denote 14 strands, each of 0.30 mm diameter. The conductor is twisted into a gentle spiral bundle, and the whole is then protected by a plastic insulator layer, normally available in a wide range of colours to permit colour-coding of the electrical system.

As might be expected, electrical cable is produced in both metric and Imperial sizes. The manufacturer will select the cable sizes used throughout the electrical system according to the loads expected in a given circuit. The cable must be able to carry the appropriate current without overheating or excessive volt-drop, both of which are affected by the conductor area. On the other hand, the larger the cable the higher will be the cost, so the manufacturer will use the smallest practical size for the job. When selecting replacement cable, or cable for additions to the system, use a size equivalent to that already fitted, or a little larger. Owners of Italian machines should note that it is not uncommon for wiring of too small a diameter to be fitted as standard. This can result in frequent electrical problems, especially serious volt-drop in the headlamp circuit. In such cases it is worth rewiring that particular circuit using heavier cable. A summary of the commonly-available Imperial and metric cables sizes is shown below.

Size	Inch cables current rating (A)	Volt drop (V/ft/A)
23/0.0076	5.75	0.00836
9/0.012	5.75	0.00840
14/0.010	6.00	0.00778
36/0.0076	8.75	0.00534
14/0.012	8.75	0.00540
28/0.012	17.5	0.00770
35/0.012	21.75	0.00216
44/0.012	27.5	0.00172
65/0.012	35	0.00116
97/0.012	50	0.0008
120/0.012	60	0.00064
60/0.018	70	0.00057

Size	Metric cables current rating (A)	Volt drop (V/m/A)
16/0.20	4.25	0.0371
9/0.30	5.5	0.02935
14/0.25	6.0	0.02715
14/0.30	8.5	0.01884
21/0.30	12.75	0.01257
28/0.30	17.0	0.00942
35/0.30	21.0	0.00754
44/0.30	25.5	0.006
65/0.30	31.0	0.00406
84/0.30	41.5	0.00374
97/0.30	48.0	0.00272
120/0.30	55.5	0.0022
80/0.40	70.0	0.00182

It should be noted that various factors will have an effect on the cable's capacity. It is especially important to consider the rate at which heat can be dissipated from the cable; a single exposed cable being much better than cables bunched together in a wiring loom in this respect. Where the cable is included in a loom and thus less able to shed unwanted heat, reduce the effective current rating to 60% of that shown in the above table. A good rule-of-thumb is to use a size or two larger than that indicated by the expected current rating; this will obviate any overheating problems, and will ensure minimal volt-drop in the circuit.

Electrical connectors

A wide variety of electrical connectors are in use on motor vehicles in general, though on most motorcycles you will encounter only a few of these. Irrespective of the type of terminal used, a sound electrical connection is the main requirement. It is always preferable to use soldered joints. Done properly, these ensure a good electrical and mechanical connection between the conductor and its terminal, and will not give rise to problems later. Soldered joints are less common than crimped joints these days, the latter being much quicker and easier to carry out. If done carefully, and assuming that the wire end and terminal are clean prior to assembly, this type of connection is nearly as good as a soldered joint, though there remains the possibility of corrosion occurring in the joint unless it is well protected.

Soldering techniques are discussed in some detail in Chapter 3 of this manual, and if you are not familiar with the process it is worth using a few terminals for practice before any real repairs or alterations are undertaken. Crimping is fairly straightforward to carry out, requiring the cable insulation to be stripped back by the specified amount before the terminal is crimped in place using special pliers. With most crimp-fitting terminals, two crimped joints are made, one to secure the conductor, and a second to hold the insulation. It is possible to purchase terminal kits which include a selection of terminals and a crimping tool in a partitioned box. This is a good idea as an initial purchase; you can add extra terminals later if you find that you need them.

Flat blade (Lucar) connectors are commonly-used connectors also known as spade terminals, and are either crimped or soldered to the end of a cable. Available in a range of sizes, these connectors are produced in male and female patterns, allowing cables to be connected together or to electrical components having the corresponding connector half. In most cases, the terminal will be insulated with a plastic sleeve which slides over the exposed metal part to prevent short circuits.

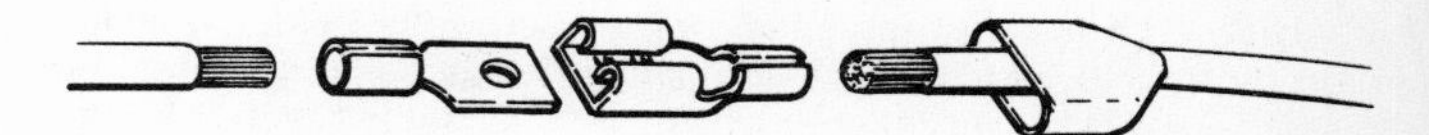

Bullet connectors are an alternative type to the flat blade connectors and are cylindrical in shape. They will often be encountered on motorcycles other than where multi-pin connectors are used. The 'bullet' terminal is fitted by soldering or crimping it to the bared cable end, the connection being made by pushing it into an insulated tubular connector.

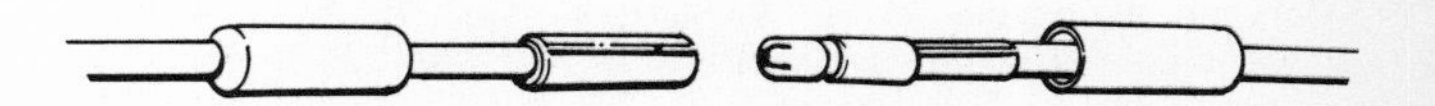

Eyelet and forked terminals will be found where components or cables from the harness need to be connected to earth (ground). The terminal is crimped or soldered to the wire end, and then secured by a bolt or screw to a sound earth point. Often a shakeproof washer will be found under the fastener head; this bites into the component paintwork, thus improving electrical contact.

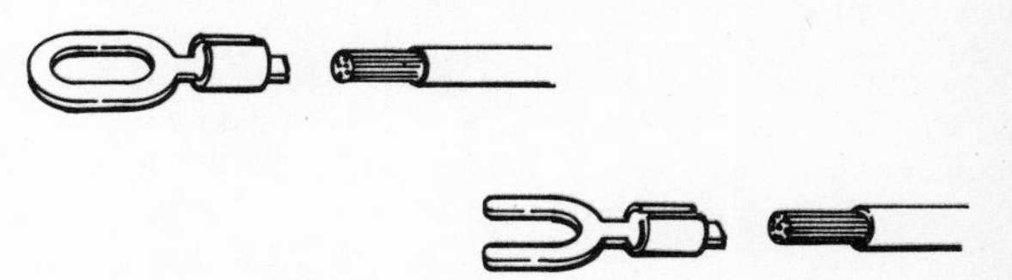

Multi-pin connectors Most connections to the main harness are made using multi-pin connector blocks, and these are usually moulded in such a way that it is impossible to fit the two halves incorrectly. Each connector will carry two or more wires relating to a certain circuit or component. Inside the plastic connector halves will be found a number of flat-blade or bullet-type connectors. These are frequently held in place with a small sprung peg or tang, and close examination of the connector will usually reveal exactly how the terminal is secured. It is normally possible to depress the tang or peg to allow the terminal to be withdrawn from the connector half, and this can be invaluable where it is necessary to remove corrosion. It is a good idea to pack the inside of each connector half with silicone grease. This excludes moisture and air, preventing subsequent electrical faults, and is especially useful on off-road machines.

Unless serious corrosion has taken place, it is not normally necessary to renew these connectors, and you will encounter some difficulty should the need arise. The connectors are normally sold complete with the switch or component to which they are wired, and it is unlikely that you will be able to locate a supplier of the connectors or terminals. The usual method of dealing with this type of problem is to remove the affected wires from the connector block and to fit a single bullet-type connector to bypass the damaged terminals.

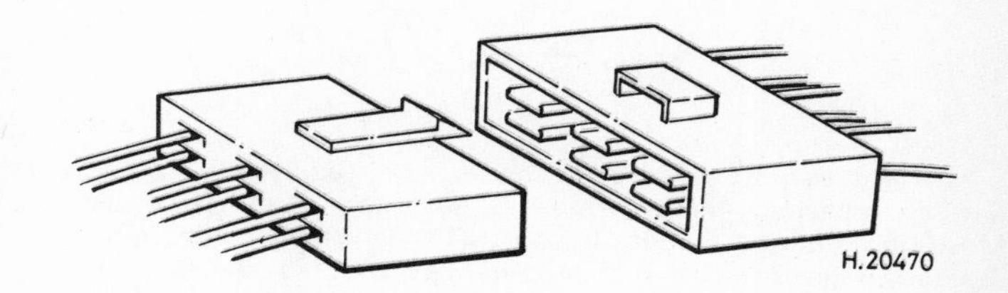

Fuses

Just about every machine uses at least one fuse to protect the electrical equipment from damage in the event of a fault such as a short circuit. The fuse acts as an intentional 'weak link' and is designed to fail well before damage can be caused to the circuit concerned. On small machines and mopeds, a single main fuse is usually all that is fitted, whilst on larger models, individual circuits are normally fused separately. The fuses are plugged into a fuse box or holder which is normally located within easy reach behind a side panel or below the seat. On many larger machines provision is made in the fuse box for the fitting of additional circuits for electrical accessories.

Glass cartridge fuses are used on many machines and comprise a short length of glass tubing inside which a fuse wire of the required rating is connected between metal end caps. The fuse is either fitted in an individual fuse holder, or more commonly, clipped between terminals in a fuse box. This type of fuse can sometimes be difficult to identify as having failed if its wire has broken or blown near the metal end caps; check by substitution, or carry out a continuity test if you are in any doubt.

Note that there are two basic types of glass cartridge fuse; cone end types which are 25.4 mm (1.00 in) long, and flat end types which are 29.4 mm (1.16 in) long. Make sure that you purchase the correct length for your machine!

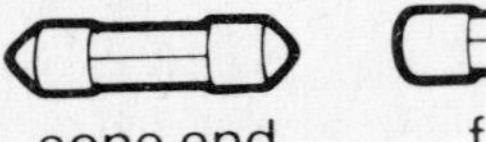

Flat bladed fuses will be found on more recent machines, and comprise a coloured plastic holder from which projects two flat blade terminals. The fuse element lies between the terminals. These fuses are simple and positive in use, though where a large number of fuses are fitted close together you may need to grasp them with a special tool or a pair of pliers to remove and fit them. As the accompanying line drawing and table indicate, the fuse rating is denoted by colour-coding, though it is normally printed on the end of the plastic holder as well.

Identification	*Rating*
Purple	3
Pink	4
Orange	5
Brown	7.5
Red	10
Blue	15
Yellow	20
White	25
Green	30

Ceramic fuses are most often found on European machines where they take the place of the glass cartridge type. The coloured ceramic body supports the one-piece fuse element and end caps, the whole assembly being clipped between terminals. On Italian machines in particular, be wary of problems caused by the soft terminals failing to grip the fuse securely; this can often cause erratic faults and can be difficult to track down unless you are aware of this possibility.

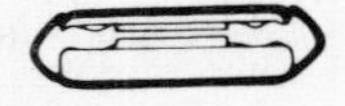

Ribbon fuses are less likely to be encountered than the more common types described above, but may be found on a few models. The fuse element consists of a flat metal strip, usually secured at each end by a small screw. When fitting a new fuse of this type be sure to handle them carefully – the soft element is easily torn or broken if twisted.

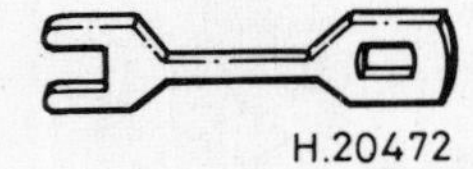

Circuit breakers are occasionally used in place of a main fuse, and have the advantage that they may be reset after a fault has occurred. Most types consist of contacts operated by a bi-metal strip, and are designed to break the circuit if the current exceeds the rated figure. The unit has a reset button mounted on the case, and this can be used to reset the circuit after the fault has been resolved and the unit allowed to cool down. The circuit breaker is a sealed unit, and if it develops an internal fault it must be renewed.

Spark plugs

Spark plugs are produced in a wide range of sizes and grades to suit various applications, and it should be noted that only one of these is likely to be suitable for a particular engine. The correct type will be recommended by the motorcycle manufacturer, and this information will be found in the owners handbook supplied with the machine, or in the Haynes Owners Workshop Manual for that model. It is unlikely that you will need to depart from the standard grade of plug, and this should never be done without first checking with the manufacturer, importer or an authorised dealer for your model. Changes of plug grade are only normally needed where the machine is used in unusual climatic conditions, or where the type of use the machine is subject to indicates that a change might be appropriate. One exception to this rule is where the motorcycle manufacturer lists alternative grades of plug for normal riding and sustained high speed use, or for frequent short journeys.

Each spark plug manufacturer employs a combination of letters and numbers to denote the thread diameter, reach, type and heat range of their plugs, as well as any special characteristics of a given plug. Although of some academic interest, space precludes the listing of the available types and sizes from all manufacturers, and in practice such information is of little consequence. The system used is specific to one manufacturer, and there is often no direct equivalent grade available from a rival company. Despite this, each spark plug manufacturer will have their own recommendation for a particular model, and spark plug stockists will have this information in the form of a comprehensive catalogue of applications. This can be useful where the make of spark plug recommended by the motorcycle manufacturer cannot be obtained locally, though as a rule it is best to adhere to the original specification.

With a few exceptions, most motorcycle engines use plugs with a 14 mm thread/20.6 mm hexagon or 12 mm thread/18 mm hexagon. A few four-stroke moped engines employ the smaller 10 mm thread/16 mm hexagon type, where fitting space is limited. The reach (thread length) of plugs varies considerably, and it is important that the correct type is chosen; too short a plug will result in masking of the electrodes, carbon build-up in the exposed threads in the cylinder head and possible misfiring. Too long a plug may result in it contacting the piston, with consequent engine damage.

The heat range of the plug is not obvious from examination, but is of great importance in use. The heat range is chosen to suit the operating temperature range of a specific engine. In most applications, the electrode area of the plug should be kept at around 450° – 950°C (840° – 1740°F). If the plug runs too cool, carbon and oil fouling of the electrodes will occur, whilst at higher temperatures there is the risk of oxide fouling, burnt electrodes and pre-ignition damage. The plug electrode temperature can be controlled by regulating the rate at which the plug can shed engine heat, and in practice this is accomplished by altering the effective length of the ceramic insulator nose inside the plug body.

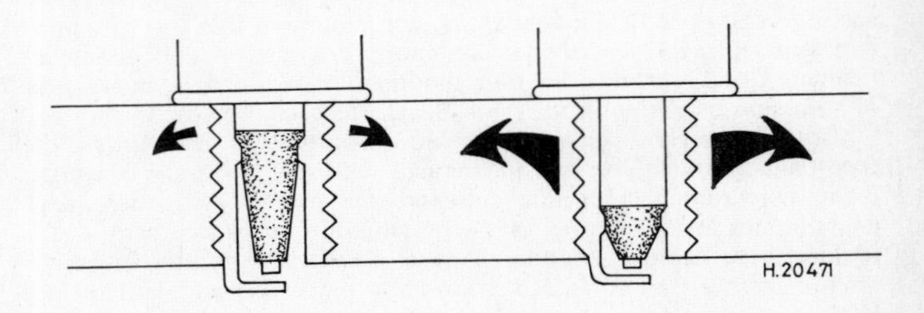

A 'hot' or 'soft' plug
The long insulator nose limits the rate at which heat is lost from the electrodes

A 'cold' or 'hard' plug
The shorter insulator nose allows heat to escape quicker, keeping the electrodes cool

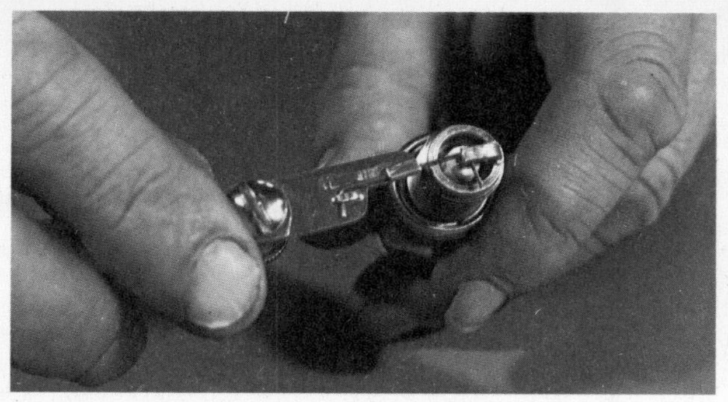

Spark plug maintenance: Checking plug gap with feeler gauges

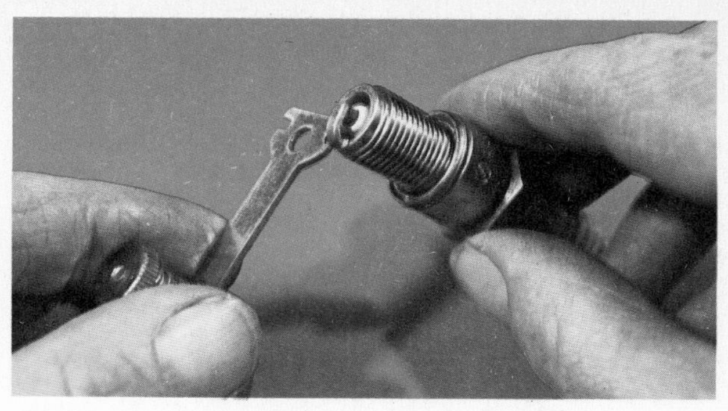

Altering the plug gap. Note use of correct tool

Spark plug conditions: A brown, tan or grey firing end is indicative of correct engine running conditions and the selection of the appropriate heat rating plug

White deposits have accumulated from excessive amounts of oil in the combustion chamber or through the use of low quality oil. Remove deposits or a hot spot may form

Black sooty deposits indicate an over-rich fuel/air mixture, or a malfunctioning ignition system. If no improvement is obtained, try one grade hotter plug

Wet, oily carbon deposits form an electrical leakage path along the insulator nose, resulting in a misfire. The cause may be a badly worn engine or a malfunctioning ignition system

A blistered white insulator or melted electrode indicates over-advanced ignition timing or a malfunctioning cooling system. If correction does not prove effective, try a colder grade plug

A worn spark plug not only wastes fuel but also overloads the whole ignition system because the increased gap requires higher voltage to initiate the spark. This condition can also affect air pollution

The accompanying colour photographs illustrate the main problems likely to be suffered by spark plugs, and this can form a useful basis for diagnosing running faults. Note that if the condition of the plug taken from your engine indicates that the plug grade is incorrect, check this carefully before considering a change of grade. It is more usual that abnormal plug condition is reflecting some other engine fault.

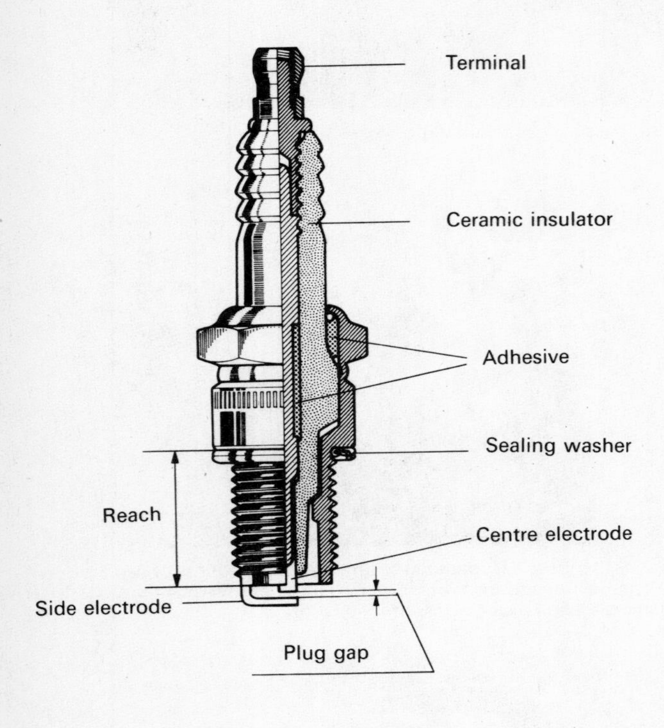

Preparation for the MOT test

In the UK, all machines more than three years old are subject to an annual test to ensure that they meet minimum safety requirements. A current test certificate must be issued before a machine can be used on public roads, and is required before a road fund licence can be issued. Riding without a current test certificate will also invalidate your insurance.

For most owners, the MOT test is an annual cause for anxiety, and this is largely due to owners not being sure what needs to be checked prior to submitting the machine for testing. The simple answer is that a fully roadworthy machine will have no difficulty in passing the Ministry test.

One common complaint is that the test is not objective, and to some extent this is true. It is not possible within the current definition of the test to give specific measurements and clearances for all aspects of every machine in use in the UK. This means that the tester must assess each machine on an individual basis, following guidelines issued by the Department of Transport. Whilst this can lead to 'grey areas' on some occasions, it is preferable to tolerate this approach in preference to the expensive and time consuming testing procedures which exist in some other countries.

In this section, the MOT test procedure is outlined, together with suggestions as to how the owner can check his or her own machine prior to submitting it for the Ministry test. It must be stressed that this is by no means an exhaustive treatment of the subject, and for owners requiring definitive information, it is suggested that a copy of *The MOT Inspection Manual, Motor Cycle Testing,* is studied. This is available from HMSO, or through bookshops.

The suggested inspection routine

The following inspection sequence is that recommended in the MOT Inspection Manual for motorcycle tests, and is similar to the procedure adopted by most testing stations. Note that it is not compulsory, and so individual testers may work in another sequence. It does, however, give a good indication of the areas to be checked.

Solo machines

1 Sit on machine: inspect and operate all controls, switches and horn. Test handlebars, head bearings, forks and front suspension
2 Move to front of machine: check all front lights and indicators. Check front brake master cylinder (if fitted)
3 Place machine on stand with front wheel raised: check or inspect steering, head bearings, front forks, wheel (including bearings), tyre, front brake
4 Proceed down offside of machine: inspect frame, seat, footrests, rear suspension and final drive and exhaust system, tyre, wheel and brake
5 Proceed to rear of machine: check rear and stop lamp, reflector and indicators
6 Repeat 4 and 5 above on nearside of machine
7 Remove machine from stand and test rear suspension
8 Check wheel alignment
9 Conduct headlamp aim check
10 Conduct brake performance check

Sidecars

1 Tyre, wheel, brake (if fitted) and suspension
2 Attachment points and structure
3 Swivel joints on leanable sidecars
4 Wheel alignment

Note: Generally the sidecar examination would be performed towards the end of the inspection of the solo machine but before conducting a headlamp alignment or brake performance check.

Preparation for the test

The above checklist, taken from the MOT Inspection Manual, gives a fairly clear indication of the areas to be checked. Space precludes a full list of detail items to be assessed, and in any case this is largely a matter of interpretation on the tester's part. Some of the more important areas of preparation, checks and settings are outlined below.

Always clean the machine before taking it for testing. This gives a better impression and allows you and the tester to see clearly the various areas to be tested. It should also be noted that the tester can refuse to accept the machine for testing if, in his opinion, it is so dirty that an inspection cannot reasonably be carried out.

It is advisable to carry out a normal service operation prior to the test. This ensures that all normal adjustments are correct, and will bring to your attention any areas which might be suspect. If you follow carefully the recommended full service procedure outlined in the owners handbook or the Haynes Owners Workshop Manual for your model, it is unlikely that the machine will fail.

You must also be able to provide proof of the age and identification of the machine, if this is in doubt. The machine must have sufficient fuel and oil, and be in a safe condition before the test can be conducted; if it is obviously unroadworthy, then taking it for a test will be a waste of time and money. Once started, the tester may choose to discontinue the test if it is felt that there is a risk to the safety of himself, the test equipment or the machine.

In addition to the normal service checks and adjustments, look carefully at the following areas. These show the most likely causes of failure.

Lights must all be in working order, clean, and correctly adjusted. The machine will be failed if any of the lights flicker or operate erratically, if the lens is damaged or discoloured, if the headlamp reflector silvering is discoloured or peeling or if a switch is faulty. Note that almost all machines must have a stop lamp operated from at least one switch (two on machines registered from 1st April 1986 onwards), and that if a stop lamp system is fitted, it must work.

The test allows for off-road machines with a clause exempting from the lighting test those machines 'used only in daylight which either have no front or rear position lamps at all or have such lamps permanently disconnected, painted over or masked. Do not expect to get away with taping over a defective headlamp on a normal road machine.

Do not forget to check the rear reflector. This is a small point, and easily overlooked. It can, nevertheless, mean failure if it is missing or damaged.

Turn signals (direction indicators) should work properly and be clearly visible. The flash rate must be between 60 and 120 flashes per minute, and the warning lamp must be working and visible.

Check that the horn works properly, noting that the note must be continuous and uniform. Note that gongs, bells and sirens are not permitted, nor are alternating two-tone horns.

Headlamp beam alignment is checked by one of a number of methods, according to the headlamp type (Symmetrical main beam type, Asymmetric dipped beam, European type and Asymmetric dipped beam, British American type). The test method is also dependent on the type of equipment used at the testing station.

It would not be easy to approximate the official test procedures at home with any guarantee of accuracy, but if the following procedure is followed, a reasonably accurate setting should be obtained. If you do not feel confident that the beam setting is correct, ask the garage or dealer who is to carry out the MOT examination to adjust it for you **before** the test commences; if the setting is incorrect during the test, this could result in a failure and the need for a re-test. Note that details of the headlamp adjusting screws will be found in the owner's handbook or the Haynes Owners Workshop Manual for your model.

1) Find an area of level ground with a flat wall or a similar surface at one end. Draw a line at 90° to the wall so that the machine can be set up perpendicular to the wall surface. Continue the line vertically up the wall for about four feet. Position the machine on the line with the front surface of the headlamp 3.84 metres (12.5 feet) from the wall.
2) Measure the distance between the ground and the centre of the headlamp lens, with the machine resting on its wheels and an assistant seated normally to hold the machine upright. Draw a horizontal line on the wall at this height, and intersecting the vertical line.
3) Check that the motorcycle wheels are in alignment and facing straight ahead along the line on the ground. With the assistant still seated normally, switch on the headlamp and note the position of the beam in relation to the markings on the wall.
4) In the case of symmetrical main beam type headlamps, set the dip switch to the main (high) beam position. The centre point of maximum intensity on the wall should coincide with the intersection of the vertical and horizontal lines.
5) With European or British American type asymmetric dipped beam headlamps, arrange the area of maximum intensity to the left of the vertical line, and just below the horizontal line.

Suspension and steering must work smoothly and without undue stiffness or freeplay. Worn forks, steering head bearings or rear suspension pivots will all result in failure, as will obvious oil leakage from the forks or rear suspension unit(s).

Controls and cables must be in good condition and correctly adjusted. Seized or worn control pivots will mean a failure, as will damaged or frayed control cables.

Wheels and tyres must be checked carefully. Tyres should be in good condition, of the correct type and size, within tread wear limits, correctly inflated and undamaged. Wire spoked wheels must have no loose, broken or rusty spokes. Aluminium rims will be failed if they have been repaired. All types of wheel must be free from buckles in the rim or excessive runout, and there must not be free play in the wheel bearings. Check that the wheels are correctly aligned and that final drive chain tension (where applicable) is correctly adjusted.

Brakes should work smoothly and progressively, as well as efficiently. Worn pads or shoes, grabbing, judder, wear or damage in the controls may result in failure. In the case of hydraulic systems there should be no signs of leakage.

Exhaust system must be of the correct type and in good condition. Severe rusting, holes or splits, damaged mountings or excessive noise will all mean failure. Repairs are acceptable, provided that they are secure and of a permanent type; exhaust bandage is unlikely to qualify in this sense.

Frame structure should be checked for damage or serious corrosion. Any indications of stress or accident-damage fracturing will mean a failure, as will extensive rusting. Rusting is likely to be more of a problem on pressed-steel frames of the type used on mopeds and scooters than on tubular frames. Footrests, seats and handlebars should be secure and in good condition.

Arranging the test

Once you are satisfied that the machine is in good order, contact the local MOT testing station and book a test. Note that this is essential if the old certificate has expired. You are allowed to use the vehicle on a public road without a current test certificate only when: taking the machine to a testing station for a test **booked in advance**, bringing it away from a testing station after it has failed the test, taking it to or bringing it from a place where by previous arrangement repairs are to be or have been done to remedy the defects for which the vehicle was failed.

Note that even in the case of the above exceptions, the machine must be roadworthy and comply with the various Construction and Use regulations. Note also that your insurance cover may not be operative if the MOT certificate has expired.

Glossary of technical terminology

A

Accelerator pump A carburettor device for temporarily increasing the amount of fuel delivered (i.e. for richening the mixture) so as to improve acceleration.

Aeration Mixing of air and (usually) oil to form an undesirably frothy mixture.

Air filter Usually a paper, fabric, felt or gauze element through which the engine draws its air, and intended to trap wear-creating particles of abrasive mixture.

Air-fuel ratio Proportions in which air and fuel are mixed to form a combustible gas.

Alternating current Electricity varying in polarity and potential (voltage), reversing direction regularly. The kind generated in an alternator (which see). Abbreviated to a.c. (Compare with **direct current**).

Alternator A generator of alternating current (a.c.) electricity. Often does not have brush gear. (see **alternating current**)

Anode The positive pole in a direct current situation (Compare wth **cathode**).

Arc weld To join material by the use of a heavy electric current so that an arc is struck to generate heat.

Argon arc To electrically weld material using a shield of inert argon gas to avoid oxidation of the parent metal. Often used in the welding of aluminium.

Armature That part of an electrical apparatus such as a solenoid, dynamo or magnet, which comprises the electrical windings in which a current flow or a magnetic field is generated or excited. (See **magnet**).

Aspect ratio With a tyre, the ratio of the section's depth to its width. Old tyres had an aspect ratio of 100% (as fat as they were deep) but modern motorcycle tyres are flatter with an aspect ratio of say 80%.

B

Belt drive Drive by a belt, originally often of rubber-fabric composition, but today may incorporate terylene or nylon and have a toothed construction.

Bevel gear Gear with slanted teeth, a pair of such gears turning the drive through ninety degrees.

Big-end The larger of two bearings of a connecting rod and the one mounted on the crankpin.

Bi-metallic Made of two metals.

Blow-back Mixture blown back out of the carburettor against the normal direction of air flow into an engine.

Bobweight A countershaft weight on a crankshaft offsetting piston and con-rod mass.

Bore Diameter of a cylinder. Slang for the cylinder itself. In some senses, the surface of a hole.

Bore: stroke ratio The ratio of cylinder diameter to stroke. When these are equal the engine is said to be square (which see). May be given as a ratio or a percentage.

Boss A raised area on a component, the thickness being provided for more strength.

C

Calipers Leg-like measuring instrument (pair of), hinged in the middle and used to gauge gaps, bores or external sizes. In brakes, the component spanning the disc and carrying one or two pistons and pads. (See **swing caliper**).

Cam follower A component with intimate contact against the camshaft (which see) so as to take up the cam's eccentricity in a non-destructive manner, transmitting the resultant linear (or non-linear) motion to the rest of the valve gear. Followers may have flat or curved feet, sometimes fitted with rollers.

Camplate A flat (or slightly bowed) plate in which are formed slots in which pegs may move for converting complete rotary motion into a sliding mode for the operation of, say, control gear (such as selector forks for controlling gearbox ratio choice).

Camshaft A rotary shaft equipped with lobes (which see) for converting rotary into linear movement, generally for the operation of valve gear in poppet valve engines.

Cantilever A bracketing arrangement notable for its overhang and absence of support at one end. Some sub-frames are cantilever.

Capacitor Strictly, a condenser (which see). But, by convention, often one of considerably larger capacity than a normal condenser and able to perform a smoothing role in battery-less current generation.

Carburettor Mechanism for mixing air and fuel to form a combustible mixture and (ideally) automatically correcting for speed, load and other variables.

Catalyst An ingredient present in trace quantities in glass fibre mixes to bring about setting. Not to be confused with accelerator, which controls the rate of setting.

Cathode The negative pole in a direct current situation. Sometimes spelt kathode. (Compare with **anode**).

Centre of gravity The point from which a mass could be suspended so that it would be in 'all round balance' and would remain in any attitude in which it was placed. The 'centre of its mass' so to say.

Centrifugal To be thrown outwards. A force. The opposite – the tending inwards – is centripetal.

Clearance volume The space inside the combustion chamber when the piston is at the top of its stroke, and extending half way up the sparking plug threads. Sometimes called the trapped volume.

Clutch A device for engaging or disengaging the engine from the driving wheel and (except for some specialised applications or in historic form) so designed that connection may be engaged smoothly and progressively at any time.

Coefficient The reduction of a characteristic to a numeral value and related to basic units (e.g. per degree Celsius).

Collet A ring-shaped device, usually divided into two segments, for wedging a component on to a rod, shaft, spindle etc. Especially to be found on valve and suspension units to enable the spring retainer to lock itself against the valve stem or damper rod.

Commutator Part of a rotating armature (which see), against which the pick-up brushes rub, so that electricity generated in the spinning armature may be collected in the proper cyclic manner for delivery to the main circuitry.

Compression Squeeze smaller, particularly a fresh charge of mixture in the cylinder by the rising piston.

Compression ratio The extent to which the contents of the cylinder are squeezed by the rising piston. The ratio of the swept volume (cubic capacity) plus the clearance volume (cylinder head space) in relation to the clearance volume alone (which see).

Concave Curved inwards. Hollow or cave-like. Opposite to convex (which see).

Concentric Tending to a common centre. Also a proprietary name of an Amal carburettor in which the float chamber is concentric with the mixing chamber's main jet, thus reducing mixture errors caused by swirl.

Condenser Electrical device able to store electricity and particularly to release it very rapidly. Can assist in the control of arcing in a make-and-break system. Properly called a capacitor.

Con-rod or connecting rod The rod in a reciprocating engine connecting the piston to the crankshaft via big and small ends (which see).

Constant rate A spring is this when each increment in load produces an equal change in length. (Contrast with **multi-rate** and **progressive rate**).

Contact breaker Abbreviated to c.b. An electrical switch designed to permit a field-producing current to flow in an HT coil and then to abruptly cut the current so that the rapid inward collapse of the magnetic flux produces a strong, high voltage current in the secondary windings.

Convoluted Coiled, sinuous. Convoluted hose is alternately tapered and expanded along its length.

Coupling gears Gears that couple two crankshafts together

Cradle A support, usually designed to embrace components. A type of frame in which the bottom tubes embrace the power unit.

Crankcases The structurally-strong chamber in which is carried the crankshaft (which see). The singular and plural forms are used indiscriminately as, in much motorcycle design, this component is made in two non-mirror images to form a pair.

Crankshaft A contrivance, using the principle of the eccentric (crank) for converting the reciprocating piston engine's linear power pulse into rotary motion.

Cross-ply Form of tyre construction, almost universal in motorcycling, in which the wraps of fabric in the tyre carcase are laid over each other diagonally instead of radially. (see **radial ply**).

Cross valve A rotary valve placed above the cylinder and handling exhaust as well as inlet gases on the four-stroke cycle. Permits high compression ratios with a good resistance to detonation and yields excellent fuel consumption.

Crownwheel The larger of the two gear wheels in the reduction (or final drive) pair at the axle of a shaft-drive motorcycle. The smaller is called the pinion.

Cruciform Cross-shaped.

Cush drive A shock-absorbing type of transmission system, often involving rubber.

Cush hub A rear wheel, provided at the hub with a means of separating wheel from sprocket in respect of transmission shocks.

Cycle Slang term for C.E.I. thread favoured by the cycle industry as it is easily rolled rather than cut. Has close pitched fine threads usually 26 t.p.i. in small sizes and 20 t.p.i. in the larger.

Cylinder A parallel-sided circular cavity, usually containing a piston.

Cylinder head End piece closing of the blind end of a cylinder.

D

Damper A device for controlling and perhaps eliminating unwanted movement. In suspension systems, for quickly arresting oscillations, and for absorbing unwanted energy to release it as heat.

Decarbonise To remove accumulated carbon and other deposits from the combustion chamber, inlet tract and exhaust system.

Decompressor A small, manually-operated valve on a two-stroke to release above-piston compression. Used to stop the engine and also, on a trials machine, as an aid to descending steep hills. (Compare with **valve lifter**).

Deflector crown A raised part or hump on the piston crown of some two-strokes to deflect the incoming fresh charge away from the exhaust port.

Density Solidity or heaviness. Also the ratio of the substance in question's mass to that of an equal volume of pure water, and termed relative density or specific gravity. (see latter).

Depression A concavity, say in a panel. Also an engine term for the amount of partial vacuum in the inlet manifold, measured in pounds per square inch at atmospheric pressure or in inches of barometric mercury.

Desmodromic A method of operating poppet valves so that they are positively closed as well as opened. Cams, and sometimes rockers and fingers, are often used. Supplementary return valve springs may or may not be featured. The purpose is to maintain full control at high rpm, with savage valve events also specified.

Detent (spring or plunger A mechanical device to lock a movement and in particular the selector system of a gearbox.

Detonation Explosion of mixture in the combustion chamber, instead of controlled burning. May cause a tinkling noise under an open throttle. Intensely destructive. May be confined to the end gases only or may involve the whole charge.

Diode An electrical device which allows a current to flow in one direction only. Often used as a simple rectifier on mopeds etc. See zener diode.

Direct current Electricity of constant polarity (direction) which may or may not fluctuate in potential (voltage). The kind of electricity produced in a dynamo or stored in a battery. Abbreviated to d.c. (Compare with **alternating current** and **alternator**).

Displacement The amount of volume displaced by the piston of an engine on rising from its lowest position to its highest. In some cases may be marginally different from the cubic capacity calculated from the bore and the crankshaft's eccentricity (throw).

Distillation The process of boiling a liquid and then condensing the vapour to yield a pure form of that liquid.

Dog A projection from a moving part, mating with another dog or a slot, on another part, so that the two components may be locked together or left free of each other. Much used in gearboxes to connect pinions or a pinion to a shaft.

Downdraught Downward inclination of the induction tract, usually the carburettor too.

Drum brake One with a rotating chamber (of drum shape) attached to a wheel and inside which are placed stationary pieces of friction material which may be forced outwards against the inner periphery of the chamber to retard motion.

Dry liner Cylinder liner not in contact with water (see **liner**).

Dry sump Four-stroke lubrication system in which the oil is carried in a separate container and not in the sump. Drainage into the sump is removed by a scavenge pump so that the sump is kept dry.

Duplex Two. A duplex frame has two front down tubes. A duplex chain has two rows of rollers (a simple chain has but one).

Dwell That period of rotation of a valve or contact-breaker cam in which events, as it were, stand still.

Dykes ring A piston ring of 'L' section, the long upright in contact with the cylinder wall and the base of the 'L' inserted in the narrowed groove in the piston. The purpose is to get the top ring as close to the piston crown as possible to protect the top land.

Dynamic Moving, in action – the opposite of lifeless or static (which see).

Dynamo A generator of direct current (d.c.) electricity.

E

Earth The grounded pole of a battery. A connection to earth (or ground). By definition, of zero potential (voltage).

Eccentric Not central. An offset pin used to drive or be driven.

Electrode A conductor with an end, from that end being taken the electricity.

Electrolyte The liquid in a battery, usually an acid but sometimes an alkali.

Electro-magnet A magnet, strongly excited by an electric current, used to create a local field by means of flowing the current in windings. Has the quality of losing virtually all magnetism (which see) practically instantaneously the moment the current flow ceases.

Energy transfer A system of ignition in which closed contact-breaker points allow energy to build up in the alternator windings, point opening resulting in a rush of current to an external ht coil which transforms its low voltage into high voltage for the sparking plug. Correct ignition timing is vital to spark strength.

Expander ring An auxiliary ring, inserted behind a scraper ring, and intended to raise ring-to-wall pressure to the desired value (see scraper ring).

F

Ferrous Iron. (Steel is, of course, an iron).

Filament Electrical resistance wire incandescing (glowing) when made to pass an adequately heavy current and thus yielding light.

Finning A thin but wide plate-like projection, usually arranged in multiples, and generally functioning for the dissipation of heat. Fins are sometimes used for strengthening and often for appearance.

Flat head An engine with a flat cylinder head instead of curved internal contours. The valve arrangement may be sv or ohv.

Flat top Piston with a flat top, in contrast to one with a concave (hollow) or convex (domed) crown.

Flat twin (or four) An engine with horizontally-opposed cylinders and having a flat configuration (also called a pancake).

Float A buoyant object in, say, a carburettor, used to operate another component such as a fuel-admission needle. Also used in a fuel gauge.

Float chamber A carburettor component used to stabilise the fuel level regardless of the head of fuel supplied by gravity flow from an overhead tank or by a pump. It uses a float to operate a cut-off valve.

Float needle The needle part of the fuel cut-off and admission valve in a carburettor float chamber (which see). It works in conjunction with a float to raise it against a needle seat.

Flywheel A rotating mass of considerable weight and radius, used to smooth out power impulses and to store energy to assist clutch engagement 'nicety'. It is also an aid to idling, low-speed running when in gear, and delays changing down on hills.

Free play Clearance in a mechanism, say, a clutch control, introduced to avoid any destructive tightness.

Friction The resistance set up in two bodies moving relatively to each other. (Compare with **stiction**).

Fulcrum The point (or sometimes edge, where line contact is involved) about which a leverage system pivots. (See **lever**).

Full-wave Half of the wave of alternating current is positive and half negative. Full-wave means from maximum plus to maximum minus, or 'all of the electricity'. But half-wave is from either maximum to just the neutral mid-point. (See **rectifier**)

G

Gaiter A tubular, usually corrugated and always flexible, shroud round a sliding or otherwise moving joint, and used for protection of working components and/or for appearance.

Gassing The giving-off of gas from the cells of a battery due to excessive charging. Explosive hydrogen-oxygen mixture is released. Gassing does not commence prior to the achievement of full charge. Also flooding of an engine with fuel so that it ceases to run or fails to start because of a hopelessly rich mixture.

Gas weld To join material with the aid of heat from gaseous combustion. In the workshop, acetylene and oxygen are frequently employed, though propane and compressed air is now much used in industry.

Gear A component, often circular, with projections for the positive transmission of movement to a companion gear which may, or may not be, of the same shape and size.

Gearbox An assembly containing the transmission components used in varying the ratio of the gearing. Even when this is effected by short chains and sprockets, and other methods of ratio variation, the term gearbox is still used.

Gear ratio The ratio of turning speeds of any pair of gears or sprockets being the arithmetical product of the number of their teeth. Particularly the total drive ratio of each set of gears in a gearbox or the overally transmission ratio(s) of a vehicle.

Grease A stabilised mixture of a metallic soap and a lubricating oil. Lime (calcium) and lithium are both used as base soaps.

Gudgeon pin The pin, usually made of hardened steel, linking the piston to the small end of the connecting rod. Possibly the most high stressed bearing of an engine.

H

Hairpin A type of spring, once much used for valve gear, shaped like a hairpin (but sometimes with one or more coils at the 'closed' end).

Halogen lighting See quartz iodide. The filament regeneration cycle at work in a quartz iodine lamp will, in fact, function with other halogen gases of which bromine is also coming into commercial usage.

Heat sink The ability of, say, a brake drum to store heat until it can later be shed.

Helicoil Proprietary system of strengthening female threads formed in weak metal by making a rude thread of a special form and lining it with a tough wire insert that itself forms the true thread. Also used for reclaiming stripped threads (in holes only).

Hertz A measurement of frequency. A Hertz is a movement of one cycle per second. For large values, the Kilo-Hertz is a convenient unit, being 1000 cycles per second.

High tensile Material of high tensile (or 'stretch' strength). Tough.

Homogeneous Thoroughly mixed. The same all through. Applied to fuel/air mixture.

Honing Achieving a good finish and precision control of size, to better than one ten thousandth part of an inch is, say, cylinders, by a slowly-proceeding abrasion process. Similar to grinding, but done slowly.

Horizontally-opposed A type of engine in which the cylinders are opposite to each other with the crankshaft in between.

Hub The centre part of a wheel, sometimes called a nave.

Hub centre steering Motorcycle steering modified to car practice so that the lock to lock axis lies within the hub itself.

Hydrocarbon Hydrogen and carbon compound forming the basis of all lubricants and oils formed from crude oil.

Hydrometer A device for measuring the specific gravity (S.G.) of a liquid, and of battery acid in particular so to assess the state of charge. It consists of a small float giving a reading against its containing glass cylinder's graduations, liquid being raised into the cylinder by means of a squeeze bulb.

Hygroscopic Water absorbing. (Silica Gel is a commonly-used, extremely hygroscopic substance).

Hypoid oil An extreme-pressure oil formulated to stand up to severe and unique conditions in hypoid transmission gears.

I

Idler Gear interposed between two others to avoid using overlarge proper gears. An idler does not alter the ratio between the proper gears.

Impeller A powered device used to impel coolant through an engine and radiator to assist natural thermo-syphon action (which see). Usually a rotary, vane-type pump.

Incombustible Unburnable. The opposite of combustible.

Inertia The property of matter by which it wants to continue at rest, or in motion, without change of direction or velocity (See **momentum**).

Injector Equipment for squirting. Used for both fuel and oil. Also the proprietary name of the Wal Phillips fuel injector, a system of metering fuel to a petrol engine using gravity supply, a rotary variable jet and a throttle butterfly.

Insulator Substance or component for handicapping the transfer of heat or entirely preventing the transmission of electricity.

Interference fit Two parts so sized that the inner is slightly larger than the outer. When forced together they jam in place, grasping each other to obstruct separation.

In-unit Engine and gearbox, manufactured as separate components, but fastened together to form one whole. (Compare with **unit**).

Isolastic Proprietary name registered by Norton-Villiers for a frame in which the engine, transmission and rear wheel are isolated in respect of their vibration by means of rubber mountings.

J

Jet A hole through which passes air, petrol or oil, the size determining the quantity.

Jockey A wheel placed between the centres of a belt or chain, engaging with one run of it, and used to adjust tension.

K

Kickstarter A crank, operated by foot, for starting an engine.

Kilovolt One thousand volts, abbreviated to Kv.

Kinetic energy The energy of motion, and not that of position.

L

Land The raised portion between two grooves. However, the 'raised' part may not actually be proud of the main material but merely superior to the 'lower' section(s).

Latent heat The amount of heat input needed to change a solid to its liquid state, or a liquid to a gas, without change of temperature. A fuel with a high latent heat, such as methanol, has a considerable cooling effect on an engine.

Layshaft A gearbox shaft parallel to the mainshaft and carrying the laygears with which the mainshaft gears mesh to achieve ratio change.

Leading link A form of front suspension using a pivoting link – approximately horizontal – with the axle in front of the pivot. In the short link the pivot is reasonably close to the axle. In the long link design the pivot is behind the wheel. The same distinctions apply to the trailing link (which see) but are rarely applied.

Leading shoe A brake shoe which has its cam end 'reached' by any given spot on a revolving brake drum before that spot gets to the shoe's pivot end. (Compare with **trailing shoe**).

Lean out Found on sidecar outfits. The machine's steering head is leant out of the vertical (the top away from the sidecar) to combat road camber.

Liner A detachable insert in a component used either to reduce size or to provide a better working surface or to restore one. (see **sleeve**).

Lines of force The imaginary lines tracing out the pattern of a magnetic system and 'along which magnetism flows'. Shown by placing a card over a magnet, sprinkling on it some iron filings, and tapping the card. (See **magnetism**).

Live The pole of a battery other than the earth one. A lead carrying current 'above' ground potential. (See **earth**).

Lobe The total part of a cam that is eccentric to its centre, the part not on the base circle.

Long reach Sparking plug term for a plug hole of three-quarters-inch depth. (See **reach** and compare with **short reach**).

Long stroke Undersquare (which see), but to a significant degree. (See also **bore:stroke ratio**).

Lubricant A substance interposed between rubbing surfaces to decrease friction. Friction cannot be eliminated entirely and so-called frictionless bearings do not exist.

M

Magnetism A force invested in magnetic situations or substances, having the quality of attraction and repulsion depending on polarity, and of some similarity to electricity.

Magneto A self-contained ignition spark generating instrument featuring primary and secondary (ht) windings and requiring no external power source. Ones with fixed magnets are called rotating armature magnetos (or just magnetos); those of the fixed winding sort are called either rotating magnet or polar-inductor magnetos.

Main bearing The principal bearing(s) on which a component is carried but usually reserved exclusively for the crankshaft.

Mainshaft A principal shaft, as in an engine or a gearbox.

Master cylinder The operator end of an hydraulic control system, so called because (on cars) it operates several slave cylinders (which see). A large piston, in a cylinder, is depressed manually to expel fluid down a pipe(s) to the slave cylinder(s).

Megaphone An outwardly tapering chamber in which exhaust gas expands, sometimes fitted with a small reverse cone (which see). The original purpose was to extend a plain pipe rearward without altering its effective length. Later it was found that a megaphone could increase power if blended with other engine characteristics such as valve timing.

Mesh The closeness of fit of the teeth of gears and similar articles. Also the grid-like formation, often woven if made from fabric, but sometimes made of metal or moulded from plastic; mesh then refers to the closeness of the weave as in tight mesh and loose mesh.

Mixing chamber That part of a carburettor distinct from the float chamber both in function and layout and in which the air and the fuel mix as they meet. (See **concentric**).

Momentum The desire of a moving object to continue in motion. (See **Inertia**).

Monobloc Made in one. Also a proprietary name used by Amal (and then taking a capital M) to differentiate a carburettor featuring mixing and float chambers formed in one piece from an earlier layout in which these items were separate.

Monograde oil An oil, the viscosity of which falls within the limits set for a single SAE number. (See **multigrade, viscosity** and **SAE.** Compare with **straight oil**).

Multigrade oil An oil the viscosity and temperature characteristics of which are such that the low and high temperature viscosities fall within the limits of different SAE numbers. (See **Viscosity, SAE** and **monograde**).

Multi-rate A spring which changes length unequally for equal increments of load. (Contrast with **constant rate** and **progressive rate**).

N

Needle roller A roller, usually of hardened steel, having its length greatly superior to its diameter.

Negative earth Connecting the negative or minus pole of the battery to earth.

Non-ferrous Not iron.

O

Octane A measure of the knock resistance of a fuel on a scale of zero to 100 with higher values extrapolated. Most British petrol lies in the band 92 to 101, the larger the number, the more knock resistant the fuel.

Odometer A mileage recorder.

OHC Abbreviation for overhead camshaft (which see).

OHV Abbreviation for overhead valve (which see).

Oilbath Invariably for a chain and intended to provide lubrication by partial submergence, as well as to protect from dirt.

Oil cooling The use of oil as a cooling medium to transfer heat from a hot component to the environment (atmosphere or even a cooler part of the machinery).

Oil thrower A specially-shaped ring or plate designed to throw oil away from a protected site. May take the form of a reverse scroll (which see).

Oldham coupling A tongue and groove sliding joint, much used for vertical shafts (which see), having the ability to handle modest amounts of misalignment of shafts.

Otto cycle The cycle of operation of an engine featuring charge compression as a prime element, and invariably applied to four-stroke engines only.

Overhead valve A four-stroke engine with the poppet valves in the cylinder head and not at the side of the cylinder (side valve). Abbreviated to ohv.

Overhead cam An ohv four-stroke with the cams mounted in or on the cylinder head.

Overlap The duration of crankshaft rotation during which the inlet and exhaust valve are open at the same time. Equally split overlap means the inlet opens as much before TDC as the exhaust closes after.

Over-square Bore greater than the stroke (see **short stroke**).

P

Pawl A catch to mesh with a ratchet wheel, usually to prevent reverse motion.

Periphery Round the outside. Circumference.

Permanent magnet A magnet made of very retentive steel alloy which holds its magnetism well, in contrast to soft iron and electro-magnets (which see).

Petroil Lubricant system for two-strokes using oil mixed with the petrol prior to the petroil mixture's admission to the carburettor.

Petroleum jelly A semi-solid formed mainly from a wax but having a high oil content and refined from crude petroleum. Often used on battery terminals.

Phosphor bronze An alloy of copper and tin plus a trace of phosphorus, having excellent bearing qualities, e.g. small-end bushes.

Pinion In transmission terms, strictly the smaller of a pair of gears but colloquially any gear. The larger gear (or sprocket) is termed the wheel.

Pinking Detonation (which see). A tinkling noise from an overloaded engine.

Piston A moving plunger in a cylinder, intended to seal the cylinder and to accept or deliver thrust.

Piston boss The material below the piston crown (and also joined to the skirt) which carries the gudgeon pin.

Pitch The nominal distance between two specified points such as gear teeth, spring coils or chain rollers.

Planetary A system of gears in which two or more wheels orbit round a central sun wheel. Found in some transmission systems.

Plug cap A cover over the top of a sparking plug fulfilling several purposes such as neatness, protection, a convenient method of attaching the H.T. lead, and for incorporating the legally-required suppressor (which see).

Plug lead The wire carrying the high tension current from the spark creation apparatus to the sparking plug. Such a wire is very heavily insulated. In some cases the actual wire conductor is replaced by graphited cord (in slang, string H.T. leads).

Plunger pump An oil pump consisting of a reciprocating plunger in a chamber, and provided with ports.

Pneumatic Utilising or pertaining to air or another gas. Nitrogen is sometimes used instead of compressed air to inflate tyres.

Polycarbonate A lightweight, rigid yet resilient, shock-absorbing plastic, easily moulded and therefore chosen for the shells of safety helmets.

Poppet valve One which pops open and shut and in its crudest form like a disc on the end of a rod, the rod guiding and the disc opening or closing an associated hole.

Porous Permitting the passage of gas or liquid through the interstices in the 'structure' of the material. The opposite is impermeable.

Port Strictly, a hole or opening but now also applied to the passageway leading to the actual window.

Positive earth Connecting the positive or plus pole of the battery to earth.

Power band The band of rpm in which the engine produces really useful power in contrast to the speeds outside of it in which disproportionately much less power is available.

Pre-ignition Auto-ignition taking place before the desired moment and happening, not by sparking, but by incandescence. Invariably extremely damaging.

Pre-load Compression applied to a spring, even when it is at its maximum permitted length. Not in a free state. Suspension springs are often pre-loaded and all valve springs must be.

Pressure The exerting of a pushing force. Expressed in Britain as pounds per square inch.

Pre-unit Engine and gearbox as separate entities and produced prior to a later but similar design featuring the two built as a whole and single unit.

Primary chain That joining engine to gearbox, being the first chain the power passes through on its way to the tyre.

Primary current The low voltage current (also called low tension or L.T.) in an ignition circuit. That handled by the contact-breaker (which see).

Primary gears The pair of gears connecting engine to gearbox and providing the primary (or first) reduction.

Printed circuit A route for electricity to flow impressed on to insulating material, instead of actual wires being used to join the terminals involved. Very often, the circuit is, in fact, printed.

Progressive rate A spring that builds up resistance to deflection as the load in it increases (see **Constant rate** as **Multi-rate**).

P.S.I. Abbreviation for pounds per square inch, a unit of pressure.

P.T.F.E. Abbreviation for poly-tetra-fluoro-ethylene, a very low friction plastic often used for bushes.

Pump Slang term concerning hydraulics. Several rapid

applications of the operating lever or pedal 'overcharges' the system so that sponginess or lost movement is temporarily overcome by excessive master cylinder action sending more than the proper amount of fluid down the slave cylinder feed pipe.

Pushrod A stout rod used to transmit a push as in clutch or valve operation.

Q

Quadrant A selector piece, usually provided with gear-like teeth for driving purposes, and strictly occupying a fourth part of a circle (namely 90°). Found in kickstarters and gearchange mechanisms and often without any suggestion of a quarter part.

Quartz iodide A type of incandescent filament light bulb utilising a reversible action based on quartz glass and iodine gas whereby material evaporated from the filament is later returned to the filament. The higher temperatures thereby possible (which would destroy an ordinary tungsten lamp) produce high outputs.

Quill A tube-like component, often tapering, generally used for oil injection into a rotating crankshaft.

R

Radial ply Form of tyre construction, only occasionally used in motorcycling in which the wraps of fabric in the tyre carcase are laid over each other radially (or bandage fashion), and not diagonally. (See **cross-ply.**)

Radiator Device for losing heat. An input via a cooling fluid is transferred to the atmosphere via a large surface area.

Rake The slope of the stanchions of front forks relative to the vertical. Also the slope of the steering head column in some usages.

Ratchet A wheel or quadrant with inclined or castellated teeth into which a pawl can notch to prevent reverse movement or to achieve one-way drive with an over-running capacity as in a kickstarter.

Ratio The proportion of one thing to another, in terms of quantity. Often reduced to a comparison against unity (one) as a base figure. (See **gear ratio** and **compression ratio**).

Reach The depth or height or length to which something extends. Particularly used in conjunction with the effective depth of sparking plug holes (See **long reach** and **short reach**).

Rebore A cylinder equipped with a new working surface by removing a worn one. Such a bore is, of course, always therefore larger.

Rebound A bouncing back, particularly of suspension or of spring-controlled poppet valves.

Reciprocate Backwards and forwards. A reciprocating engine has a piston that goes 'to and fro', coming to a standstill at each reversal.

Reciprocating weight The mass of parts that reciprocate. In the case of a piston and con-rod assembly, all of the piston, ring and gudgeon pin mass, plus half of the mass of the rod as determined by laying the rod horizontal and weighing the small-end eye, the big-end being free to pivot on a 'frictionless' mount placed away from the balance pin.

Recoil The bouncing backwards, towards its static position, of a spring as it asserts itself.

Rectifier Electrical device passing current in one direction only (and thus a wave), used to convert alternating current into direct current. Rectifiers may be of full-wave or half-wave types.

Recuperation Hydraulic term for the action which goes on when a column of fluid is returned to the slave-cylinders.

Reed valves A two-stroke valve functioning like a reed with pressure itself causing the 'flap' to open or close. Capable of working at extremely high speeds.

Relay An operational device to reduce the load on switchgear and/or cabling. The relay functions as the main current-carrying switch while the control switch is nothing more than a trigger carrying a tiny current. Relays may be used to switch heavy currents for horns and headlamps.

Reverse flow Control of the fresh charge into the two-stroke combustion space by oblique and upward angling of the transfer ports, so directing the new charge that it drives the burnt gases out before it.

Rim The edge, margin or periphery. In the case of a wheel, the part that carries the tyre.

Ring flutter Literally, flutter of a piston ring in its groove. Taper rings suffer less from this than conventional rings.

Rivet A headed pin of which the unheaded end is subsequently clenched over so as to fasten components together which it sandwiches between its ends.

Rocker A device pivoting between its ends and transmitting a push on one end in the opposite direction at the other. Rockers can also handle pulling forces. Some rockers (e.g. in some valve gear) lack the true rocking action and are pivoted at one end and so do not reverse motion.

Rocking couple The tendency of some kinds of reciprocating engine to generate a rocking effect on the machine.

Roller bearing One containing rollers as the support medium, and not balls. The rollers run in specially-prepared tracks and may be kept clear of each other by a cage. Rollers are usually of hardened steel.

Rotary Capable of rotation. Spinning. An engine, the principal components of which spin instead of reciprocate.

Rotary valve A valve for two-stroke or four-stroke which, by rotation, opens and closes gas passageways at the appropriate times and usually disc, conical or cylindrical in shape. Normally found on inlet systems.

Runout Wobble. Out of truth. Total runout is the full measurement from one extreme to the other. Sometimes referred to as total indicator reading. Confusion occurs when runout is measured from the midpoint which is, of course, half the total runout.

S

S.A.E. Abbreviation for Society of Automotive Engineers. S.A.E. numbers form a system for classifying lubricating oils into viscosity ranges at prescribed temperatures.

Scavenge To clear away, particularly exhaust gas from the cylinder and oil from a dry sump (which see).

Schnurle Loop scavenging with the attendant flat top (or deflectorless) piston. The incoming charge via the transfer port loops up the cylinder, across the head, and down towards the exhaust port, propelling burnt gas residues before it.

Schrader A proprietary air valve for an inner tube (see valve core).

Seat angle The angle at which valve seats are formed, usually 45° but sometimes 30°, the latter giving greater effective open gap at low lifts. In some engines the valve face angle is up to one degree smaller than that of the seat for manufacturing and bedding-in reasons.

Secondary current That flowing, at high voltage (or **high tension, H.T.** which see), in the multi-turn windings, in the plug lead, and across the plug points. Could be of the order of 20,000 volts, though usually less.

Security bolt A reinforced rubber pad, put inside a tyre, and pulled down on to the tyre beads (by an external nut on its bolt) so as to hold the cover in position on the rim. If the cover moves during acceleration or braking, it may otherwise wrench the inner tube's valve enough to cause a puncture.

Seizure A binding together through pressure, temperature or lack of lubrication, and often all three. Also called freezing up.

Selector fork A forked-shaped prong, mating with the track of a gearbox pinion or dog, for the purpose of sliding that component from side to side, under the instructions of the camplate or cam drum, so as to change ratio.

Semi-conductor Material passing electricity freely in one direction but not at all in the other.

Serrated washer A washer which is made with tiny 'teeth' round the rim. Also called shakeproof. An internal shakeproof has a regular outer edge and the 'teeth' on the inside.

Shim A piece of very thin tough metal used to adjust the position or effective thickness of a component during assembly.

Shock absorber A mechanical or rubber device for ironing out minute irregularities in power delivery to smooth the transmission of power. Also, on a three-wheeler, an alternative term for a suspension damper.

Shoe A rigid component able to press against another, sometimes, as in many brake shoes, being faced with a friction material.

Short reach Sparking plug term for a plug hole of half-inch depth. (See **reach** and compare with **long reach**).

Short stroke Oversquare (which see) but to a significant degree. (See also **bore:stroke ratio**).

Siamese Two joined into one, particularly two exhaust pipes joining together and entering a common silencer.

Side valve An engine having its valve gear at the side of the cylinder and not overhead. Some designs place the valves at the front or rear but these are still called side valves.

Sidewall The part of a tyre between the bead and the tread. On this part is moulded the maker's name, sizing etc.

Silencer Device to quieten. Normally applied to the exhaust side of an engine, but inlet silencers also exist.

Silentbloc A proprietary part consisting of a rubber insert bonded to metal to provide shock-insulating support.

Silicon Next to oxygen, the most abundant element (such as sand), and a valuable ingredient in aluminium alloys.

Silver solder One containing a high proportion of silver and having a moderate melting point, good capillary migration, and giving a strong joint.

Single leading shoe A brake with only one shoe leading and the other trailing, a very common type. Abbreviated to 1LS or SLS.

Sintered metal Finely divided metal powder subject to immense pressure (and sometimes heated) in a mould, producing an article with a porous structure and so accurately sized and shaped that little or no machining is required. Sintered metal usually holds oil like a sponge.

Skimming Machining operation involving the removal of the minimum amount of metal for the purpose of straightening or flattening. Operations involving stock removal are often erroneously called skimming.

Skirt A hanging down. On a piston that part below the gudgeon pin and ring belt areas.

Slave cylinder The equipment end of an hydraulic control system (e.g. brake or clutch end). A small piston in a cylinder is forced outward by hydraulic pressure and so operates the equipment concerned. (Compare with master cylinder).

Sleeve Much as liner (which see) but more in the sense of restoration of size.

Slider A part that moves up and down. On tele forks, the moving bottom leg, which slides on the fixed stanchion (which see).

Small-end The smaller bearing of the two on a connecting rod (which see) and through which the piston is attached. Usually fitted with a plain bearing or a needle-roller assembly.

Solenoid An electrical operating device consisting of a soft iron core drawn into an electro-magnetic field by magnetic suction, the displacing core being able to push on or pull at some mechanism it is desired to operate by remote control.

Sparking plug Device for arcing an electric current, as a spark, between two electrodes inserted in the combustion space, so as to initiate burning, the entire assembly being detachable and mounted in a threaded hole.

Specific gravity The density or 'weight' of a substance compared with an equal volume of pure water as unity (1.0). Thus common steel is 7.830 and ice 0.918.

Spindle The fixed rod about which an article turns or perhaps swings in an arc.

Spine Backbone. A spine-type frame consists of and derives all its strength from its upper member (though the engine unit may also play some minor role).

Splayed head A cylinder head of a four-stroke twin having its inlet and/or exhaust ports boldly angled to each other in plan view. The 'opposite' of the parallel conditions of port routing.

Spoke A wire rod, hooked at one end and threaded at the other, uniting a wheel rim to the hub. A sturdy, integral part of a wheel, joining centre to periphery.

Sprag A jamming device.

Spring A deformable component used to permit a movement but to provide a positive return. Springing mediums may be metal, rubber or even gas.

Sprocket Toothed wheel used in chain drive.

Stainless steel A tough durable steel that does not rust and is highly stain resistant. Usually incorporates at least 25 percent of the alloying metal, chromium. The 18/8 Austenitic grade is non-magnetic and totally resistant to salt water. The S80 Ferritic is magnetic and partly salt resistant, but extremely stubborn to machine.

Stanchion A strong, rigid, structural member. In a telescopic fork, that tubular part attached to the fork crowns and on which travels the moving slider (which see).

Steering head The top of the front forks of a motorcycle.

Stem A narrow, lengthy projection from the main body of a component.

Stiction Initial resistance to movement. When once overcome, the item moves more readily. Mainly used in respect of suspension systems. Not to be confused with friction – a rubbing resistance – stiction exists only when movement is absent.

Stoichometric Air: fuel ratio calculated in accordance with the pure mathematics of molecular exchanges. An engine cruising will run a little weaker than this and under power a bit richer.

Stove enamel A painted finish hardened and dried by heat, extremely durable and possessing a high gloss. In slang, called stoving.

Straight oil A mineral oil without additives (See **Monograde**).

Stroke The linear travel of a component. In a reciprocating engine, the distance between the highest and lowest points of the piston. Usually, it is numerically twice the crankshaft throw though many piston engines exist where this is marginally untrue.

Stud A rod threaded at both ends (but not necessarily with the same threads or even having the same diameters). Usually screwed into one component, there to become a permanent or semi-permanent fixture, for another component to be nutted up to it.

Subframe The rear part of a motorcycle frame which carries the seat and upper abutments for the suspension. However, if the frame proper is extended rearward or so designed to fill this function, there is no subframe.

Sulphuric acid A strongly corrosive liquid, acid in nature, used in a lead battery. It is a registered poison.

Sump A well, hollow or reservoir. Sumps of engines may be oilholding (wet) or scavenged free of draining-down oil (dry).

Supercharger A rotating pump for increasing the quantity of mixture delivered to an engine. In slang, a blower.

Suppressor An electrical resistance of 5 to 15 thousand ohms value inserted into the H.T. side of an ignition system to suppress interference with TV and radio reception in houses. Also, a resistance, perhaps of one ohm, placed in series in the L.T. circuit or across a dynamo to eliminate interference in a receiver carried on the same vehicle.

Surging A wave-like action in the centre of a spring, often destructive.

Swept volume The volume of an engine as swept by the piston.

Swing caliper A kind of brake caliper featuring only one moving piston and obtaining compensation between the two pads by its whole body being free to swing about the disc. Otherwise calipers are fixed.

Swinging arm A radius arm which sweeps out an arc and used to carry the wheel mounted in its free end to provide suspension. An arm, strictly is one-sided though general usage has resulted in a swinging fork also being called, erroneously, a swinging arm.

Swirl Rotary or swirling motion given to a charge mixture as it enters the cylinder by offsetting the inlet tract.

Switch A device for making or breaking an electrical circuit, often mechanical. However, electrical relays are but semi-mechanical and perform switching functions as can solid state devices which have no moving parts at all.

Synthetic A paint based on artificial materials and not on organic ones such as cellulose. Half-hour synthetic is a trade term for such a paint, on which further work may be done in 30 minutes.

T

Taper A narrowing width along the length.

Taper pin A tapered pin of metal driven through two or more components until it jams in the hole thus locking all together.

Taper roller A hardened steel roller, being tapered instead of cylindrical, and able to take heavy axial ('length-ways') as well as radial loads.

Tachometer Rev-counter or tacho. An instrument for measuring engine rpm.

Tappet A reciprocating cylinder placed between a cam and a valve stem or pushrod to absorb wiping loads and functioning as a cam follower (which see). May have provision for valve clearance adjustment.

Telescopic Two tubes, one fitting snugly inside the other, which are able to slide in and out like a telescope. Such form an effective front suspension (slang: tele) which may or may not feature hydraulic damping.

Terminal Literally an end. The fitting to which an electrical connection is made. On a battery usually called terminal post.

Thermal efficiency The ratio of useful work available from an engine to the heat supplied from the fuel in question. Values vary from 10 percent for a steam engine, through 35 percent for a petrol motor, to 45 percent for a high-speed diesel.

Thermo-syphon Natural cooling, utilising the fact that two columns of liquid at different temperatures possess natural circulation because the hotter weighs less on account of its lower density. Thus the hot engine must be lowermost and the cool radiator must be sited at a higher level (see **impeller**).

Thermostat A controlling device changing operational conditions according to temperature rise or fall. It may take the form of a rod or strip producing more or less push according to how much distortion is created by its heat. Frequently used to control the temperature of engine coolant so stabilising engine temperature.

Throttle A variable impediment in the inlet tract, designed to provide control of power output by limiting the amount of fresh mixture induced. Literally, a throttling.

Throw The amount the crankshaft is eccentric from the crankshaft's rotational centre and, in a conventional engine, equal to half the stroke.

Throwback The tendency of a 'suddenly' compressed spring to throw back beyond the position it should assume had the load been applied infinitely gently.

Thrust face A working surface of a piston, bearing, shim etc, which takes the thrust and any rubbing action. The 'active' face in contrast to the 'passive' one.

Timing The opening and closing points of valves and breathers, and in the moment of ignition. Usually recorded in degrees of crankshaft rotation or in linear piston movement in millimetres or inches, in relation to dead centre positions.

Timing chain The chain driving the valve gear and/or ignition equipment.

Timing clearance The use of extra clearance to the valves for timing purposes only, so as to give unambiguous indication of the opening and closing points. Some 20 thou is often used.

Timing gears Those involved in driving the valve mechanism and/or ignition equipment.

Timing light Electrical device, featuring a bulb, which gives visual indication of the moment the ignition equipment makes its spark, and used for ignition timing in the workshop.

Toe-in The turning inwards of a sidecar wheel so that it points slightly across the bows of the outfit. Intended to combat the slewing effect created by an offset driving wheel. But note that some racing outfits feature toe-out, the reverse.

Tolerance Permissible variation in manufacturing limits. Do not confuse with clearance.

Tongue A narrow projection, for example in an Oldham coupling (which see).

Tooth An accurately shaped projection on a component so that, in conjunction with others, motion may be positively transmitted from one toothed component to another. Gears, sprockets and racks may all have teeth.

Toothed belt A flexible belt, used for driving overhead camshafts, superchargers etc, which has a toothlike formation on it able to engage positively with matching pulleys. Also called a cogged belt.

Torque A twisting force, measured in pound feet, tending to cause rotation.

Torque converter Device, such as a gearbox, for varying the relative speeds of input and output shafts, thus varying the torque (which see) but not the power.

Torsion Twist or turning.

Torsion bar A spring in the form of a rod deriving its springiness from being twisted along its length.

Total loss A system of lubrication in which the oil is lost after its one and only delivery to the working surfaces. Thus two-stroke oiling is total loss.

Track The distance apart, measured sideways, of two wheels, as the rear and sidecar wheels of an outfit. (See wheelbase).

Trail The castor action of a steerable front wheel, causing it to take up a natural straight ahead position during running. With positive trail, the centre of the contact path of the tyre lies to the rear of the projected line of the steering head. With negative trail, used in sidecar racing, there is a reverse condition.

Trailing link A form of front suspension using a pivoting link – approximately horizontal – with the axle behind the pivot. (See **leading link**).

Trailing shoe A brake shoe of the opposite kind to a leading shoe (which see).

Transfer port The port (or passageway) through which the fresh mixture, in a two-stroke, is transferred from below the piston to above the piston.

Transistor An electrical device with zero warm-up time which magnifies the strength of a small input signal provided it is also supplied with a 'bulk' supply of current to draw on. Used in transistorised ignition systems (which see).

Transistorised ignition A system in which all switching is done by transistorised circuiting and a magnet plus magnet detector produces the triggering instead of contact-breaker points. (see **transistor**).

Tread The part of a tyre intended to grip on the road. A plain tread or treadless tyre has no pattern on it.

Triangulation Achieving rigidity and stiffness by making a structure triangular or introducing triangles into it. A triangle cannot be collapsed without bending one of its sides whereas a square, say, can – it then being said to lozenge.

Trickle charger A small, mains-powered charger used to boost a battery. Provision is often made for 6 and 12 volt charging and sometimes for varying the rate (amperage). If 1/2 to 2 amps is delivered, this is said to be a trickle. More is categorised as a fast rate of charging.

Triplex Three. A triplex chain has three rows of rollers (see also **duplex**).

Trunnion A component permitting the marriage of pivoting and linear movement. Found often in plunger oil pump drives and in some forms of three-wheeler suspension.

Tufnol Proprietary material used for timing gears and a resin reinforced fabric.

Tungsten A rare metal used as an alloy with tough steels and as an electric filament in a conventional light bulb. The term tungsten lighting is deliberately used to distinguish it from halogen (which see).

Turbulence Agitation in a liquid or gas and especially in the fresh charge inside a cylinder. Adequate turbulence may assist good combustion.

Twirl A twirling motion given to a charge entering a cylinder. Also called swirl.

Twistgrip Rotary throttle control on the right handlebar, operated by twisting. But twistgrip ignition controls on the left handlebar also exist.

Two leading shoe A brake possessing two leading shoes. A normal brake has but one. Abbreviated to 2LS.

Two-stroke An operating cycle for an internal combustion engine, devised in 1880 by Sir Douglas Clerk, in which combustion takes place on every ascent of the piston, that is the cycle is complete in two strokes of the piston (see four-stroke).

U

Undersquare Stroke greater than the bore (see **long-stroke**).

Unit Engine and gearbox manufactured as one single unit and not capable of being separated.

Unsprung weight That part of the suspension which lies the road side of the springs and partially the weight thereof.

V

Valve bounce A poppet valve crashed on to its seat too hard for the spring to hold it down on that seat.

Valve cap Either the safety or dust cap on a tyre valve, or the now rare hardened and loose cap on the end of an inlet or exhaust valve protecting the valve from wear when struck by the rocker.

Valve core A detachable inner half to a valve in a pneumatic tyre, mating with a seat formed integrally in the valve body formed as part of the tube.

Valve lifter A manually-operated lever to raise a four-stroke's exhaust valve off its seat independently of the normal operating mechanism. Used to make the engine easy to kick over, to help an electric starter, to stop the engine, and on a trials machine, to assist in the descent of steep hills (compare with **decompressor**).

Valve seat That part of a valve, or the part of the cylinder head against which it seats, which contacts the matching part. Much usage centres on the cylinder head half rather than the valve itself. Many seats are renewable. Some seats are formed in other components (e.g. side valve barrels).

Variable transmission A system of gearing which adjusts itself, within limits, to load and speed, and without steps. Also called stepless or infinitely variable.

Vee belt A flexible belt, usually of rubber and canvas, having a vee-like section.

Vee engine A motor with its cylinder axes arranged, not in parallel, but in vee formation.

Velocity Speed, gait, rate of movement.

Venturi A narrowing down of a gas passage intended to cause a pressure reduction. Found in carburettors and used to create the suction needed to lift fuel from a jet.

Vernier gauge A parallel jawed sliding caliper able to measure to precision exactitude by a vernier system.

Vibration Shaking. The higher the frequency, the faster the vibration.

V.I. Improver Viscosity index improver. An oil additive helping an oil to keep its viscosity (thickness) as it thins out when heated. (See **viscosity and multigrade**).

Viscosity Thickness. Indicated by an S.A.E. number. The higher the numerical figure the thicker (or more viscous) the fluid.

Viscosity index The rate of change of viscosity with temperature. A low V.I. means a small temperature change produces a large change in viscosity. A typical quality straight mineral oil might have a V.I. of 95 and a multigrade (with its concomitant V.I. improver) could have a value of 120. (see **viscosity**).

Volt Unit of electrical pressure or tension.

Voltmeter Electrical instrument for measuring voltage.

Volume Space occupied by gas or liquid, usually measured in cubic centimetres and abbreviated to cc, cl or cm³.

Vortex A whirling of gas or liquid, the rotation creating a central cavity.

Vulcanising Hot curing of a repair to a rubber article such as a tube. The basic process was discovered by Charles Goodyear of Massachusetts, in 1839.

W

Wankel A type of engine, invented by Felix Wankel, containing an inner rotor in a specially shaped chamber and which, in its usual form, possesses three compression and combustion spaces to each rotor.

Washer A piece of material, often steel and circular, used to spread the load over a larger area and to absorb the scouring action of the nut, bolt head or screw. It may also have a sealing role to play. The hole pierced is invariably circular.

Watercooling Indirect transmission of heat from a rejection surface to the atmosphere, water being used as an intermediate carrier (See **radiator**).

Water pump See impeller and thermo-syphon.

Watt Unit of electrical 'size' and the product of volts multiplied by amperes.

Wave A cyclic variation in direction, polarity and/or intensity in electrical current. In gas, a cyclic variation in pressure, often with reference to the speed of sound.

Weld To join materials by heat and sometimes by pressure. Metals and plastics are two groups of material commonly welded.

Wet liner A liner inserted into a cylinder block so that cooling water has direct contact with the liner for much of the liner's area. (See **liner**).

Wet sump An engine lubrication system, sometimes called car-type, in which the oil is carried in a tray below the crankshaft. Such engines rely on gravity drainage of circulated oil and need no return pump. (Compare with **dry sump**).

Wheelbase The distance, measured lengthwise, between the centres of front and rear wheels. Not to be confused with track.

Windings Coils of wire for generating a magnetic field in which electricity is generated, and wound in an orderly manner round a former or coil.

Woodruff key Slip of metal fitting simultaneously into grooves in a shaft and in a component thereon, for location. Often supposed, erroneously, to be able to drive.

X

Xenon An inert gas naturally present in the air in small quantities and in its radioactive form sometimes used for air-leak detection in pressurised vessels such as tyres.

Y

Yoke A component that connects two or more others.

Z

Zener diode An electrical component allowing a controlled leak to earth above a specified voltage, surplus current appearing as heat. It prevents overcharging with an alternator system.

Other titles in this series

Motorcycle Carburettor Manual Book no. **603**

There can be few motorcyclists who have not been confronted by carburation problems in one form or another. The number of symptoms which can be attributed to some sort of carburettor malfunction is just short of infinite, and this often leads to some tentative screwdriver-twiddling on the part of the owner, in the vain hope that this will resolve the problem. Unfortunately, unless he is very lucky indeed, this will only add another discrepancy, and will often mask the true cause of the original problem. Given that many machines possess four or more carburettors, this can lead to a very interesting tangle, and can take hours to sort out.

This manual sets out to overcome such problems, first by giving the reader a sound knowledge of the simple principles which govern the way a carburettor functions, and then by examining in detail the more practical aspects of tuning and the correction of maladjustments. In addition to this the overhaul of the main types of carburettors is discussed in detail as is the practical use of tuning aids and equipment by which greater accuracy can be obtained.

The manual is divided into thirteen Chapters covering the theory of carburation, development history, carburettor types, overhaul, tuning and also includes a fault diagnosis section.

Motorcycle Basics Manual Book no. **1083**

Quite a large proportion of motorcycles, mopeds and scooters in everyday use are ridden by owners who have little understanding of how their machine works. Many are quite content to remain in this state of blissful ignorance, having their machine serviced at regular intervals as and when the need occurs, or repairs carried out when problems necessitate taking the machine off the road. Not everyone has the ability or inclination to look after these matters themselves and it is well that they are aware of their own limitations.

If, however, the correct mental approach is adopted it will often be found that a complicated-looking assembly is nothing more than a number of relatively simple basic units bolted together to make up a single compact unit. These individual units are described in ten Chapters accompanied by well over 150 line drawings. Complicated technical terminology has been kept to a minimum and used only where considered necessary, but to help readers even further, a basic glossary of terms is included.

Many of the topics covered are: basic and improved two- and four-stroke engine tyes, engine designs and layouts, fuel and exhaust systems, ignition systems, transmission, lubrication and cooling, wheels, brakes, tyres, suspension, steering, frames and electrical systems.

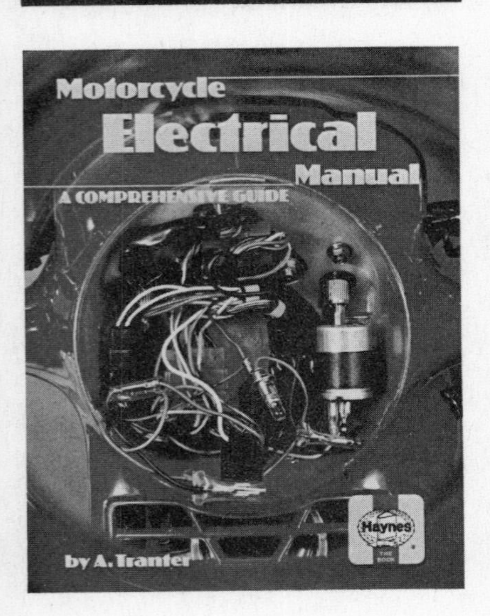

Motorcycle Electrical Manual Book no. **446**

The modern motorcycle is a far cry from those of twenty years ago and not least have been the changes in electrical equipment. Electrical loading now demands generators of greatly increased output while ignition and control units use electronic circuits of sophisticated designs. No mechanic can afford to ignore electrical matters and armed with the knowledge contained in this informative manual, should have no difficulty in working on the most complex machines of today.

The manual is well illustrated, with over 200 line drawings, and is divided into twelve Chapters. The first part of the manual is devoted to explaining the basic principles of ignition, electricity and magnetism, subsequent Chapters going into greater detail on the various system types and operation. These are grouped together as follows: coil and battery ignition, flywheel magneto ignition, electronic ignition, spark plugs, dc generators, alternators, lighting and signalling, starter motors, batteries and wiring.

The author, Tony Tranter, pioneered the first successful Motorcycle Engineering course in the UK, the City and Guilds 389 Motorcycle Mechanics Certificate course. He is a Chartered Electrical Engineer with an industrial background in vehicle electrics and electronics, has a long teaching experience and is the former Principal of the Merton Technical College, London.

MOTORCYCLE MANUALS

MOTORCYCLE MANUALS	No
BMW	
Twins	249
K100 & 75	1373
BSA	
Bantam	117
Unit Singles	127
Pre-unit Singles	326
A7 & A10 Twins	121
A50 & A65 Twins	155
Rocket 3 – see Triumph Trident	
BULTACO	
Competition Bikes	219
CZ	
125 & 175 Singles	185
DUCATI	
V-Twins	259
GARELLI	
Mopeds	189
HARLEY DAVIDSON	
Sportsters	702
Glides	703
HONDA	
Four Stroke Mopeds	317
SS50 Sports Mopeds	167
PA50 Camino	644
NC50 & NA50 Express	453
SH50 City Express	1597
NB, ND, NP & NS50 Melody	622
NE & NB50 Vision	1278
MB, MBX, MT, MTX50	731
C50 LA Automatic	1334
C50, C70 & C90	324
ATC70, 90, 110, 185 & 200	565
XR75 Dirt Bikes	287
XL/XR 80-200 2-valve Models	566
CB/CL100 & 125 Singles	188
CB100N & CB125N	569
H100 & H100S Singles	734
CB/CD125T & CM125C Twins	571
CG125	433
CB125, 160, 175, 200 & CD175 Twins	067
MBX/MTX125 & MTX200	1132
CD/CM185, 200T & CM250C 2-valve Twins	572
ATC250R (US)	798
XL/XR 250 & 500	567
CB250RS Singles	732
CB250T, CB400 T & A Twins	429
CB250 & CB400N Super Dreams	540
CB400 & CB550 Fours	262
CB350 & CB500 Fours	132
500 & 450 Twins	211
CX/GL500 & 650 V-Twins	442
CBX550 Four	940
CBR600 & 1000 Fours	1730
CB650 Fours	665
CB750 sohc Four	131

MOTORCYCLE MANUALS	No
Sabre (VF750S) & Magna V-Fours (US)	820
CB750 & CB900 dohc Fours	535
GL1000 Gold Wing	309
GL1100 Gold Wing	669
KAWASAKI	
AE/AR50 & 80	1007
KC, KE & KH100	1371
AR125	1006
Z200 & KL250 Singles	438
250, 350 & 400 Triples	134
400 & 440 Twins	281
400, 500 & 550 Fours	910
500 & 750 3 cyl Models	325
ZX600 Fours	1780
650 Four	373
750 Air-cooled Fours	574
900 & 1000 Fours	222
ZX900, 1000 & 1100 Liquid-cooled Fours	1681
MOBYLETTE	
Mopeds	258
MOTO GUZZI	
750, 850 & 1000 V-Twins	339
MZ	
TS125	1270
ES, ETS, TS150 & 250	253
ETZ Models	1680
NORTON	
500, 600, 650 & 750 Twins	187
Commando	125
NVT	
(BSA) Easy Rider Mopeds	457
PUCH	
Maxi Mopeds	107
Sports Mopeds	318
SUZUKI	
A50P, A50 and AS50	328
CL50 Love	1084
CS50 & 80 Roadie	941
FZ50 Suzy	575
FR50, 70 & 80	801
GT, ZR & TS50	799
TS50 X	1599
Trail Bikes ('71 to '79)	218
Air-cooled Trail bikes ('79 to '89)	797
A100	434
GP100 & 125 Singles	576
GS & DR125 Singles	888
GT125 & 185 Twins	301
250 & 350 Twins	120
GT250X7, GT200X5 & SB200 Twins	469
GS/GSX250, 400 & 450 Twins	736
GT380 & 550 Triples	216
GS550 & 750 Fours	363
GS/GSX550 4-valve Fours	1133
GT750 Three Cylinder Models	302
GS850 Fours	536
GS1000 Four	484

MOTORCYCLE MANUALS	No
GSX/GS1000, 1100 & 1150 4-valve Fours	737
TOMOS	
A3K, A3M, A3MS & A3ML Mopeds	1062
TRIUMPH	
Tiger Cub & Terrier	414
350 & 500 Unit Twins	137
Pre-Unit Twins	251
650 & 750 2-valve Unit Twins	122
Trident & BSA Rocket 3	136
VELOCETTE	
Singles	186
VESPA	
P/PX125, 150 & 200 Scooters	707
Scooters ('59 to '78)	126
Ciao & Bravo Mopeds	374
YAMAHA	
FS1E, FS1 & FS1M	166
RD50 & 80	1255
V50, V75, V80 & V90	332
SA50 Passola	733
DT50 & 80 Trail bikes	800
T80 Townmate	1247
YB100 Singles	474
100, 125 & 175 Trail bikes	210
RS/RXS100 & 125 Singles	331
RD125 Twins	327
RD & DT125LC	887
TZR125 & DT125R	1655
TY50, 80, 125 & 175	464
XT & SR125 Singles	1021
200 Twins	156
250 & 350 Twins	040
XS250, 360 & 400 sohc Twins	378
250, 360 & 400 Trail Bikes	263
RD250 & 350LC Twins	803
RD350 YPVS Twins	1158
RD400 Twin	333
500 Twin	308
XT, TT & SR500	342
XZ550 Vision V-Twins	821
650 Twins	341
XJ650 & 750 Fours	738
Yamaha XS750 & 850 Triples	340
Yamaha XV750, XV920 & TR1 V-Twins	802
XS1100 Four	483

Index